Advanced Robotics and Intelligent Automation in Manufacturing

Maki K. Habib
The American University in Cairo, Egypt

A volume in the Advances in Computational Intelligence and Robotics (ACIR) Book Series

IGI Global
DISSEMINATOR OF KNOWLEDGE

Published in the United States of America by
IGI Global
Engineering Science Reference (an imprint of IGI Global)
701 E. Chocolate Avenue
Hershey PA, USA 17033
Tel: 717-533-8845
Fax: 717-533-8661
E-mail: cust@igi-global.com
Web site: http://www.igi-global.com

Library of Congress Cataloging-in-Publication Data

Names: Habib, Maki K., 1955- editor.
Title: Advanced robotics and intelligent automation in manufacturing / Maki K. Habib, editor.
Description: Hershey, PA : Engineering Science Reference, an imprint of IGI Global, 2019. | Includes bibliographical references and index. | Summary: “This book examines recent advances and state-of-the-art technologies in the field of robotics engineering and mechatronics for advanced intelligent manufacturing”-- provided by publisher.
Identifiers: LCCN 2019027722 (print) | LCCN 2019027723 (ebook) | ISBN 9781799813828 (hardcover) | ISBN 9781799813835 (ebook) | ISBN(softcover) 9781799823223
Subjects: LCSH: Manufacturing processes--Automation--Manuals, etc. | Robots, Industrial--Manuals, etc.
Classification: LCC TS183 .H35945 2019 (print) | LCC TS183 (ebook) | DDC 670.42/72--dc23
LC record available at https://lccn.loc.gov/2019027722
LC ebook record available at https://lccn.loc.gov/2019027723

This book is published in the IGI Global book series Advances in Computational Intelligence and Robotics (ACIR) (ISSN: 2327-0411; eISSN: 2327-042X)

British Cataloguing in Publication Data
A Cataloguing in Publication record for this book is available from the British Library.

All work contributed to this book is new, previously-unpublished material.
The views expressed in this book are those of the authors, but not necessarily of the publisher.

For electronic access to this publication, please contact: eresources@igi-global.com.

Advances in Computational Intelligence and Robotics (ACIR) Book Series

ISSN:2327-0411
EISSN:2327-042X

Editor-in-Chief: Ivan Giannoccaro, University of Salento, Italy

MISSION

While intelligence is traditionally a term applied to humans and human cognition, technology has progressed in such a way to allow for the development of intelligent systems able to simulate many human traits. With this new era of simulated and artificial intelligence, much research is needed in order to continue to advance the field and also to evaluate the ethical and societal concerns of the existence of artificial life and machine learning.

The **Advances in Computational Intelligence and Robotics (ACIR) Book Series** encourages scholarly discourse on all topics pertaining to evolutionary computing, artificial life, computational intelligence, machine learning, and robotics. ACIR presents the latest research being conducted on diverse topics in intelligence technologies with the goal of advancing knowledge and applications in this rapidly evolving field.

COVERAGE

- Cyborgs
- Computational Intelligence
- Artificial Life
- Intelligent control
- Agent technologies
- Automated Reasoning
- Machine Learning
- Algorithmic Learning
- Pattern Recognition
- Natural Language Processing

The Advances in Computational Intelligence and Robotics (ACIR) Book Series (ISSN 2327-0411) is published by IGI Global, 701 E. Chocolate Avenue, Hershey, PA 17033-1240, USA, www.igi-global.com. This series is composed of titles available for purchase individually; each title is edited to be contextually exclusive from any other title within the series. For pricing and ordering information please visit http://www.igi-global.com/book-series/advances-computational-intelligence-robotics/73674. Postmaster: Send all address changes to above address.

Titles in this Series

For a list of additional titles in this series, please visit:
https://www.igi-global.com/book-series/advances-computational-intelligence-robotics/73674

Handbook of Research on the Internet of Things Applications in Robotics and utomation
Rajesh Singh (Lovely Professional University, India) Anita Gehlot (Lovely Professional University, India) Vishal Jain (Bharati Vidyapeeth's Institute of Computer Applications and Management (BVICAM), New Delhi, India) and Praveen Kumar Malik (Lovely Professional University, India)
Engineering Science Reference • ©2020 • 433pp • H/C (ISBN: 9781522595748) • US $295.00

Handbook of Research on Applications and Implementations of Machine Learning Techniques
Sathiyamoorthi Velayutham (Sona College of Technology, India)
Engineering Science Reference • ©2020 • 461pp • H/C (ISBN: 9781522599029) • US $295.00

Handbook of Research on Advanced Mechatronic Systems and Intelligent Robotics
Maki K. Habib (The American University in Cairo, Egypt)
Engineering Science Reference • ©2020 • 466pp • H/C (ISBN: 9781799801375) • US $295.00

Edge Computing and Computational Intelligence Paradigms for the IoT
G. Nagarajan (Sathyabama Institute of Science and Technology, India) and R.I. Minu (SRM Institute of Science and Technology, India)
Engineering Science Reference • ©2019 • 347pp • H/C (ISBN: 9781522585558) • US $285.00

Semiotic Perspectives in Evolutionary Psychology, Artificial Intelligence, and the Study of Mind Emerging Research and Opportunities
Marcel Danesi (University of Toronto, Canada)
Information Science Reference • ©2019 • 205pp • H/C (ISBN: 9781522589242) • US $175.00

Handbook of Research on Human-Computer Interfaces and New Modes of Interactivity
Katherine Blashki (Victorian Institute of Technology, Australia) and Pedro Isaías (The University of Queensland, Australia)
Engineering Science Reference • ©2019 • 488pp • H/C (ISBN: 9781522590699) • US $275.00

For an entire list of titles in this series, please visit:
https://www.igi-global.com/book-series/advances-computational-intelligence-robotics/73674

IGI Global
DISSEMINATOR OF KNOWLEDGE

701 East Chocolate Avenue, Hershey, PA 17033, USA
Tel: 717-533-8845 x100 • Fax: 717-533-8661
E-Mail: cust@igi-global.com • www.igi-global.com

Table of Contents

Detailed Table of Contents

This chapter describes the development of a robotic CAM system for six-DOFs articulated industrial robot to generate online motion supported by cutter location source (CLS) data and numerical control (NC) data. The robotic CAM system realizes a practical data interface between industrial robots with open architecture controllers and commercially available CAD/CAM systems, and it includes functions that generate minute position and orientation components for real time motion control from CLS data and NC data without the need for teaching. The design principles of the developed robotic CAM system, and the experimental results on three real robots, RV1A, VS068, and FANUC R2000iC, are presented and demonstrated using both CLS and NC data generated through the developed CAM system.

The chapter approaches the issues of modeling the process of load lifting by a person while wearing an exoskeleton. The classification of existing gravitational compensation systems for industrial exoskeletons is shown, as well as examples of its use. A mathematical model of lifting a person's load in the exoskeleton is presented, as well as numerical parameters are calculated. It is shown that the introduction of an elastic element reduces the level of energy consumption during work, and can also facilitate the level of the worker. Industrial exoskeleton prototype design is presented. A particular focus is given to studying the influence of the gravity compensator on the magnitude of the moments generated by the electric drives of the hip and knee joints. It is shown that the use of gravity compensators enables to reduce significantly the load on electric drives.

This chapter deals with cable-driven robots when applied in physical rehabilitation. In general, neurorehabilitation is limited to physical therapy that is delivered by clinicians and potentially augmented by robotic tools to facilitate neurorehabilitation and to reduce the consequences of central nervous system injury. Among the robotic tools for rehabilitation can be considered the cable-driven manipulators. First, this chapter presents the upper and lower human limbs movements. The main rehabilitation robots are presented as exoskeletons and cable-driven manipulators. After, the cable-driven manipulators theory is introduced focusing on considerations for robot design in rehabilitation and control with safe human-machine interaction. Experimental examples with different cable-driven robot's structures are presented so that this chapter suggests that these structures can be used as a complement to conventional therapies and not as a substitute. Finally, this chapter presents the clinical evidence in cable-driven robots when applied in physical rehabilitation.

The chapter presents different case studies involving technology upgrading involving Industry 4.0 technologies and artificial intelligence. The work analyzes four cases of study of industry projects related to manufacturing process of kitchen, tank production, pasta production, and electronic welding check. All the cases of study concern the analysis of engineered processes and the inline implementation of image vision techniques. The chapter discusses other topics involved in the production process such as augmented reality, quality prediction and predictive maintenance. The classic methodologies to map production processes are matched with innovative technologies of image segmentation and data mining predicting defects, machine failures, and product quality. The goal of the chapter is to prove how the combination of image processing techniques, data mining approaches, process simulation, chart process modeling, and process reengineering can constitute a scientific research project in industry research.

This chapter proposes a framework to make the programming of cobots faster, user-friendly and flexible for assembly tasks. The work focusses on an industrial case of a small (10kg) air compressor and investigates the technologies that can be used to automate this task with human-robot collaboration. To this end, the framework takes a radically different approach at the motion stack level and integrates the cobot with a constraint-based robot programming paradigm that enhances the robot programming possibilities. Additionally, the framework takes inputs from the operator via speech recognition and computer vision to increase the intuitiveness of the programing process. An implementation is made with focus on industrial robustness and the results show that this framework is a promising approach for the overall goal of achieving flexible assembly in the factories by making robot programming faster and intuitive.

The use of robotics in the industrial environment has, in general, very similar goals. Because of productivity requirements, or due to reliability, industries have been constantly equipping their floor with robots. In that sense, the chapter observed—in a fiberglass company—the chance of using a robot to execute a boring and repetitive task. The task mentioned is, actually, the manufacturing of fiberglass reinforced plastic (FRP) molded grating. To confirm the possibility of using a robot to this job, a cost and time analysis was made about the whole molded gratings manufacturing process. Afterward, research about robotics was taken in parallel with the conception of the robot (named "roving-robot"). Calculations were made to the mechanical project of the robot. Applying computer-aided design (CAD), technical drawing and bill of materials were generated to permit the robot assembling. All of these project steps are presented in this chapter.

Designing a two-wheeled self-balancing scooter involves in the synergistic approach of multidisciplinary engineering fields with mutual relationships of power transmission, mass transmission, and information transmission. The scooter consists of several subsystems and forms a large-scale system. The mathematical models are in the complex algebraic and differential equations in the form of high dimension. The complexity of its controller renders difficulties in its realization due to the limit of iteration period of real time control. Routh model reduction technique is employed to convert the original high-dimensional mathematical model into a simplified lower dimensional form. The modeling is derived using a unified variational method for both mechanical and electrical subsystems of the scooter, and for the electronic components equivalent circuit method is adopted. Simulations of the system response are based on the reduced model and its control design. A prototype is developed and realized with Matlab-Labview simulation and control environment.

In this chapter, effective learning approach of inverse kinematics using neural networks with efficient weights update ability has been presented for a serial link structure and industrial robot. Generally, in making neural networks learn a relation among multi inputs and outputs, a desired training data set prepared in advance is used. The training data set consists of multiple pairs of input and output vectors. The input layer receives each input vector for forward computation, and it is compared with the yielded vector from the output layer. The time required for the learning process of the neural networks depends on the number of total weights in the neural networks and that of the input-output pairs in the training data set.

This chapter presents the theoretical foundations and methodology to develop a bioinspired hybrid control architecture for a biped robotic device that reproduces gait and human motor control strategies with the ability to adapt the trajectory to environmental conditions. The objective is to design robotic devices (such as exoskeletons), through the functional integration of hybrid dynamic system modeling (event-driven and continuous dynamics) with efficient and robust conventional control techniques and bioinspired control algorithms, with a near-natural human gait pattern. The human gait cycle is modeled as a hybrid dynamic using a finite state machine (FSM). The gait trajectories are to be generated in such a way that they will be capable of adapting to disturbances in the path followed by the robotic device; this will be achieved using a neuronal oscillator that simulates the behavior of a central pattern generator (CPG).

This chapter describes solving multi-objective reinforcement learning (MORL) problems where there are multiple conflicting objectives with unknown weights. Previous model-free MORL methods take large number of calculations to collect a Pareto optimal set for each V/Q-value vector. In contrast, model-based MORL can reduce such a calculation cost than model-free MORLs. However, previous model-based MORL method is for only deterministic environments. To solve them, this chapter proposes a novel model-based MORL method by a reward occurrence probability (ROP) vector with unknown weights. The experimental results are reported under the stochastic learning environments with up to 10 states, 3 actions,

Preface

Robotics, automation, intelligence and autonomy together with the evolution of industry 4.0 (I4.0) are opening new dimensions to develop strategies that are shaping the way to develop new customized products. Speed, quality, capabilities, and opportunities all play crucial roles in establishing competitive advantages.

Many studies pointed to the manufacturing sector's need for technology innovation and to identify novel science and technology solutions with the need to develop technical capabilities for advanced manufacturing. The new rising of manufacturing is associated with the rise of I4.0 era where automation is going to be heavily used and integrated with robotics and heavy data exchange processes that include mainly Internet of Things (IoT), cloud computing, cyber physical systems (CPSs), new form of human-machine interaction in real time, sustainability and intelligent decision making, etc. This facilitate the way to the development of smart factories that could provide new levels of production quality and speed, new capabilities, customer tailored products and broader opportunities for products that could not be realized with current manufacturing methods. However, these technologies are not easy to develop or to adapt broadly across industries. For this reason and others, advanced robots and intelligent automation that represent greater agility, flexibility, and autonomy has been identified as a viable solution to achieving these objectives. Robotics with greater mobility and environment interaction is the driver for developments in different dimensions and concurrency. Multi robotic systems will allow tasks to be decomposed in ways that exceed human capabilities through seamless communications and synchronizations; human-robot interaction; cooperation between multiple robots achieves a common goal by carrying out the task together or by decomposition. Cooperation can be achieved without explicit representation of actions and will use skill-based or learning-based automation.

This new book presents the challenge in the field of robotics, automation, intelligence and production processes by introducing major multiple research and knowledge areas in the field that are quickly growing and have potential in the evolution of new technologies. This book aims to highlight the state-of-the-art technologies and try to maximize the insights of the readers in understanding the recent advancements in robotics engineering and Mechatronics for advanced intelligent manufacturing. In addition, this book aims to enable engineers, researchers, scientist, industrial managers and the industry to have the latest knowledge and development in the subject field supported by theory, innovative design and applications.

This book includes 11 chapters that have the following titles with brief on their focus.

Chapter 1, "Development of Robotic CAM System that Generates Online Motion Supported by CLS and NC Data," by Fusaomi Nagata, Maki Habib, Takamasa Kusano, and Keigo Watanabe, describes the development of a robotic CAM system for six-DOFs articulated industrial robot to generate online motion supported by cutter location source (CLS) data and numerical control (NC) data. The robotic CAM system realizes a practical data interface between industrial robots with open architecture controllers and commercially available CAD/CAM systems.

Chapter 2, "Industrial Exoskeletons With Gravity Compensation Elements," by Sergey Jatsun and Andrey Yatsun, South-West State University, Russian Federation covers the issues related to the mathematical modeling of load lifting by a person while wearing an exoskeleton. The classification of existing gravitational compensation systems for industrial exoskeletons is shown, as well as examples of its use.

Chapter 3, "Cable-Driven Robots in Physical Rehabilitation: From Theory to Practice," by Rogério Gonçalves, Thiago Alves, Giuseppe Carbone, and Marco Ceccarelli, deals with cable-driven robots when applied in physical rehabilitation. In general, neurorehabilitation is limited to physical therapy that is delivered by clinicians and potentially augmented by robotic tools, to facilitate neurorehabilitation and to reduce the consequences of central nervous system injury.

Chapter 4, "Intelligent Processes in Automated Production Involving Industry 4.0 Technologies and Artificial Intelligence: Industry 4.0," presents and Intelligence, by Alessandro Massaro, Nicola Contuzzi, and Angelo Galiano proposes different case studies related to technology upgrading involving Industry 4.0 technologies and artificial intelligence. The work analyzes four cases of study of industry projects related to manufacturing process of kitchen, tank production, pasta production, and electronic welding check.

Chapter 5, "Framework for Quick & Intuitive Programming of Robot Applications," by Asad Tirmizi, Patricia Leconte, Karel Janssen, and Jean Hoyos, Maarten Witters, proposes a framework to make the programming of cobots faster, user-friendly and flexible for assembly tasks. The work focusses on an industrial case of a small

(<10kg) air compressor and investigates the technologies that can be used to automate this task with Human-Robot collaboration.

Chapter 6, "Designing a Robot for Manufacturing Fiberglass Reinforced Plastic (FRP) Molded Grating," by Marcos Carnevale, and Armando Carlos Pina Filho aims to confirm the possibility of using a robot to manufacture fiberglass reinforced plastic (FRP), by introducing cost and time analysis about the whole molded gratings manufacturing process. Research on robotizing the process was taken in parallel with the conception of the robot (named "roving-robot"). Applying computer aided design (CAD), technical drawing and bill of materials were generated, to permit the robot assembling.

Chapter 7, "Modeling, Model Reduction and Control of a Hands-free Two Wheeled Self-Balancing Scooter," by Qiong Li, Wangling Yu, and H. Henry Zhang, focuses on designing a two wheeled self-balancing scooter involves in the synergistic approach of multidisciplinary engineering fields with mutual relationships of power transmission, mass transmission and information transmission. The scooter consists of several subsystems and forms a large-scale system.

Chapter 8, "An Efficient Learning of Neural Networks to Acquire Inverse Kinematics Model," by Fusaomi Nagata, Maki K. Habib, and Keigo Watanabe, presents an effective learning approach of inverse kinematics using neural networks with efficient weights update ability has been presented for a serial link structure and industrial robot.

Chapter 9, "Hybrid Dynamic Modelling and Bioinspired Control Based on Central Pattern Generator of Biped Robotic Gait," by João Maurício Rosário, Luis Izquierdo-Córdoba, and Darío Amaya Hurtado, introduces the theoretical foundations and the methodology to develop bioinspired hybrid control architecture for a biped robotic device that reproduces gait and human motor control strategies with the ability to adapt the robot trajectory to environmental conditions.

Chapter 10, "Model-Based Multi-Objective Reinforcement Learning by a Reward Occurrence Probability Vector," by Tomohiro Yamaguchi, Shota Nagahama, Yoshihiro Ichikawa, Yoshimichi Honma, and Keiki Takadama, describes solving multi-objective reinforcement learning (MORL) problems where there are multiple conflicting objectives with unknown weights. Previous model-free MORL methods take large number of calculations to collect a Pareto optimal set for each V/Q-value vector. In contrast, model-based MORL can reduce such a calculation cost than model-free MORLs.

Chapter 11, "Autonomous Surgical Robotics at Task and Subtask Levels," by Tamás Nagy, and Tamás Haidegger, introduces the current research directions of subtask automation in surgery. In addition, it discusses the recent advances in motion planning, perception and human—machine interaction, along with the limitations of the task-level autonomy.

The development and the advancement in the field of Robotics, Intelligence and manufacturing and its integration with industry 4.0 toward smart factories help to motivate graduate students, researchers and scientist in all different engineering disciplines who are looking to broaden their knowledge and expertise to develop innovative, intelligent, interdisciplinary and sustainable intelligent robotics and manufacturing systems. The book widens the opportunities and open new roads to the future.

Maki K. Habib
The American University in Cairo, Egypt

Chapter 1
Development of Robotic CAM System That Generates Online Motion Supported by CLS and NC Data

Fusaomi Nagata
Sanyo-Onoda City University, Japan

Maki K. Habib
https://orcid.org/0000-0001-8088-8043
The American University in Cairo, Egypt

Takamasa Kusano
SOLIC Co. Ltd., Japan

Keigo Watanabe
Okayama University, Japan

ABSTRACT

This chapter describes the development of a robotic CAM system for six-DOFs articulated industrial robot to generate online motion supported by cutter location source (CLS) data and numerical control (NC) data. The robotic CAM system realizes a practical data interface between industrial robots with open architecture controllers and commercially available CAD/CAM systems, and it includes functions that generate minute position and orientation components for real time motion control from CLS data and NC data without the need for teaching. The design principles of the developed robotic CAM system, and the experimental results on three real robots, RV1A, VS068, and FANUC R2000iC, are presented and demonstrated using both CLS and NC data generated through the developed CAM system.

DOI: 10.4018/978-1-7998-1382-8.ch001

1. INTRODUCTION

Mechatronics deals with development, design, practice and applications associated with modern systems, control techniques and intelligence aiming to solve engineering problems and fulfilling needs. Habib (2007) showed time line evolution of interdisciplinary mechatronics as a field of synergy since 1970s. In addition, he introduced the knowledge space paradigm of mechatronics, while exploring the importance and the prospect of the future development of mechatronics both at the research development and education levels.

One of representative mechatronics systems is the industrial robotics. Robotics represents a well-developed example of interdisciplinary class of mechatronic systems, which has been remarkably advanced and applied to several tasks such as welding, handling, painting, polishing, etc. In addition, due to improvement of robot accuracy, load handling, repeatability and speed, the robotics integrated with artificial intelligence and learning capabilities is evolving to play a key role in our current manufacturing and future industry and in human life as well.

It is important to mention that the relation between the current CAD/CAM systems and the NC machine tools are well established and widely adopted in manufacturing industries compared to that between the CAD/CAM systems and the industrial robots that is not well established yet. A model generated using CAD and verified through computer aided engineering (CAE) can be introduced to the CAM software in which its outcome controls the machine tool. Generally, the CAD/CAM system main-processor generates CLS data according to each model's shape and machining conditions. Then, the post-processor produces the NC data that match the target actual NC machine tool that will use this data. The controller of the NC machine tool will use the generated NC date sequentially and accurately controls the positions of main head and the angles of other axes. On the other hand, however, the CAM system for industrial robots has not been sufficiently considered and as such it is not well developed yet. Due to this circumstance, generally a teaching pendant is used to obtain sequence of positions and orientations data of the arm tip through teaching before going to execute the trajectory connecting the taught points. Hence, it is necessary to develop innovative software that can integrate the simulation at the design level and the manufacturing at the robot level to increase flexibility and utilizes robotics capabilities effectively in solving problems. In addition, such software should be capable to receive the output of a CAM system and generate the suitable machining data code understandable by the robot side. Having the capability to share and exchange data between different CAD/CAM systems and different robot controllers will lead to fulfil the standardization demand is such evolution. The demand to develop innovative software that can integrate the design process and the manufacturing requirements increases. As a result of this many researchers in the field started their research toward achieving such goal developing software that realizes a robot driven machining process by having CAM robotics programing solution.

Kamisetty & McDermott (1992) examined the issues involved in the development of a translator that interface between CAD/CAM software and IBM 7535 SCARA industrial robot. However, the translator was not involve real time operation rather than generating offline the translated CAM output in terms of set of position, process parameters and relevant instructions compatible to the robot's native language, and then downloads the results into the robot. Following this, Ge et al. (1993) showed a basic transformation from CAD data to position and orientation vectors for a polishing robot. Then, Sugitani et al. (1996) developed welding robots which were successfully controlled by teaching-less CAD/CAM system, in which there were 26 sets of arc welding robot for steel bridge panel fabrication. Ahn & Lee (2000) also proposed an off-line automatic teaching method using vision information for robotic assembly task. Also, Kushida et al. (2001) proposed a method of force-free control for an industrial articulated robot arm. The control method was applied to the direct teaching of the robot arms, in which the robot arm was directly moved by human force. Nagata et al. (2001, 2006) developed a joystick teaching based system for polishing robot to safely obtain desirable orientation data of a sanding tool attached to the tip of robot arm. Maeda et al. (2002) proposed a simple teaching method for industrial robots through human demonstration. The proposed automated camera calibration enabled labor-saving teaching and compensates the absolute positional error of the industrial robot. Further, Sugita et al. (2003) developed two kinds of teaching support devices, i.e., a three-wire type and an arm type dedicated for a deburring and finishing robot. The validity the proposed devices were verified through experiments by using an industrial robot. Also, CAD-based off-line teaching system was proposed by Neto et al. (2010), which allow users with basic CAD skills to generate robot programs off-line, without stopping the production by using a robot. Nagata et al (2012) developed and implemented a CAM System integrated with the Industrial Robot RV1A. Then Nagata et al. (2016a and 2016b) introduced the integration of a robotic CAM system was integrated with the industrial robot controller to realize teaching-less operation and a fuzzy reasoning method is developed to process radius of curvature to skillfully regulate endmill's feed rate. The presented development used fuzzy feed rate controller and demonstrates the mapping obtained by the fuzzy feed rate to support machining on foamed polystyrene materials and later the authors replaced the mapping controller by a multi-layered neural network. In addition Nagata et al group (2016a, 2016b and 2018) continued the development of a post-processor for the industrial robot FANUC R2000iC is introduced. In order to help avoiding the conventional teaching process, the post-processor regularly generates a rigorous robotic program in real time from the cutter location source data called

CLS data. The CLS data are easily made through the main process of CAD/CAM, so that a promising data interface is realized between the industrial robot and CAD/CAM. The effectiveness and usefulness of the developed system are demonstrated through actual machining experiments.

This chapter describes, from the view point of robotic servo controller, the development of a robotic CAM system for an articulated industrial robot with six-DOFs that can generate online motion supported by cutter location source (CLS) data and numerical control (NC) data. The robotic CAM system includes an important function that generates minute position and orientation components for real time control from CLS data and NC data. The robotic CAM system realizes a practical data interface between the industrial robot and commercially available CAD/CAM systems such as Creo and Unigraphics, so that the robot can move in real time along a desired trajectory generated by CAD/CAM systems without the need for teaching. A reverse post-processor is used with the robotic CAM system to generate in real time CLS data for the industrial robot from NC data provided by the CAM system for a five-axis NC machine tool with a tilting head, in which orientation components of 4^{th}(B)-axis and 5^{th}(C)-axis in NC data are transformed into normal direction vector in CLS data. The transformation accuracy about orientation components in CLS data is tested and evaluated. The developed CAM system has a high applicability for other industrial robots with an open architecture controller whose servo system is technically opened to end-users, and also works as a straightforward interface between commercially available general CAD/CAM systems and the industrial robot. The design principle of the developed robotic CAM system and the experimental results on actual robots RV1A, VS068 and FANUC R2000iC are presented and demonstrated using both CLS and NC data generated from the developed CAM system.

Figure 1a and 1b severally show two real articulated-type industrial robots with open architecture controllers, which are used to evaluate the effectiveness of the proposed robotic CAM system. In Fig. 2, the proposed robotic CAM system is shown in comparison with the conventional CAM process using an NC machine tool. Furthermore, Fig. 3 presents the proposed robotic CAM system compared with the conventional off-line teaching process for an industrial robot. As can be seen from the figures, the industrial robot with the proposed CAM system can work based on both NC data and CL data without conducting a teaching task and without using robot language to program the robot.

Figure 1a. Shows articulated industrial robot with an open architecture controller

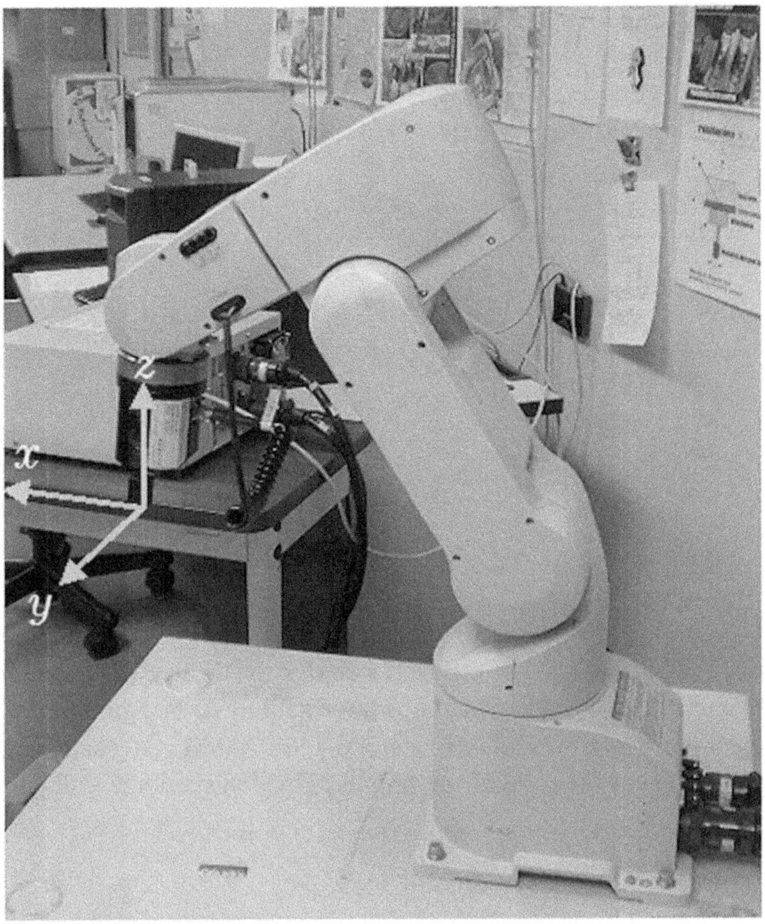

2. GENERATING A DESIRED TRAJECTORY

2.1 Main-Processor of General CAD/CAM System

Various kinds of CAD/CAM system such as Catia, Unigraphics, Pro/Engineer, etc. are widely used in manufacturing industries. The main-processor of each CAM can generate CLS data consisting of position and orientation components along a 3D model. This section introduces the details of how to calculate a desired position and orientation for robotic servo system. For example, robotic sanding task needs a desired trajectory so that the sanding tool attached to the tip of the robot arm can follow the desired object's surface, keeping contact with the surface from normal direction. If the object is fortunately designed by a CAD/CAM system and manufactured by an

Figure 1b. Presents machining scene of an impeller model using FANUC R2000iC industrial robot
(SOLIC Co. Ltd., Japan)

NC machine tool, then CLS data can be referred as the desired trajectory consisting of position and orientation elements. It is important to show a guideline for raising the relationship between the main-processor of CAD/CAM system and an industrial robot, which is one of the roles of the propose CAM system that will be presented in the next subsection.

Figure 2. Introduces the outline of the proposed robotic CAM system compared to that of the conventional CAM process using an NC machine tool

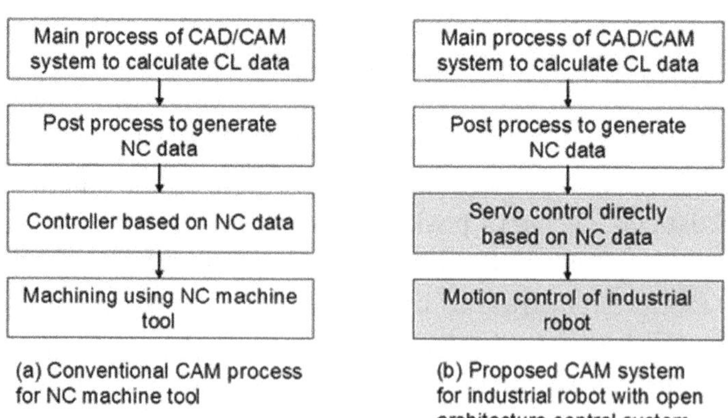

Figure 3. Introduces the outline of the proposed robotic CAM system compared to that of the conventional off-line teaching process for an industrial robot

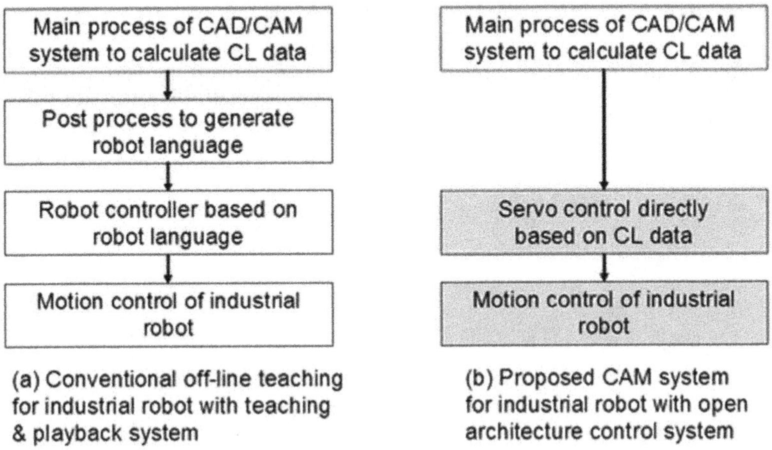

2.2 Position and Orientation Components for Discrete-Time Control System

In order to realize non-taught operation, we have already proposed a generalized trajectory generator (Nagata et al., 2001) using CLS data, which yields desired trajectory ${}^{w}\boldsymbol{r}(k)$ at the discrete time k given by

$$ {}^{w}\boldsymbol{r}(k) = [{}^{w}\boldsymbol{x}_d^T(k)\,{}^{w}\boldsymbol{o}_d^T(k)]^T \tag{1} $$

where the superscript w denotes the work coordinate system,

$$ {}^{w}\boldsymbol{x}_d(k) = [{}^{w}x_d(k)\,{}^{w}y_d(k)\,{}^{w}z_d(k)]^T $$

and

$$ {}^{w}\boldsymbol{o}_d(k) = [{}^{w}o_{dx}(k)\,{}^{w}o_{dy}(k)\,{}^{w}o_{dz}(k)]^T $$

are the position and orientation components, respectively. ${}^{w}\boldsymbol{o}_d(k)$ is the normal vector at the position ${}^{w}\boldsymbol{x}_d(k)$. In the following, we explain in detail how to make ${}^{w}\boldsymbol{r}(k)$ using the CLS data.

A target workpiece with curved surface is generally designed by a CAD/CAM system, so that CLS data can be calculated by the main-processor. The CLS data consist of sequential points along the model surface given by a zigzag path or a whirl path. In this approach, the desired trajectory ${}^{w}\boldsymbol{r}(k)$ is generated along the CLS data. The CLS data are usually calculated with a linear approximation along the model surface. The i-th step is written by

$$\boldsymbol{cl}(i) = [p_x(i) p_y(i) p_z(i) n_x(i) n_y(i) n_z(i)]^T \tag{2}$$

$$\{n_x(i)\}^2 + \{n_y(i)\}^2 + \{n_z(i)\}^2 = 1 \tag{3}$$

where

$$\boldsymbol{p}(i) = [p_x(i) p_y(i) p_z(i)]^T$$

and

$$\boldsymbol{n}(i) = [n_x(i) n_y(i) n_z(i)]^T$$

are position and orientation vectors viewed from the origin ${}^{w}O$, respectively. ${}^{w}\boldsymbol{r}(k)$ is obtained by using both linear equations and a tangential velocity scalar v_t called the feed rate.

A relation between $\boldsymbol{cl}(i)$ and ${}^{w}\boldsymbol{r}(k)$ is shown in Fig. 4. In this case, assuming

$${}^{w}\boldsymbol{r}(k) \in [\boldsymbol{cl}(i-1), \boldsymbol{cl}(i)]$$

we obtain ${}^{w}\boldsymbol{r}(k)$ through the following procedure. First, a direction vector

$$\boldsymbol{t}(i) = [t_x(i) t_y(i) t_z(i)]^T$$

is given by

$$\boldsymbol{t}(i) = \boldsymbol{p}(i) - \boldsymbol{p}(i-1) \tag{4}$$

so that v_t is decomposed into x-, y- and z-components in work coordinate system as written by

$$v_{tj} = v_t \frac{t_j(i)}{\left\| \boldsymbol{t}(i) \right\|}(j = x, y, z) \tag{5}$$

Using a sampling width Δt, each component of the desired position ${}^{w}\boldsymbol{x}_d(k)$ is represented by

$${}^{w}x_d(k) = {}^{w}x_d(k-1) + v_{tx}\Delta t \tag{6}$$

$${}^{w}y_d(k) = {}^{w}y_d(k-1) + v_{ty}\Delta t \tag{7}$$

$${}^{w}z_d(k) = {}^{w}z_d(k-1) + v_{tz}\Delta t \tag{8}$$

Next, how to calculate the desired orientation ${}^{w}\boldsymbol{o}_d(k)$ is considered. By using the orientation components of two adjacent steps in CLS data, a rotational direction vector

$$\boldsymbol{t}_r(i) = [t_{rx}(i) t_{ry}(i) t_{rz}(i)]^T$$

is defined as

$$\boldsymbol{t}_r(i) = \boldsymbol{n}(i) - \boldsymbol{n}(i-1) \tag{9}$$

Each component of desired orientation vector can be linearly calculated with $\boldsymbol{t}_r(i)$ as

$${}^{w}o_{dj}(k) = n_j(i-1) + t_{rj}(i) \frac{\left\| \boldsymbol{x}_d(k) - \boldsymbol{p}(i-1) \right\|}{\left\| \boldsymbol{t}(i) \right\|}(j = x, y, z) \tag{10}$$

${}^{w}\boldsymbol{x}_{d}(k)$ and ${}^{w}\boldsymbol{o}_{d}(k)$ shown above are directly obtained from the CLS data without either any conventional complicated teaching process or recently proposed off-line teaching methods. The desired position and orientation in the discrete time domain are very important to control the tip of an industrial robot in real time, i.e., to design a feedback control system.

If the linear approximation is applied when CLS data are generated by the main-processor of a CAD/CAM system, the CLS data forming a curved line are composed of continuous minute lines such as $\boldsymbol{p}(i)-\boldsymbol{p}(i-1)$ and $\boldsymbol{p}(i+1)-\boldsymbol{p}(i)$ shown in Fig. 5. In this case, it should be noted that each position vector in CLS data such as $\boldsymbol{p}(i)$ and $\boldsymbol{p}(i+1)$ have to be carefully dealt with in order to be accurately followed along the CLS data. For example, ${}^{w}\boldsymbol{x}_{d}(k+1)$, ${}^{w}\boldsymbol{x}_{d}(k+5)$ and ${}^{w}\boldsymbol{x}_{d}(k+9)$ are not calculated by using Eqs. (6), (7) and (8) but have to be directly set with $\boldsymbol{p}(i)$, $\boldsymbol{p}(i+1)$ and $\boldsymbol{p}(i+2)$, respectively, just before the feed direction changes. $\boldsymbol{p}(i)-{}^{w}\boldsymbol{x}_{d}(k)$,

$$\boldsymbol{p}(i+1)-{}^{w}\boldsymbol{x}_{d}(k+4)$$

and

$$\boldsymbol{p}(i+2)-{}^{w}\boldsymbol{x}_{d}(k+8)$$

are called the fraction vectors.

3. DEVELOPMENT AND IMPLEMENTATION THE INDUSTRIAL ROBOT RV1A

3.1 Communication With an Open Architecture Controller of an Industrial Robot

A Windows PC and an open architecture controller of RV1A are connected with Ethernet as shown in Fig. 6. The servo system in Cartesian coordinate system of the RV1A is technically opened to users, so that absolute coordinate vectors of position and orientation can be given to the reference of the servo system. The servo rate of the robot is fixed to 7.1 ms by the robot maker. Figure 7 illustrates the communication scheme by using UDP packet, in which sampling period is set to 10 ms. The data size in a UDP packet is 196 bytes. The packet transmitted by 'sendto()' includes values of desired position

Figure 4. Illustrates the relation between CLS data $\mathbf{cl}(i)$ and the desired trajectory $^{w}\mathbf{r}(k)$, in which ^{w}O is the origin of the work coordinate system

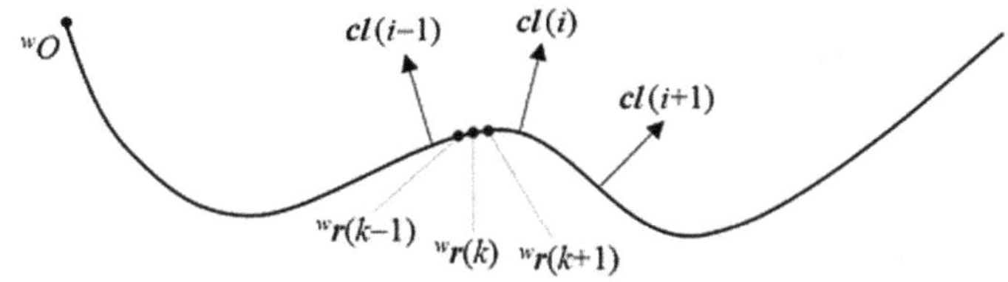

$$\boldsymbol{X}_d(k) = [X_d(k)\, Y_d(k)\, Z_d(k)]^T \text{ [mm]}$$

and desired orientation

$$\boldsymbol{O}_d(k) = [\varphi_d(k)\, \theta_d(k)\, \psi_d(k)]^T \text{ [rad]}$$

in robot absolute coordinate system. $\varphi_d(k)$, $\theta_d(k)$ and $\psi_d(k)$ are the rotational angles around *x*-, *y*- and *z*-axes, respectively, which are called X-Y-Z fixed angles or roll, pitch and yaw angles. The desired position and the desired orientation are set in a UDP packet as the reference of arm tip for the Cartesian servo system. Also, the packet received by 'recvfrom()' includes values of current position

Figure 5. Demonstrates the relation between position component $\mathbf{p}(i)$ in CLS data and the desired position $^{w}\boldsymbol{x}_d(k)$, in which $\mathbf{p}(i) - {}^{w}\boldsymbol{x}_d(k)$, $\mathbf{p}(i+1) - {}^{w}\boldsymbol{x}_d(k+4)$ and $\mathbf{p}(i+2) - {}^{w}\boldsymbol{x}_d(k+8)$ are called the fraction vectors

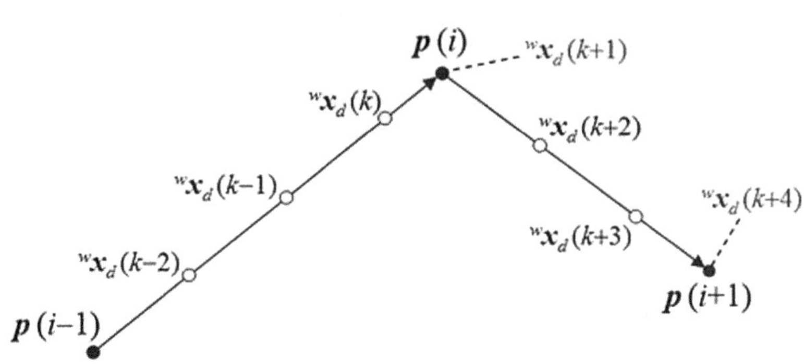

Figure 6. Shows the system communication block diagram using UDP packet between a PC and the industrial robot RV1A

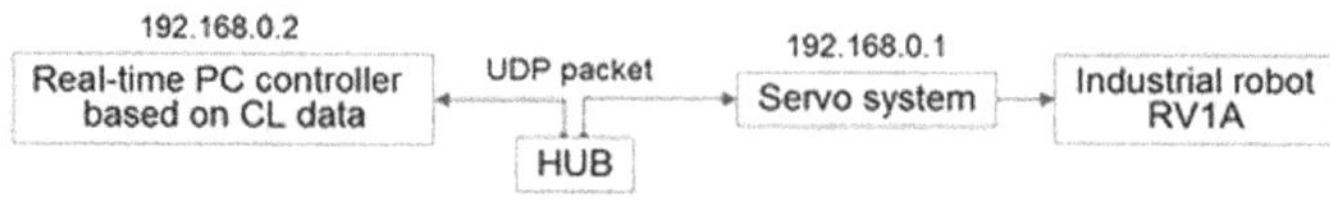

$$\boldsymbol{X}(k) = [X(k)\,Y(k)\,Z(k)]^T \text{ [mm]}$$

and current orientation

$$\boldsymbol{O}(k) = [\varphi(k)\,\theta(k)\,\psi(k)]^T \text{ [rad]}$$

in robot absolute coordinate system, which can be used for feedback quantity.

3.2 Desired Position and Orientation Based on CLS Data

How to make $\boldsymbol{X}_d(k)$ and $\boldsymbol{O}_d(k)$ based on the position and the orientation given by Eq. (1) is discussed in detail by using an example shown in Fig. 4, where the tip of the robot arm follows the trajectory keeping the orientation along normal direction to the surface. Each components of $\boldsymbol{X}_d(k)$ is represented with the initial position

$$\boldsymbol{X}_d(0) = [X_d(0)\; Y_d(0)\; Z_d(0)]^T$$

Figure 7. The communication scheme implemented on the PC using Windows socket functions for UDP, in which sampling period Δt is set to 10 ms

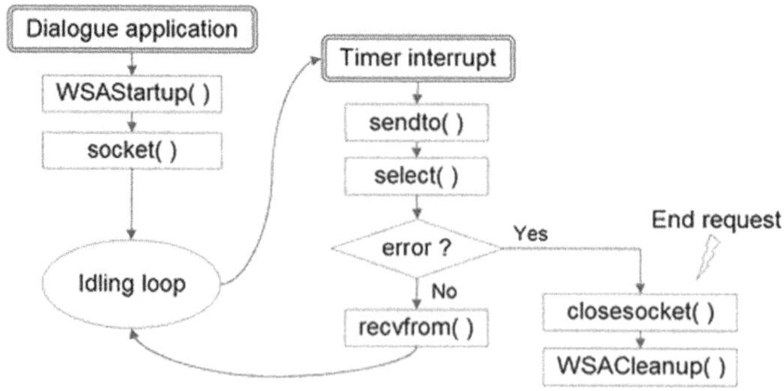

as

$$X_d(k) = X_d(0) + {}^w x_d(k) \tag{11}$$

$$Y_d(k) = Y_d(0) + {}^w y_d(k) \tag{12}$$

$$Z_d(k) = Z_d(0) + {}^w z_d(k) \tag{13}$$

where $\boldsymbol{X}_d(0)$ means the coordinate values of ${}^wO$ in robot absolute coordinate system.

Next, we consider the rotational components. Generally, rotational matrices R_X, R_Y and R_Z around x-, y- and z-axes are respectively given by

$$R_X = \begin{pmatrix} 1 & 0 & 0 \\ 0 & \cos\varphi(k) & -\sin\varphi(k) \\ 0 & \sin\varphi(k) & \cos\varphi(k) \end{pmatrix} \tag{14}$$

$$R_Y = \begin{pmatrix} \cos\theta(k) & 0 & \sin\theta(k) \\ 0 & 1 & 0 \\ -\sin\theta(k) & 0 & \cos\theta(k) \end{pmatrix} \tag{15}$$

$$R_Z = \begin{pmatrix} \cos\psi(k) & -\sin\psi(k) & 0 \\ \sin\psi(k) & \cos\psi(k) & 0 \\ 0 & 0 & 1 \end{pmatrix} \tag{16}$$

Accordingly, the rotational matrix $R_Z R_Y R_X$ with a roll angle $\varphi(k)$, a pitch angle $\theta(k)$, and a yaw angle $\psi(k)$ is given by

$$R_Z R_Y R_X = \begin{pmatrix} C_\theta C_\psi & S_\varphi S_\theta C_\psi - C_\varphi S_\psi & C_\varphi S_\theta C_\psi + S_\varphi S_\psi \\ C_\theta S_\psi & S_\varphi S_\theta S_\psi + C_\varphi C_\psi & C_\varphi S_\theta S_\psi - S_\varphi C_\psi \\ -S_\theta & S_\varphi C_\theta & C_\varphi C_\theta \end{pmatrix} \tag{17}$$

where, for example, S_θ and C_θ means $\sin\theta(k)$ and $\cos\theta(k)$, respectively. Figure 8 illustrates the trajectory following control along CLS data shown in Fig. 4. As can be seen, the direction of ${}^w\boldsymbol{o}_d(k)$ is just the inverse to that of the arm tip. When referring the orientation components in CLS data, the arm tip can be determined only with the roll angle and the pitch angle except for the yaw angle. That means that $\psi(k)$ can always be fixed to 0 [rad]. Thus, the roll angle $\varphi(k)$ and the pitch angle $\theta(k)$ at the discrete time *k* are simply obtained by solving

$$\begin{pmatrix} -{}^w o_{dx}(k) \\ -{}^w o_{dy}(k) \\ -{}^w o_{dz}(k) \end{pmatrix} = \begin{pmatrix} C_\theta & S_\varphi S_\theta & C_\varphi S_\theta \\ 0 & C_\varphi & -S_\varphi \\ -S_\theta & S_\varphi C_\theta & C_\varphi C_\theta \end{pmatrix} \begin{pmatrix} 0 \\ 0 \\ 1 \end{pmatrix} \tag{18}$$

where $[0\ 0\ 1]^T$ is the initial orientation vector in robot absolute coordinate system as shown in Fig. 8. Finally, $\varphi(k)$ and $\theta(k)$ are calculated by using inverse trigonometric functions as

$$\varphi(k) = a\tan 2\{-{}^w o_{dx}(k), -{}^w o_{dz}(k)\} \tag{19}$$

$$\theta(k) = a\sin\{{}^w o_{dy}(k)\} \tag{20}$$

If the desired yaw angle $\psi_d(k)$ is fixed to 0, then the desired roll angle $\varphi_d(k)$ and pitch angle $\theta_d(k)$ can be calculated with Eqs. (19) and (20), respectively.

4. EXPERIMENT

In this section, an experiment of trajectory following control based on CLS data is conducted to evaluate the effectiveness of the proposed CAM system. Figure 9 shows the desired trajectory generated by using the main-processor of 3D CAD/CAM Pro/Engineer, which consists of position and orientation components written by multi–lined "GOTO/" statements. It is tried that the arm tip is controlled so as

Figure 8. shows the orientation control of the arm tip based on CLS data presented in Fig. 4 with the assigned initial orientation $[\varphi(k)\,\theta(k)\,\psi(k)]^T = [0\,0\,0]^T$

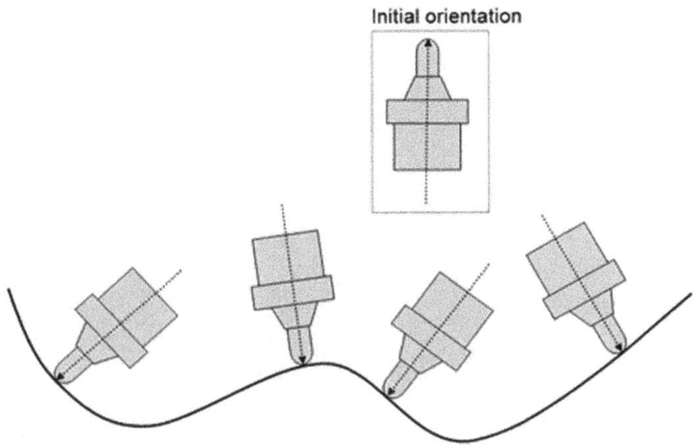

to follow the position and orientation. The value of feed rate, i.e., tangent velocity, is set to 1 mm/s; the desired values composed of $\boldsymbol{X}_d(k)$ and $\boldsymbol{O}_d(k)$ are given to the references of the servo controller in RV1A every sampling period through UDP packets.

When the desired position $\boldsymbol{X}_d(k)$ and orientation $\boldsymbol{O}_d(k)$ is transmitted to the robotic servo system in RV1A, the roll angle $\varphi_d(k)$ has to be given within the range from $-\pi$ to π. Figure 10 shows four examples of roll angle around x-axis, in which (a) is the initial angle represented by $\phi(k) = 0$, (b) is the case of $-\pi \leq \phi(k) < 0$, (c) is the case of $\phi(k) = \pm\,\pi$, and (d) is the case of $0 < \phi(k) \leq \pi$. It should be noted how to calculate the desired roll angle $\varphi_d(k)$ for an manipulated value if the orientation changes as (b)→(c)→(d) or (d)→(c)→(b), i.e., in passing through the situation (c). In such cases, the sign of $\varphi_d(k)$ has to be suddenly changed as shown in Fig. 11, i.e., (c) is the most critical orientation from the viewpoint of roll angle. In order to smoothly control the orientation of the arm tip, the following control rules were applied for the correction of the roll angle (Yoshitake et al., 2012).

$$\text{if } \phi_d(k) > 0 \text{ and } \phi(k) < 0, \text{ then } \Delta\phi(k) = K_p\{\phi_d(k) - \phi(k) - 2\pi\} \quad (21)$$

$$\text{else if } \phi_d(k) < 0 \text{ and } \phi(k) > 0, \text{ then } \Delta\phi(k) = K_p\{\phi_d(k) - \phi(k) + 2\pi\} \quad (22)$$

$$\text{else if } \sin\{\phi_d(k)\} = \sin\{\phi(k)\}, \text{ then } \Delta\phi(k) = K_p\{\phi_d(k) - \phi(k)\} \quad (23)$$

Figure 9. Presents the CLS data $\mathbf{cl}(i) = [\mathbf{p}^T(i)\mathbf{n}^T(i)]^T$ *consisting of position and orientation components that are used to describe the desired trajectory of the tip of robot arm.*

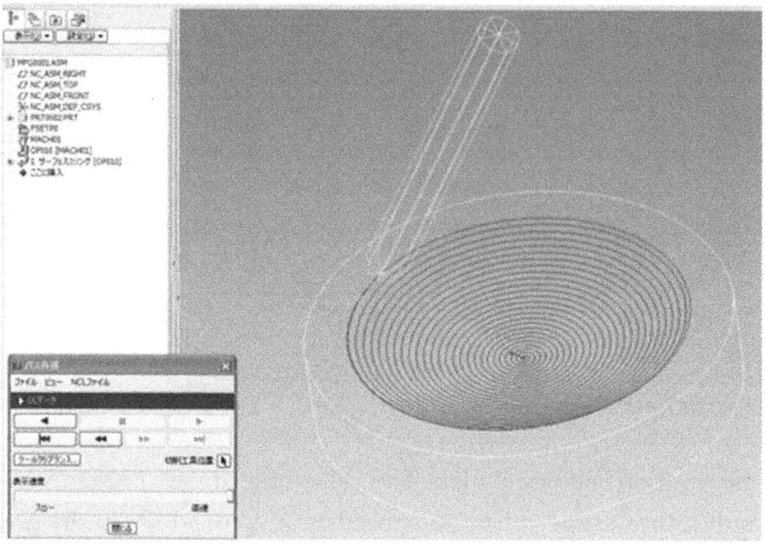

where $\varphi_d(k)$ and $\varphi(k)$ are the desired roll angle transmitted to the robotic servo controller from the PC and actual roll angle transmitted from the robotic servo controller to the PC, respectively. $\Delta\varphi(k)$ is the output of proportional control action with a gain K_p. In fact, the desired roll angle $\varphi_d(k+1)$ can be generated as

$$\phi_d(k+1) = \phi(k) + \Delta\phi(k) \tag{24}$$

It was confirmed from an experiment that desirable control results of position and orientation could be obtained as shown in Fig. 12. The arm tip could gradually move up from the bottom center along the desired spiral path shown in Fig. 9. In this case, the orientation of the arm tip was simultaneously controlled so as to be normal direction to the surface. As an example, position error $X_d(k) - X(k)$ in x-direction is shown in Fig. 13. The values of the errors are under the neighborhood of 0.15 mm that is well-known as the repetitive position accuracy officially announced by the industrial robot's maker. Also, Fig. 14 shows the position error $X_d(k) - X(k+1)$. It is observed that the one sampling period delayed x-directional position $X(k+1)$ could accurately follow the reference $X_d(k)$. The position error

Figure 10. The roll angles around x-axis, in which (a) The initial angle represented by $\phi(k) = 0$, (b) The case of $-\pi \le \phi(k) < 0$, (c) is the case of $\phi(k) = \pm\pi$, and (d) The case of $0 < \phi(k) \le \pi$

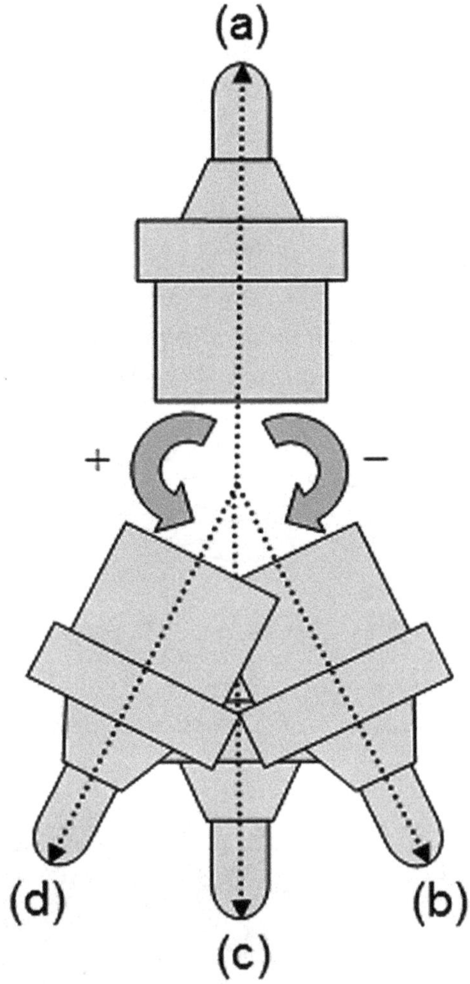

$X_d(k) - X(k)$ shown in Fig. 13 is the typical problem concerning the delay of response when a feedback control law is applied to a dynamic system.

Thus, it was successfully demonstrated that the proposed CAM system allows the tip of the robot arm to desirably follow the desired trajectory given by multi-axis CLS data without any complicated teaching tasks.

Figure 11. Initial part of the calculated desired roll angle $\phi_d(k)$ from the desired trajectory presented in Fig. 9

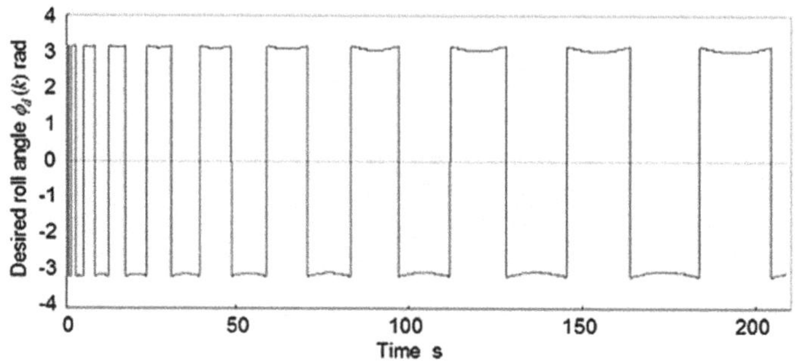

Figure 12. Illustrated the 3D view in robot absolute coordinate system, in which the initial range of the actual controlled results X(k), Y(k) and Z(k) are plotted

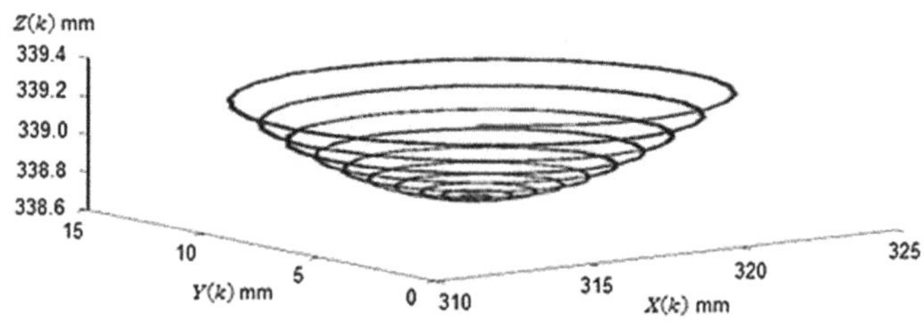

Figure 13. Presents the initial range of the position error $X_d(k) - X(k)$ in x-direction

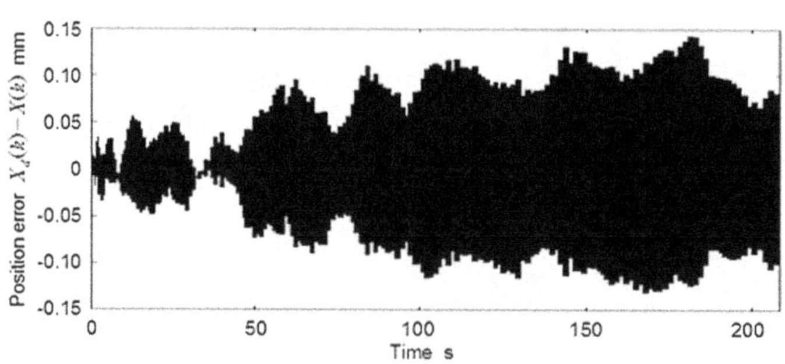

5. ROBOTIC CAM SYSTEM EXTENDED FOR DEALING WITH NC DATA

A robotic CAM system based on CLS data and the implementation to an industrial robot have been introduced up to the previous section. Generally, the main processor

Figure 14. Presents the initial range within one sampling period of the delayed position error $X_d(k) - X(k+1)$ in x-direction

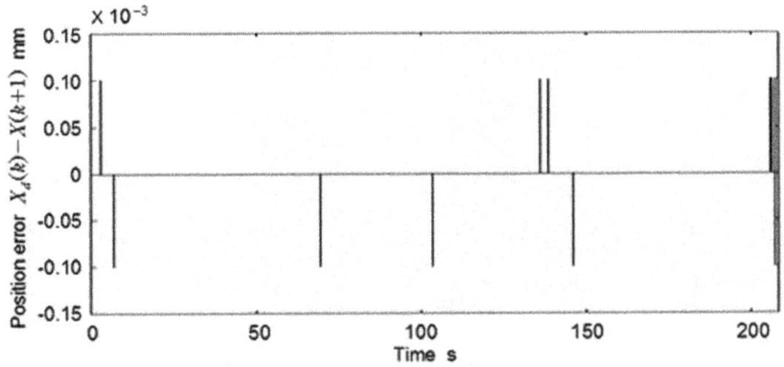

of a CAD/CAM system calculates CLS data, then the post-processor transforms the CLS data to the corresponding NC data according to various types of NC machine tools actually used. In this section, the proposed robotic CAM system is extended to deal with NC data post-processed for a five-axis NC machine tool with a tilting head (Nagata et al., 2009) as shown in Fig. 15. If the NC data can be reversely transformed to the original CLS data, it means that the proposed robotic CAM system available for CLS data allows the industrial robot to be also controlled based on the NC data. We call this transformation (NC data → CLS data) the reverse post-process as shown in Fig. 16.

In the proposed reverse post-processing shown in Fig. 17, the position components have only to be directly extracted. On the other hand, the orientation components have to be dealt with considering the geometric configuration. In the case of this five-axis NC machine tool, the tilt angle $-90 \leq B(i) \leq 90$ and the turn angle $0 \leq C(i) \leq 360$ at the i-th step in NC data can be transformed to the corresponding normal vector

$$\boldsymbol{n}_r(i) = [n_{rx}(i)\, n_{ry}(i)\, n_{rz}(i)]^T$$

in CLS data by calculating

$$n_{rx} = \sin\left\{\frac{B(i)}{180}\pi\right\}\cos\left\{\frac{C(i)}{180}\pi\right\} \tag{25}$$

$$n_{ry} = \sin\left\{\frac{B(i)}{180}\pi\right\}\sin\left\{\frac{C(i)}{180}\pi\right\} \tag{26}$$

Figure 15. Demonstrates a five-axis NC machine tool with a tiling head that is used in the advanced woodworking industry
(SOLIC Co. Ltd., JAPAN)

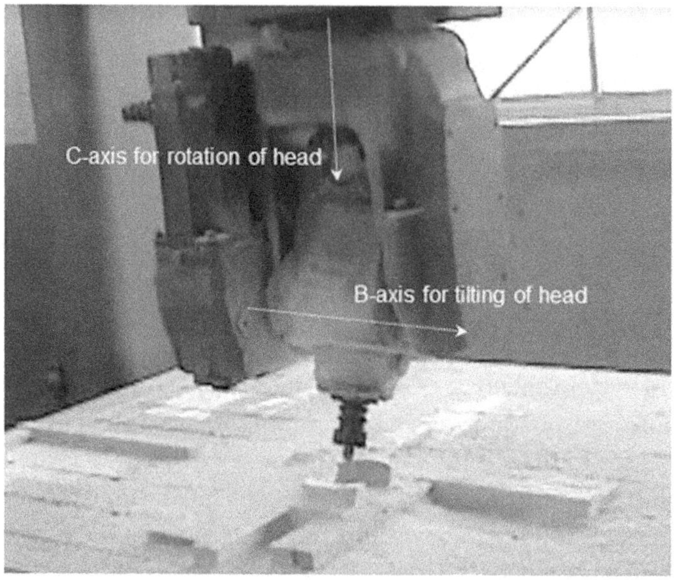

Figure 16. Shows the conversion of the tilting and rotation angles into a normal direction vector as main part of the proposed reverse post-processor

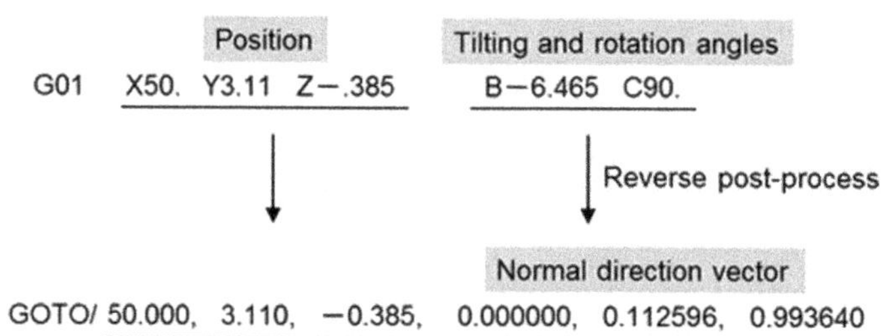

$$n_{r_z} = \cos\left\{\frac{B(i)}{180}\pi\right\} \tag{27}$$

The accuracy of the transformed $\boldsymbol{n}_r(i)$ to the original $\boldsymbol{n}(i)$ was examined by using the case shown in Fig. 17, so that it was confirmed that the errors were relatively small, i.e.,

Figure 17. Illustrates the post-process action from CLS data → NC data, and the proposed reverse post-process action from NC data → CLS data

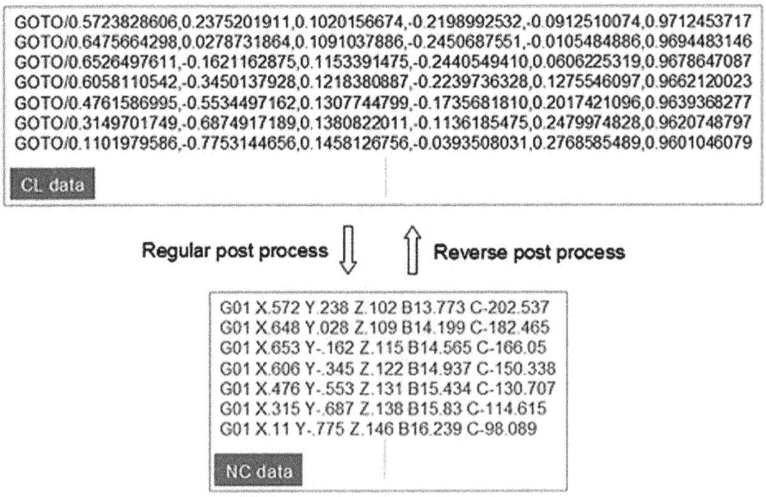

$$\max_i \left| n_{rx}(i) - n_x(i) \right| = 0.0000089,$$

$$\max_i \left| n_{ry}(i) - n_y(i) \right| = 0.0000085$$

and

$$\max_i \left| n_{rz}(i) - n_z(i) \right| = 0.0000055.$$

Here, $\boldsymbol{n}(i)$ is the normal vector included in the original CLS data in Fig. 17, which is generated from the main-processor of 3D CAD/CAM Pro/Engineer. Also, $B(i)$ and $C(i)$ included in NC data in Fig. 17 are the tilt angle and the turn angle, respectively, which are calculated by the post-processor of Pro/Engineer. Further, $\boldsymbol{n}_r(i)$ is the normal vector calculated by using the proposed reverse post-processor given by Eqs. (25), (26) and (27).

Figure 18. Outline font and its spline interpolation. The CLS file consisting of improved spline curve is applied to the machining using the industrial robot VS068

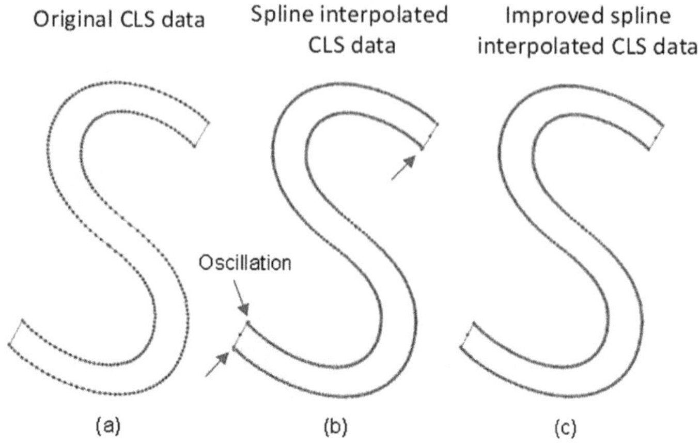

Figure 19. Machining scene of an outline font using an industrial robot VS068 with ORiN middleware, in which the desired trajectory of the small ball-end mill is a CLS file made from the outline font drawn by using Adobe Illustrator

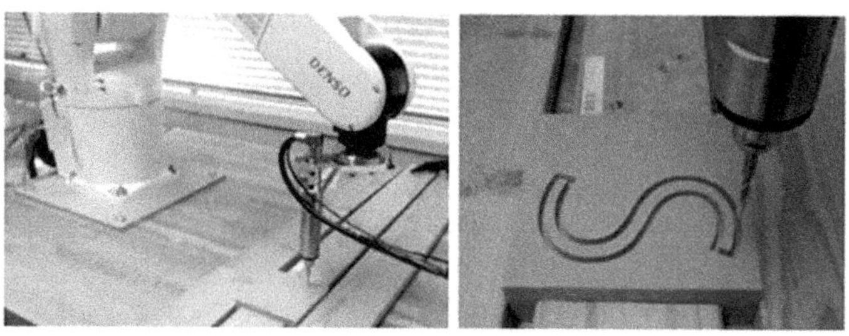

6. IMPLEMENTATION WITH THE INDUSTRIAL ROBOTS VS068 AND FANUC R2000iC

The developed CAM system was integrated and demonstrated with another two industrial robots VS068 and FANUC R2000iC (Nagata et al, 2012, 2016a, 2016b & 2018), in which CLS data are available for desired trajectory of a ball-end mill attached to the arm tip. Figure 18 introduces the font and its spline interpolation, while Fig. 19 shows the machining scene of an outline font using an industrial robot VS068 with ORiN middleware. Figure 20 illustrates the machining scene of a foamed polystyrene using an industrial robot FANUC R2000iC and Fig. 21 presents the complex CLS data for rough machining of an impeller model designed

Figure 20. Machining scene of a foamed polystyrene using an industrial robot FANUC R2000iC (SOLIC Co. Ltd., Japan), in which the desired trajectory of the ball-end mill is a CLS file made by using CAD/CAM Creo

by using CAD/CAM Creo. Figure 22 show the actual machining done using FANUC R2000iC to machine the impeller model.

Figure 21. Complex CLS data for roughly machining an impeller model designed by using CAD/CAM Creo

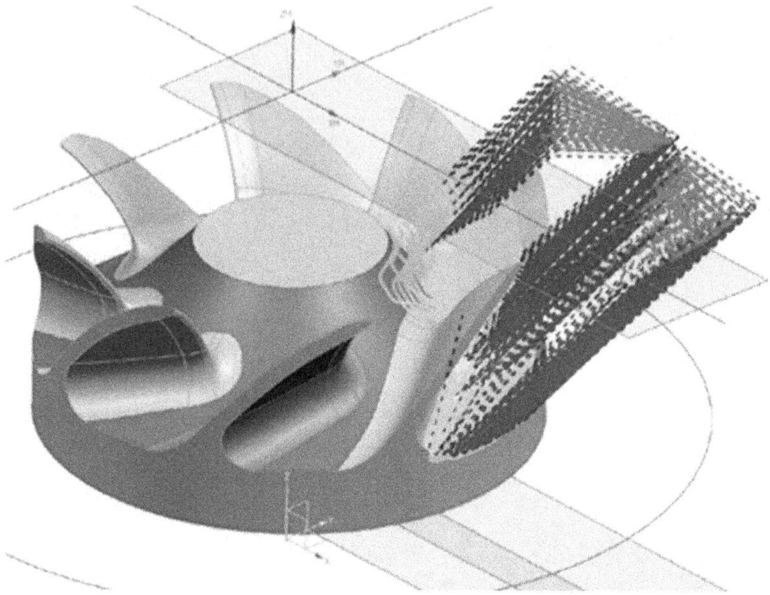

Figure 22. the output of the CAM model in Fig. 21 is applied to machining of a foamed polystyrene using an industrial robot FANUC R2000iC

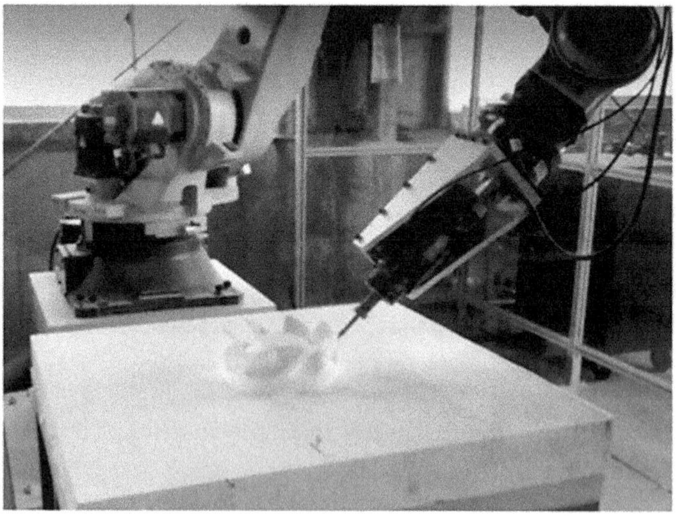

7. CONCLUSION

A robotic CAM system that can generate online motion and support both CLS and NC date for any articulated-type industrial robots supported by an open source controller is presented. The integrated system includes not only conventional main-processor of CAM but also robotic servo system and its kinematics. Further, the reverse post-processor has been developed so that the arm tip of the industrial robot can be flexibly controlled based on both CLS data and NC data. In addition, successful experiments on real industrial robots RV1A, VS068 and FANUC R2000iC were demonstrated and shown in this chapter. This development helps the integration between a commercially available CAD/CAM system and any industrial robot to support real time manufacturing processes. Finally, the presented examples, which have been numerically demonstrated, show the transformation accuracy about the orientation components in CLS data.

REFERENCES

Ahn, C. K., & Lee, M. C. (2000). An Off-Line Automatic Teaching by Vision Information for Robotic Assembly Task. *Proceedings of IEEE International Conference on Industrial Electronics, Control and Instrumentation*, 2171–2176.

Ge, D. F., Takeuchi, Y., & Asakawa, N. (1993). Automation of Polishing Work by an Industrial Robot, – 2nd report, Automatic Generation of Collision- Free Polishing Path –. [in Japanese]. *Transactions of the Japan Society of Mechanical Engineers*, *59*(561), 1574–1580. doi:10.1299/kikaic.59.1574

Habib, M. K. (2007). Mechatronics - A unifying interdisciplinary and intelligent engineering science paradigm. IEEE Indsutrial Electronics Magazine, 2, 2 – 24.

Kamisetty, K. V., & McDermott, K. J. (1992). Development of a CAD/CAM Robotic Translator for Programming the IBM 7535 SCARA Robot off-line. *Computers in Industry*, *20*(2), 219–228. doi:10.1016/0166-3615(92)90056-S

Kushida, D., Nakamura, M., Goto, S., & Kyura, N. (2001). Human Direct Teaching of Industrial Articulated Robot Arms based on Force-Free Control. *Artificial Life and Robotics*, *5*(1), 26–32. doi:10.1007/BF02481317

Maeda, Y., Ishido, N., Kikuchi, H., & Arai, T. (2002). Teaching of Grasp/Graspless Manipulation for Industrial Robots by Human Demonstration. *Proceedings IEEE/RSJ International Conference on Intelligent Robots and Systems*, 1523–1528. 10.1109/IRDS.2002.1043971

Nagata, F., Hayashi, S., Nagatmi, T., Otsuka, A., & Watanabe, K. (2016). Application of Fuzzy Reasoning and Neural Network to Feed Rate Control of a Machining Robot. *International Journal of Applied Electromagnetics and Mechanics*, *52*(3-4), 897–905. doi:10.3233/JAE-162217

Nagata, F., Kusumoto, Y., & Watanabe, K. (2009). Intelligent Machining System for the Artistic Design of Wooden Paint Rollers. *Robotics and Computer-integrated Manufacturing*, *25*(3), 680–688. doi:10.1016/j.rcim.2008.05.001

Nagata, F., Watanabe, K., & Izumi, K. (2001). Furniture Polishing Robot using a Trajectory Generator based on Cutter Location Data. *Proceedings of 2001 IEEE International Conference on Robotics and Automation*, 319–324. 10.1109/ROBOT.2001.1620978

Nagata, F., Watanabe, K., & Kiguchi, K. (2006). *Joystick Teaching System for Industrial Robots using Fuzzy Compliance Control. In Industrial Robotics: Theory, Modelling and Control* (pp. 799–812). INTECH.

Nagata, F., Yamane, Y., Okada, Y., Kusano, T., Watanabe, K. & Habib, M. K. (2018). Development of Post Processor Approach for an Industrial Robot FANUC R2000iC. *Journal of Artificial Life and Robotics*, *23*(2), 186-191.

Neto, P., Pires, J. N., & Moreira, A. P. (2010). CAD based Off-Line Robot Programming. *Proceedings of IEEE International Conference on Robotics Automation and Mechatronics*, 516–521.

Sugita, S., Itaya, T., & Takeuchi, Y. (2003). Development of Robot Teaching Support Devices to Automate Deburring and Finishing Works in Casting. *International Journal of Advanced Manufacturing Technology*, *23*(3/4), 183–189.

Sugitani, Y., Kanjo, Y., & Murayama, M. (1996). Systemization with CAD/CAM Welding Robots for Bridge Fabrication. *Proceedings of 4th International Workshop on Advanced Motion Control*, 80–85. 10.1109/AMC.1996.509384

Yoshitake, S., Nagata, F., Otsuka, A., Watanabe, K., & Habib, M. K. (2012). Proposal and Implementation of CAM System for Industrial Robot RV1A. *Proceedings of the 17th International Symposium on Artificial Life and Robotics*, 158–161.

KEY TERMS AND DEFINITIONS

CAD/CAM: Computer-aided design and computer-aided manufacturing. Here, CAD/CAM is used to make a desired trajectory of the robot arm without a robotic teaching process.

CLS Data: Cutter location source data are generated from the main-processor of CAM, which are written with multi-lined "GOTO" statements including position and orientation vectors.

Industrial Robot: Articulated-type industrial robot RV1A with six degree-of-freedoms. The servo system and communication interface are technically opened to users.

NC Data: Numerical control data according to the type of machine tool are converted from CLS data. The tool for this process is called the post-processor.

Reverse Post-Processor: Reverse post-processor is the online conversion tool from NC data to CLS data in the proposed robotic CAM system.

Robot Language: An industrial robot has an original robot language provided by the robot maker to describe a program for a robotic application. The drawback is that it is not well standardized among various types of industrial robots.

Robotic Teaching: Robotic teaching is the well-known process to make a desired trajectory by using a teaching pendant. The robotic teaching is a time-consuming task.

Servo Controller: Six servo motors are built in the corresponding six joint of the industrial robot RV1A. The joint angles can be controlled to follow reference values by the servo controller.

Chapter 2
Industrial Exoskeletons With Gravity Compensation Elements

Sergey Fedorovich Jatsun
South-West State University, Russia

Andrey Yatsun
South-West State University, Russia

ABSTRACT

The chapter approaches the issues of modeling the process of load lifting by a person while wearing an exoskeleton. The classification of existing gravitational compensation systems for industrial exoskeletons is shown, as well as examples of its use. A mathematical model of lifting a person's load in the exoskeleton is presented, as well as numerical parameters are calculated. It is shown that the introduction of an elastic element reduces the level of energy consumption during work, and can also facilitate the level of the worker. Industrial exoskeleton prototype design is presented. A particular focus is given to studying the influence of the gravity compensator on the magnitude of the moments generated by the electric drives of the hip and knee joints. It is shown that the use of gravity compensators enables to reduce significantly the load on electric drives.

DOI: 10.4018/978-1-7998-1382-8.ch002

1. INTRODUCTION

A person's manual labor can be considered as a workload performed by a person in this type of operation, and the functional stress of the body as an integral response of the body to the load. The workload is a combination of factors of the workflow, performed under certain conditions of the production environment (Pons, 2008; Anam & Al-Jumaily, 2012; RViteckova et al., 2013; Arisumi et al., 2008). The labor severity is understood as a degree of cumulative impact of working conditions on the functional state of the human body, his health and performance, on the process of reproduction of labor power and safety. The severity of work is determined by the degree of stress on the human muscular system. Since a person experiences particular stress during loading and unloading (lifting) operations, so the working conditions can be attributed to extreme ones, which lead to a decrease in a person's working capacity, causing functional changes that exceed the limits of the norm, but do not produce pathological changes. In some cases, there are super-extreme working conditions, causing pathological changes in a human body that makes impossible to complete the workload.

It is necessary to create comfortable working conditions, ensuring optimal working efficiency and maintaining a person's health, or at least such conditions, which provide a given performance and maintain the health, and at the same time do not cause inner discomfort and abnormal functional changes. One of the ways to solve this problem is the use of an exoskeleton - external skeleton that supports and protects a body. Particularly effective are industrial exoskeletons allowing to carry out complex movements of both lower and upper extremities (Gou et al., 2014; Kajita et al., 2001), so they are able to expand significantly human capabilities, including loading and unloading operations.

Currently, industrial exoskeletons Fortis (by Lockheed Martin), HAL (Cyberdyne), Atoun (Panasonic) and others are becoming more and more common in the market. They find practical application (Sellaouti & Stasse, et al., 2006), significantly enhancing human capabilities in terms of facilitating movement, weight transfer and various types of activities that require considerable effort, as is shown in Figure 1. Particularly effective are the exoskeletons that allow performing complex types of movement of both lower and upper extremities, which significantly expands the human capabilities when performing cargo handling operations.

The carbon fiber frame gives Atoun Model Y exoskeleton longevity and lightness, it can be worn outdoors and in rainy weather due to its waterproof and dustproof.

Recently, the development of exoskeletons with passive elements of gravity compensation has been embarked on. The term «gravity compensation» is used to designate the properties of individual links of the mechanism to have a statically stable position, independent of the vertical traverse of the links. For an external

Figure 1. Industrial active exoskeletons Atoun model Y and Cyberdyne HAL-LB03

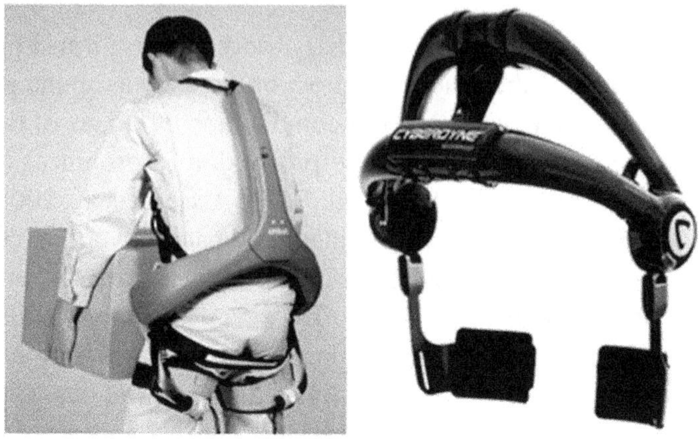

observer, the movement of these links of the mechanism looks as if the gravitational field did not act on them. Issues related to the design of mechanisms similar to those presented in this section, especially the optimal choice of parameters and the location of elastic elements, are the subject of research conducted by a number of scientists (Bosch et al., 2016; Mooney & Luke et al., 2014; Anam & A-Jumaily, 2012; Jatsun & Savin et al., 2016).

The referred works note the advantages of gravity compensation systems based on the system of elastic elements in comparison with passive balancing systems and mechanisms that use friction in joints to achieve static equilibrium. Thus, in the paper Yu et al. (2014), it is mentioned that the most of the existing mechanisms of gravity compensation use the introduction of additional links in the original mechanism. In articles Elliot et al., (2013); Shamaei et al., (2014); O'Sullivan et al., (2015) it is indicated that "perfect balancing" («perfect equilibration» in the original text of the article) – the property of the mechanism to fully compensate for the effect of gravity - can be obtained for multi-link mechanisms of a certain configuration, including four-link mechanism. At the same time, the development of methods for the optimal parametric synthesis of multi-link mechanisms with a system of elastic elements that implement gravity compensation remains an urgent scientific task.

Industrial SuitX exoskeleton to reduce the load on the lower back while holding heavy objects. LegX reduces leg load by supporting the knees and quadriceps (fig. 2).

A similar mechanism of gravity compensation is used in the exoskeleton «FORTIS Exoskeleton», developed by Lockheed Martin. In this device, gravity compensation is implemented in order to simplify operations with a heavy tool for the exoskeleton operator. Figure 2 shows a photograph of a person while wearing an exoskeleton.

Figure 2. A worker in a an FORTIS passive exoskeletons

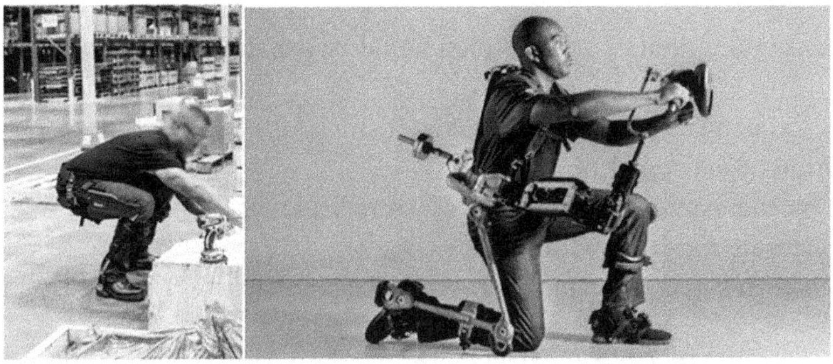

A «Reduced Gravity Exoskeleton» device, developed by the University of New Mexico (New Mexico State University), could serve an example. The device is stationary and uses a multi-link lever mechanism and a spring system to compensate for the effect of the gravitational field. Figure 3 shows the diagram of the device and a photo of the layout with a mobile robot attached to it and demonstrates separate units of the device.

In such exoskeletons, one or several of its links have statically stable positions independent of their vertical traverse (vertical traverse means hereinafter the movement along the lines of the gravitational field). In this case, it is implied that the mechanism does not require the operation of the drives to maintain this property.

Figure 3. Diagram of a device for rehabilitation of a person "Reduced Gravity Exoskeleton" and a photograph of the breadboard device

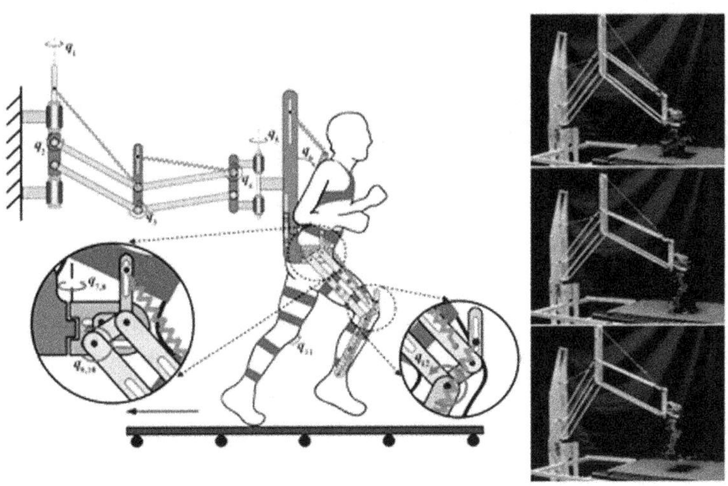

2. CLASSIFICATION OF GRAVITY COMPENSATION DEVICES

Mechanisms with gravity compensation could be separated into three main types:

- The mechanism of the exoskeleton with additional masses to change the position of the center of mass.
- The mechanism of the exoskeleton with elastic elements to create forces that compensate for gravity force.
- Mechanisms using friction in swivels and actuators.

Mechanisms related to the first type (Kajita et al., 2003) are designed in such a way that the force of gravity acting on the link for which gravity compensation is carried out passes through its point of attachment, for example, through a hinged connection with a frame. Such mechanisms are characterized by a number of advantages associated with a high operation reliability. The disadvantage of such mechanisms is an increase of the weight of the links, which leads to the increase of their inertia and has a negative impact on the work of the exoskeleton Anam, Al-Jumaily, 2012).

The mechanisms of the second type are the most promising. Such equipment is characterized by a number of advantages associated with achieving the result with a minimal increase in the weight of the design of the exoskeleton. The mechanisms belonging to the third category are also widely used in robotics. Here the static equilibrium of the position of the mechanism is achieved through the use of a controlled friction force in the joints and actuators. A disadvantage of the devices of this type is that, during the operation of the mechanism, the frictional properties of individual friction pairs may change, which leads to the loss of the properties of the gravity compensation mechanism.

3. DESCRIPTION OF AN INDUSTRIAL EXOSKELETON EXOHEAVER

Exoskeletons with the elements of gravity compensation open up the possibility of significantly expanding the capabilities of a person in performing tasks that a human cannot perform in a usual state; first of all, this concerns operations related to lifting and transporting loads. A general scheme of a person while wearing the exoskeleton is shown in Figure 4, where the following designations of the main modules are accepted:

1) A module of shoulder protection; 2) A soft back support; 3) A module of load lifting; 4) Thoracic fixings; 5) A back frame; 6) Lumbar fastenings; 7) Femoral hinge;

Figure 4. General scheme of a person with load wearing the exoskeleton

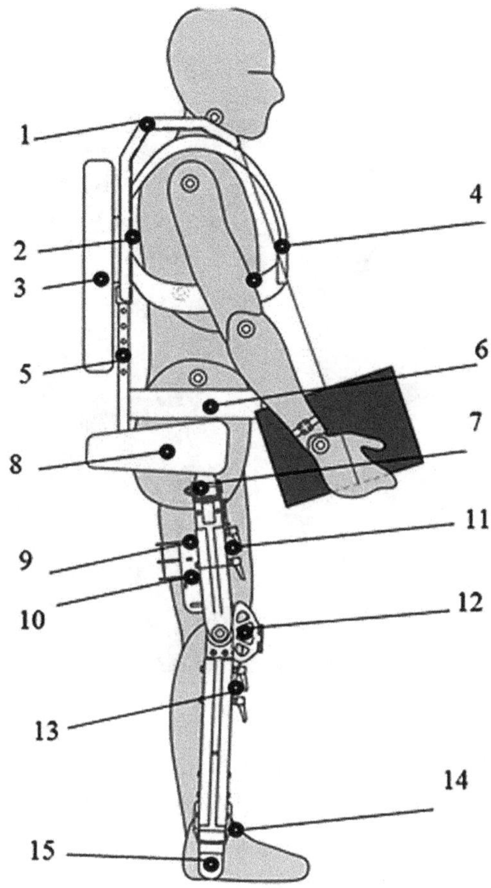

8) Drive of the femoral module; 9) Seat supports; 10) Femoral cuffs; 11) Clips to adjust the length of the thigh; 12) Protection of the knees; 13) Clips to adjust the length of the lower leg; 14) Clip for attaching the foot; 15) Elastic supports. The exoskeleton allows to reduce of the load on the legs in a vertical position through the use of the base module of the exoskeleton of the lower extremities, as well as to relieve the lumbar region, upper shoulder girdle and arms when using modules of attachments.

Based on this scheme let us describe the passive industrial exoskeleton ExoHeaver, which was designed in Mechatronics Laboratory of South-West State University (Kursk, Russia). Structurally, the exoskeleton repeats a human skeleton. The metal links of the apparatus are connected by means of hinges, each having a different number of the degrees of mobility in Figure 5. The hinge attachment of the thigh with the body, like that of a human, has three degrees of freedom. It can rotate

Figure 5. ExoHeaver exoskeleton (3D-model and photo)

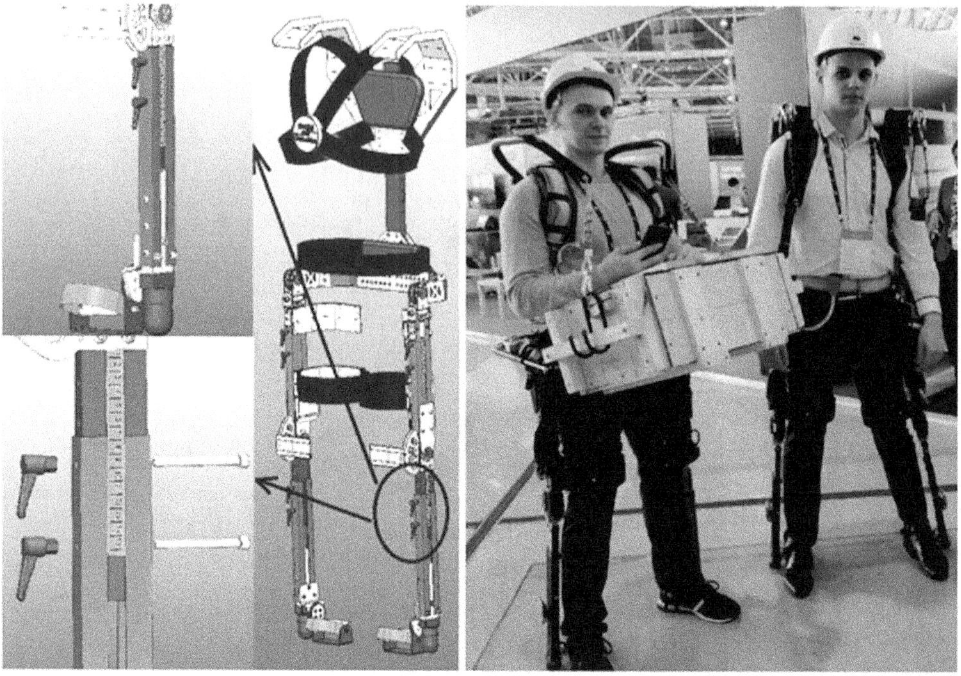

around two horizontal axes: "back and forth" and "lead - cast", and also around the vertical axis. The shin is connected to the thigh by a cylindrical hinge with a single axis of rotation. The foot is equipped with a flexible support that takes up the entire load, while reducing the risk of a person's getting ankle injuries. The frame of the exoskeleton is made of high quality materials, primarily of aluminum alloys and steel. The hinges are made with the use of modern technologies and high-quality bearings.

The main links of the exoskeleton are equipped with a number of adjustments necessary to ensure comfortable operation and the ability to easily adjust it in accordance with user parameters. All the links of the exoskeleton are adjusted in the required range, allowing people with different anthropometric parameters to use the exoskeleton.

The power frame implements a modular concept founded on the basic module of the exoskeleton, which features an exoskeleton of the lower extremities, equipped with a special attachment mounting system, such as the module "stops for sitting", the module "shoulder protection ", load lifting module, the module" third hand "and much more.

The attachment module "shoulder protection" allows holding and carrying loads, placing them on a metal frame to protect the shoulders. Thus, the spine is stress-relieved, since the load is transmitted through the frame directly to the supporting surface. The operator only needs to keep balance. Also, additional attachments, such as those described in this report, or third-party products, such as additional lighting, a radio station, etc., can be installed on the shoulder protection sections.

4. CALCULATION OF THE PHYSICAL DYNAMIC LOAD

Let us consider that physical dynamic load (PDL) is measured as external mechanical work per shift (kg-m). So, to determine the physical dynamic load it is necessary to find the mass of the load moved manually in each operation, and the path of its movement in meters. The total number of operations for the transfer of the load per shift is calculated, and the amount of external mechanical work (kg-m) per shift as a whole is summed up. According to the amount of external mechanical work per shift, depending on the type of load (total or local) and the distance of movement of the load, it is determined what class of working conditions this work belongs to. If the distances of movement of the load are different, then the total mechanical work is compared with the average distance of movement.

Dynamic work is a process of periodic contraction and relaxation of skeletal muscles, leading to the load shift, as well as the body of a person or its parts in space. Physical dynamic load is divided into local (with the participation of the muscles of the arms and shoulder girdle) and total (with the participation of the muscles of the arms, body and legs). To calculate the physical dynamic load, the mass of the load moved manually in each operation and the path of its movement in meters are determined. The total number of operations for the transfer of the load per shift is calculated and the value of external mechanical work per shift as a whole is summed up. The amount of external mechanical work per shift, depending on the type of load (local or total) and the distance of movement of the load, determine which class of working conditions the specified work belongs to.

Let us consider an example. A worker moves a spare part with weight of the load (p) 15 kg from the conveyor performs the necessary scope of operations and takes the part back to the conveyor with the load travel path (l) - 5m. The total number of spare parts processed per shift (shift target) (n) is 200 pieces;

The total distance of the load shift per turn is determined by taking into account the dual movement of the load is $L = x l n = 2000$ m. The external mechanical work expended by the worker on the entire path of shifting parts is $A = pL = 30000$ kg-m

Now we define the external mechanical work of the worker while wearing the exoskeleton. In this case, the actual weight of the load (p) - 15 kg is perceived as reduced mass (p) - 3 kg; The external mechanical work expended by the worker on the entire path of shifting parts is determined by A= pL =6000 kg-m (with the exoskeleton).

Finally, without the exoskeleton, the work is carried out under the total load, the distance of the load shift is up to 5 m, therefore, under the total load (with the participation of the muscles of the arms, body, legs): refers to the section harmful work regime(physically demanding job). With exoskeleton the work is refers to the optimal regime (light physical activity).

5. MATHEMATICAL MODEL OF LOAD LIFTING IN INDUSTRIAL EXOSKELETON

In the process of lifting loads from the point of view of a human physiology, the load rests on the spine and the musculoskeletal system of the person. The greatest danger is the load transmitted to the spine. If it is positioned in space correctly, then it takes the load quite properly, the vertebrae and the disks will not be affected. If the back is bent in irrelevant places, or has some distortions to the side, then the disks begin to receive a load in the wrong direction, and there is a risk of their deformation, which can lead to vertebral hernia. A similar situation is observed if the load is excessive. Experimental studies have shown that it is the lumbar spine that has the greatest loading moment. Therefore, it is well known that heavy things need to be raised with feet thus unloading the spine at most. That is, you need to lift so that most of the weight of the lifted object falls on the muscles of the legs, thus removing the excessive load from the back and lumbar. If you lift heavy object (for example, weighing 20-50 kg) incorrectly, you can get injured. The legs should be shoulder-width apart, one leg slightly ahead of the other. It is necessary to squat in front of the load, grab the object with your hands, keeping your back straight, and rise (see Figure 6). This technique enables to reduce the load on the spine.

The use of an exoskeleton allows one to decrease the load on the spine and simplify the load lifting process significantly.

Let us think of a simplified computational model of a human while wearing the exoskeleton, when lifting a load, in the form of a multi-link system of solid bodies connected by hinges in Figure 7. It is assumed that the lower limbs of the human and the links of the exoskeleton move in the same way and represent 7 solid bodies interconnected by cylindrical hinges equipped with electric drives. The arms are in a vertical position and hold the load with the mass m_{80}.

At this stage, we consider the gravitational mechanism of the second type. Imagine the computational model of a human while wearing the exoskeleton in the form of a multi-link system of solid bodies connected by hinges. It is assumed that the lower limbs of a person and the links of the xoskeleton move in the same way

Figure 6. Scheme of load lifting performed by a man

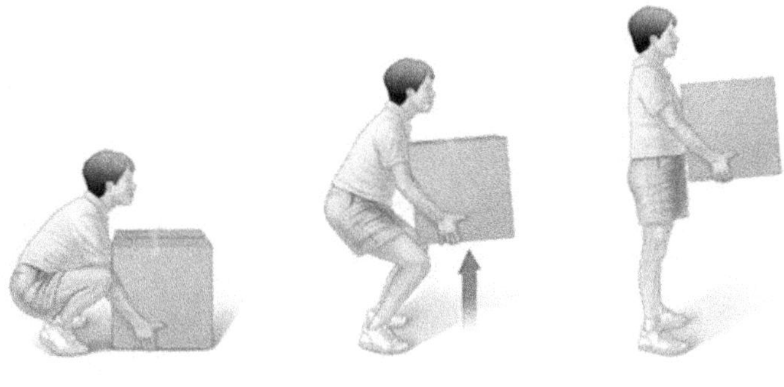

and represent seven solid bodies interconnected by cylindrical hinges equipped with electric drives. The arms are positioned vertically and hold the load with the mass m_8. The person's feet do not detach from the supporting surface.

The movement of the load occurs due to the work of the electric drives located in the zone of the ankle, knee and hip joints. Since, during the lifting of the load, the movement of the exoskeleton occurs in the sagittal plane, the position of the links of the exoskeleton in space can be determined using 6 independent parameters. Figure 8 shows the simulated position of a person in the exoskeleton at different points in time.

The vector of generalized coordinates in the general case has the form:

Figure 7. Diagram of the load separation from the surface

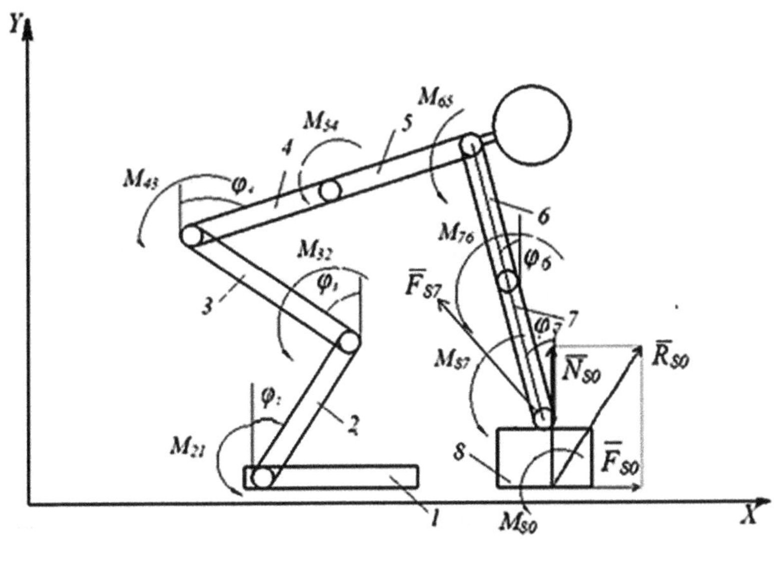

Figure 8. Calculated position of a person in an exoskeleton at various points in time

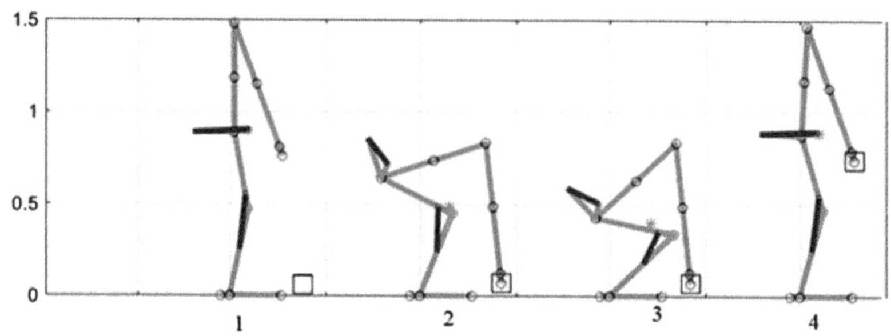

$$OH = \frac{2 \cdot \sqrt{P \cdot (P - L_{AB}) \cdot (P - L_A) \cdot (P - L_B))}}{L_{AB}}$$

It can be assumed that when lifting the load, the back remains straight, which corresponds to the condition $\overline{q} = (\phi_2, \phi_3, \phi_4, \phi_5, \phi_6, \phi_7)^T$, and the arms remain straight, that is $\phi_4 = \phi_5$. Thus, the four angles set the position of the exoskeleton during lifting

$$\phi_6 = \phi_7. \tag{1}$$

The diagram represents: $\overline{q} = (\phi_2, \phi_3, \phi_4, \phi_6)^T$ - the support surface reactions, N_{80}, F_{80}, M_{80} - the force, acting on the load from the hand side, $\overline{F}_{87}$ - the torques created by the electric drives of the hinges.

$$M_{i,i\text{-}1} \tag{2}$$

Lifting of the load will be divided into four stages (see figure 8). The first is the bent of a person to take a load; the second is the handgrip of the load, the third is the separation of the load from the surface; the fourth is lifting the load. The forces and torques acting on the load will be represented in the form of two vectors.

$$\overline{R}_{80} = \overline{N}_{80} + \overline{F}_{80} \tag{3}$$

$$\overline{\Phi}_{80} = (R_{80}, \overline{M}_{80})^T \tag{4}$$

These vectors are equal to each other $\bar{\Phi}_{87} = (F_{87}, \bar{M}_{87})^T$.

Let the vector $\bar{\Phi}_{80} = \bar{\Phi}_{87}$ determine the position of the load, both when on the surface, and in the event of separation. The torques acting in the corresponding hinges form the vector

$$\bar{S}_{80} = (x_{80} y_{80} \phi_{80})^T \tag{5}$$

This vector can be determined by the formula:

$$\bar{M} = \begin{pmatrix} M_{21} & M_{32} & M_{43} & M_{65} \end{pmatrix}^T \tag{6}$$

where

$$\bar{M}^T = J(\bar{q}) \bar{\Phi}_{80}{}^T. \tag{7}$$

The dependences of the absolute angles of rotation of the links (back, hip, shin) obtained by solving the inverse problem of kinematics are shown in Figure 9. The

Figure 9. Dependence of the absolute angles that determine the position of the exoskeleton links in time

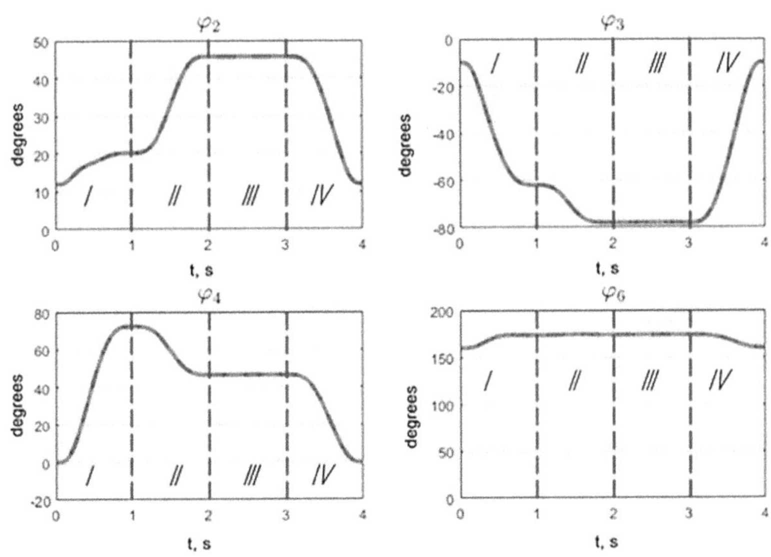

dependence of the torques on the time is shown in Figure 10. The dependence of the normal response on the time is shown in Fig. 11.

Analyzing the obtained dependencies, we conclude that the third and fourth stages associated with the separation and lifting of the load are the most load stressed. The magnitude of the torques of the ankle joint reaches $J(\overline{q}) = \dfrac{\partial \overline{S}_{80}(\overline{q})}{\partial \overline{q}}$ =250 and 300 Nm. For the knee joint, the maximum torque values equal to M_{21} =170 Nm, are observed at the end of the second stage. The load on the lumbar spine is assessed by the torque M_{32} and reach the maximum values in the second stage $M_{43}\,\mathscr{M}$, and on the third $M_{43} = 225H\ \mathscr{M}$. The moment of force in the shoulder joint reaches its maximum in the fourth stage and is estimated at $M_{43} = 275H\ \mathscr{M}$. Special attention while modeling was given to the position of the center of mass of the system exoskeleton—human - process load. Fig. 13 shows the dependence of the coordinates that determine the position of the load on the time. In this case, the projections of the center of mass are located inside the bearing surface, which is a condition ensuring the stable position of the system under consideration.

Figure 10. Dependence of torques (in Nm) on the time

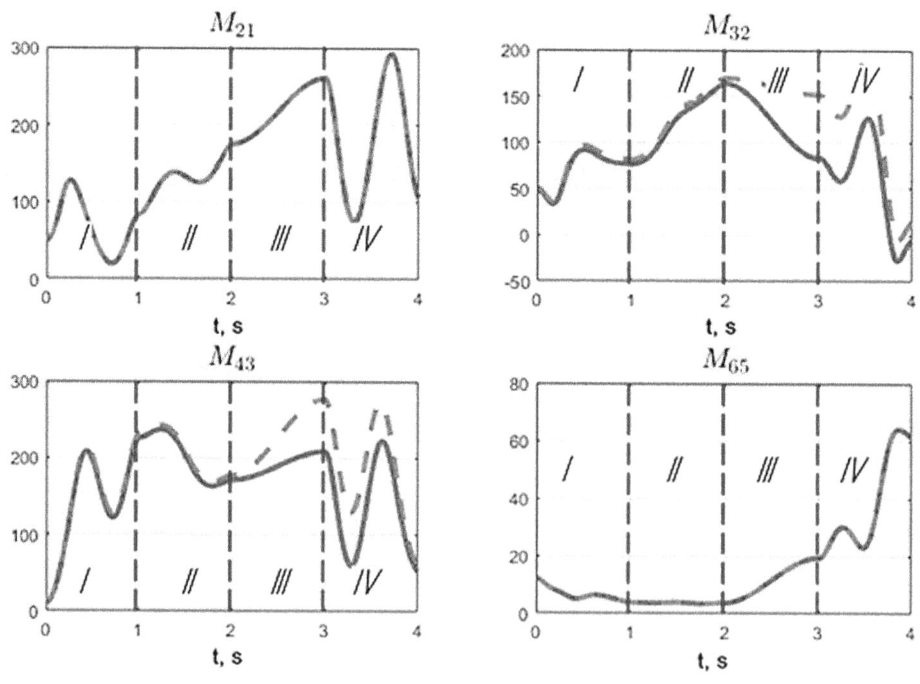

Figure 11. Dependence of the normal response force (in N) on the time

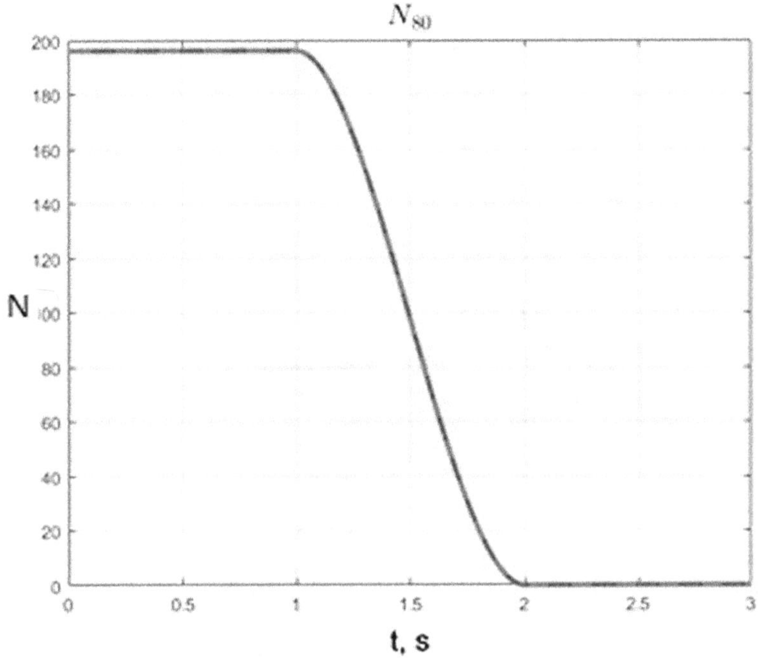

6. DIAGRAM OF THE GRAVITY COMPENSATOR

Figure 12 shows a diagram of the gravity compensator of the hip joint of the exoskeleton, which consists of an elastic element with the rigidity $M_{65} = 65H$, pivotally mounted on the hip and lower leg of the exoskeleton. It also presents the computational scheme for determining the shoulder of a linear drive on the exoskeleton, depending on the angle of the back inclination at any time period.

Let us perform following formula:

c_{32}

$$L_{AB} = \sqrt{L^2_A + L^2_B - 2 \cdot L_A \cdot L_B \cdot \cos(\beta)}$$

$$\beta = 180 - \left(\left|\phi_4\right| + \left|\phi_3\right|\right)$$

Figure 12. Computational scheme for determining the gravity compensation

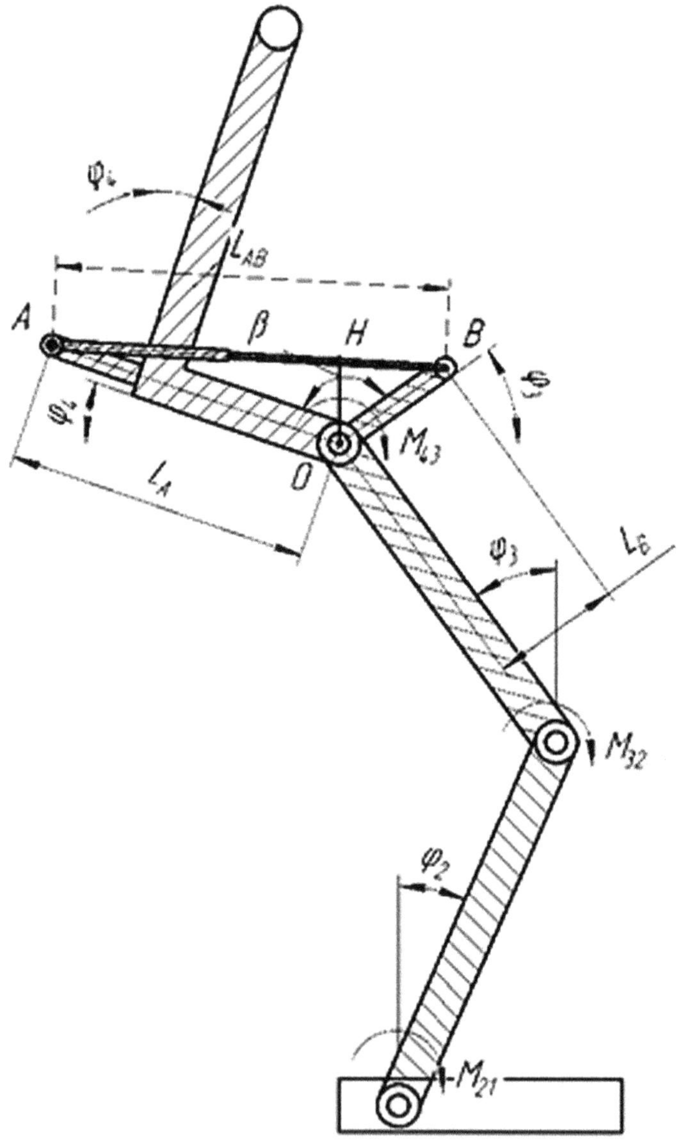

$$P = \frac{1}{2} \cdot (L_{AB} + L_A + L_B), \tag{8}$$

The dependence of the force of elasticity that occurs when the shank is rotated relative to the thigh. When a person sits, a change in the angles

$OH = \frac{2 \cdot \sqrt{P \cdot (P = L_{AB}) \cdot (P - L_A) \cdot (P - L_B)}}{L_{AB}}$ occurs and, accordingly, the relation to the angle ϕ_2, ϕ_3 this changes the length of the elastic element $\phi_3 - \phi_2 = \phi_{32}$ as a result, the potential energy

ΔL_{32}

accumulates in the elastic element. When lifting a load, this energy allows for the creation of the additional torque $\Pi = \frac{1}{2} c_{32} (\Delta L_{32})^2$, which facilitates lifting the load, since

$$M_{32}^{com}. \tag{9}$$

7. ANALYSIS OF ENERGY CONSUMPTION IN THE PROCESS OF LOAD LIFTING

The movement of a person in an exoskeleton implies, in general, energy consumption, both from the human and from the electric drives of the exoskeleton. Energy is expended on the performance of work by the forces of gravity and resistance. For example, when a person performs some task, such as lifting a load using an exoskeleton from a sitting position for the time *T*, it is necessary to perform an operation on changing the vertical coordinates of the center of mass of the human-exoskeleton-process load system. This work can be determined by the formula:

$$\Pi = -A_{32} = -\int M_{32}^{com} \dot{\phi}_{32} dt \tag{10}$$

where *A*- denotes the work on changing the position of the vertical coordinate of the center of mass of the system human-exoskeleton-process load.; *m*- stands for the weight of the system; *g*- the gravitational acceleration; V_{Cz} - the vertical velocity of the center of mass of the system; *t* – the current time point; *T*- the process execution time.

The work of the resistance forces in the hinges of the exoskeleton can be determined by the following formulas:

$$A = \int_0^T mgV_{Cz}dt \tag{11}$$

where M_{Ci}, $A_c = \sum_{i=1}^{4} M_{Ci} \cdot \dot{\phi}_i dt$ - indicate the moment of resistance and the angular velocity in the $i - th$ hinge joint.

The moment of resistance in the drive M_{Ci} depends on the magnitude and the type of friction that occurs in friction pairs, including bearings, gears and other drive elements.

For the linear viscous friction, the following formula is applied:

$$\dot{\phi}_i , \tag{12}$$

and for the dry friction:

$$M_{Ci} = \mu_i \dot{\phi}_i , \tag{13}$$

where $M_{Ci} = M_{0i} sign\dot{\phi}_i$ - indicates the coefficient of viscous friction, μ_i - the torque of the dry friction.

The work done by the electric drives is determined by the formula:

$$M_{0i} \tag{14}$$

where M_{ei}, $A_e = \sum_{i=1}^{4} M_{ei} \cdot \dot{\phi}_i dt = \eta_2 \sum_{i=1}^{4} \int_0^T uidt$ - the torque of the electric drive and the angular velocity in the $i - th$ hinge joint, u, i – denote the voltage and the current in the electric drives. $\dot{\phi}_i$ - stands for the efficiency of the exoskeleton drives. In a general case, to perform work A to move the person and the process load while wearing the exoskeleton one will need to expend the energy of human muscles E_1 and the energy of the exoskeleton drives E_2.

Then work A on lifting the load can be determined by the formula:

$$\eta_2$$

where A_1 – indicates human muscle work; A_2 – the work of the exoskeleton drives.

Dividing Eq. 14 by A we obtain the equation that determines the proportion of the energy expended by the muscles $A = A_1 + A_2$ and the exoskeleton drives ξ_1.

$$\xi_2 \tag{15}$$

where $1 = \xi_1 + \xi_2$

The parameter $\xi_1 = \frac{A_1}{A}; \xi_2 = \frac{A_2}{A}$ can be determined by measuring the voltage and the current consumed by the electric drives.

$$\xi_2$$

The parameter of the reduced energy expended by the human muscles $A_2 = \sum_{i=1}^{4} M_{ei} \cdot \dot{\phi}_i dt = \eta_2 \sum_{i=1}^{4} \int_0^T uidt$ will be determined from the condition:

$$\xi_1 . \tag{16}$$

While substituting the Eq. 10 and Eq. 11 in Eq. 16 after the corresponding transformations, we obtain an expression that determines the amount of the energy consumption of a person when performing actions related to the change of the vertical coordinate of the center of mass of the system:

$$\xi_1 = 1 - \xi_2 \tag{17}$$

Eq. 17 allows for the determination of the energy consumption by a human when lifting a load.

Having obtained by experimental or computational methods the time dependence of the current arising in the windings of electric motors, one can determine the electric power consumed by the exoskeleton electric drives. Next, you need to measure the vertical velocity of the center of mass of the system and the moments of resistance forces and, applying Eq.15, find the parameter of the reduced energy expended by human muscles $\xi_1 = 1 - \frac{\eta_2 \int_0^T uidt}{\int_0^T mgV_{Cz}dt + \int_0^T \sum_{i=1}^{m} M_{Ci}\dot{\phi}_i dt}$. Figure 13 and

Figure 13. Time dependence of the total energy value (in J) without gravity compensation

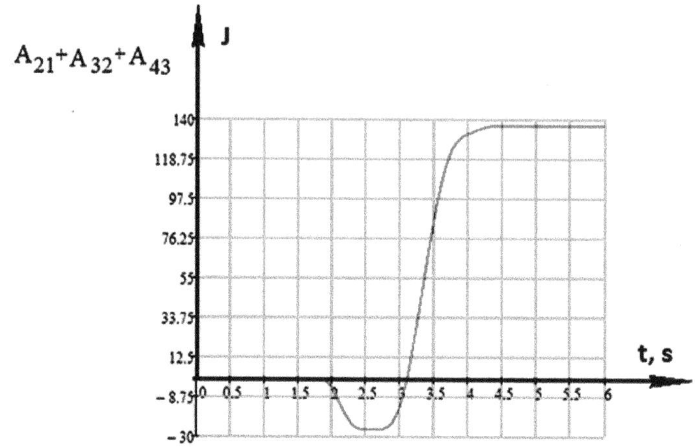

figure 14 show the time dependence of the total energy done by the exoskeleton drives during the lifting process without and with gravity compensation.

It is obvious, that elastic gravity compensator reduces the overall level of energy consumption, as well as smooths out peaks.

Let us find the dependences of the power, developed by the drives of each joint hinge, making use of the following formulas:

$$\xi_1, \; N_{21}(t) = M_{21}(t)\dot{\phi}_{21}(t) \; N_{32}(t) = M_{32}(t)\dot{\phi}_{32}(t) \tag{18}$$

Figure 14. Time dependence of the total energy (in J) with gravity compensation

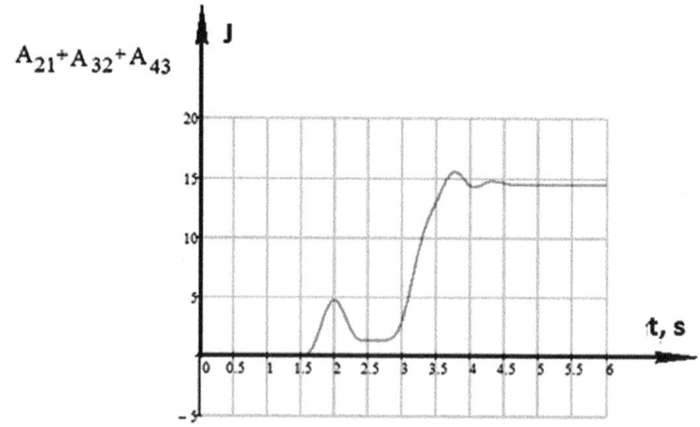

Figure 15 shows the dependence of the power $N_{43}(t) = M_{43}(t)\dot{\phi}_{43}(t)$ time dependences.

Note that the maximum power consumed by the drive of the ankle joint was less than 27 W. The power was expended in the time interval $N_{21}(t)$ s.

Figure 16 shows the power $t \in \left[3.7 \quad 4.4\right]$ time dependence.

The maximum power consumed by the knee joint drive made less than 200 W, which is much more than the maximum power expended by the ankle drive. Power was expended in the time interval $N_{32}(t)$ s.

CONCLUSION

The article has taken a close look at an industrial exoskeleton, which is applied to reduce the loads on a human when performing loading and unloading operations, a mathematical model of load lifting, executed by a person while wearing an exoskeleton, has been developed. Mathematical modeling of the process of lifting the load with the help of the electric drives of the exoskeleton has been drawn. Particular attention is paid to the study of the influence of gravity compensators on the magnitude of the torques generated by the electric drives of the femoral and knee joints. It is shown that the use of gravity compensators can significantly reduce the load on electric drives. Moreover, the analysis of energy consumption in the process of lifting goods has been conducted.

Figure 15. Time dependence of the power $N_{21}(t)$

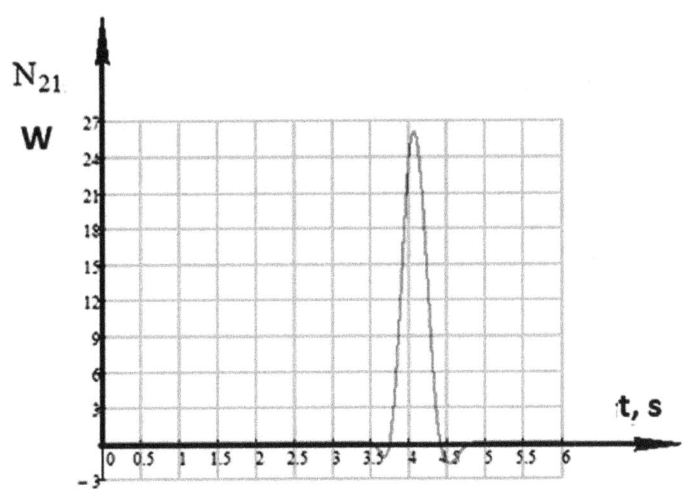

Figure 16. Time dependence of the power $N_{32}(t)$

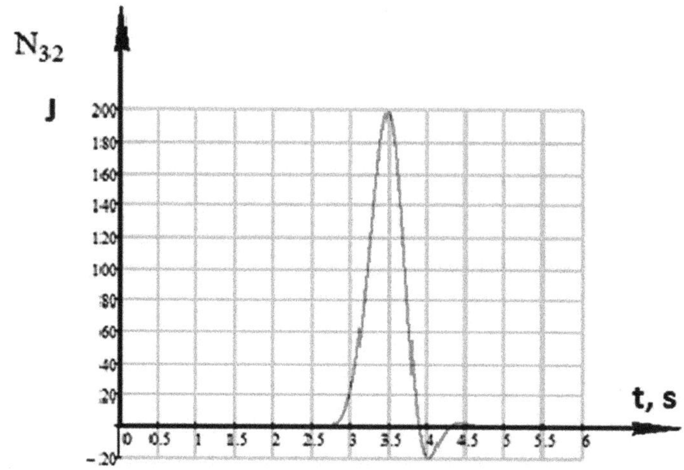

REFERENCES

Anam, K., & Al-Jumaily, A. A. (2012). Active exoskeleton control systems: State of the art. *Procedia Engineering, 41*, 988–994. doi:10.1016/j.proeng.2012.07.273

Anam, K., & Al-Jumaily, A. A. (2012). Active exoskeleton control systems: State of the art. *Procedia Engineering, 41*, 988–994. doi:10.1016/j.proeng.2012.07.273

Arisumi, H., Miossec, S., Chardonnet, J. R., & Yokoi, K. (2008, September). Dynamic lifting by whole body motion of humanoid robots. In *Intelligent Robots and Systems, 2008. IROS 2008. IEEE/RSJ International Conference on* (pp. 668-675). IEEE. 10.1109/IROS.2008.4651195

Bosch, T., van Eck, J., Knitel, K., & de Looze, M. (2016). The effects of a passive exoskeleton on muscle activity, discomfort and endurance time in forward bending work. *Applied Ergonomics, 54*, 212–217. doi:10.1016/j.apergo.2015.12.003 PMID:26851481

Chen, J., Zhang, X., & Zhu, L. (2014). *Kinematics analysis and three-dimensional simulation of the rehabilitation lower extremity exoskeleton robot.* arXiv preprint arXiv:1401.6517

Dumitru, N., Copilusi, C., Geonea, I., Tarnita, D., & Dumitrache, I. (2015). Dynamic Analysis of an Exoskeleton New Ankle Joint Mechanism. In *New Trends in Mechanism and Machine Science* (pp. 709–717). Springer International Publishing. doi:10.1007/978-3-319-09411-3_75

Elliott, G., Sawicki, G. S., Marecki, A., & Herr, H. (2013). The biomechanics and energetics of human running using an elastic knee exoskeleton. In *Rehabilitation Robotics (ICORR), 2013 IEEE International Conference on*, (pp. 1-6). IEEE. 10.1109/ICORR.2013.6650418

Gou, H., Wang, J., Wu, H., Wang, C., Yan, L., & Xiao, J. (2014). Design of information acquisition and control system for the exoskeleton robot. *Journal of Electrical and Computer Engineering, 2014*, 2. doi:10.1155/2014/309814

Harada, K., Kajita, S., Kanehiro, F., Fujiwara, K., Kaneko, K., Yokoi, K., & Hirukawa, H. (2007). Real-time planning of humanoid robot's gait for force-controlled manipulation. *Mechatronics, IEEE/ASME Transactions on, 12*(1), 53-62

Jatsun, S., Savin, S., & Yatsun, A. (2016, October). Study of controlled motion of an exoskeleton performing obstacle avoidance during a single support walking phase. In *System Theory, Control and Computing (ICSTCC), 2016 20th International Conference on* (pp. 113-118). IEEE.

Jatsun, S., Savin, S., & Yatsun, A. (n.d.). Modelling characteristics of human-robot interaction in an exoskeleton system with elastic elements. *Lecture Notes in Computer Science, 11097*, 85.

Jatsun, S., & Yatsun, A. (2018). Investigation of Human Cargo Handling in Industrial Exoskeleton. In *Proceedings - 2018 Global Smart Industry Conference*. IEEE.

Kajita, S., Kanehiro, F., Kaneko, K., Fujiwara, K., Harada, K., Yokoi, K., & Hirukawa, H. (2003, September). Biped walking pattern generation by using preview control of zero-moment point. In *Robotics and Automation, 2003. Proceedings. ICRA'03. IEEE International Conference on* (Vol. 2, pp. 1620-1626). IEEE. 10.1109/ROBOT.2003.1241826

Kajita, S., Kanehiro, F., Kaneko, K., Yokoi, K., & Hirukawa, H. (2001). The 3D Linear Inverted Pendulum Mode: A simple modeling for a biped walking pattern generation. In *Intelligent Robots and Systems, 2001. Proceedings. 2001 IEEE/RSJ International Conference on* (Vol. 1, pp. 239-246). IEEE.

Li, Yan, Qian, Wu, Wu, & Men. (2015). Review on Lower Extremity Exoskeleton Robot. *Open Automation and Control Systems Journal, 7*, 441–453.

Martelli, D., Vannetti, F., Cortese, M., Tropea, P., Giovacchini, F., Micera, S., ... Vitiello, N. (2014). The effects on biomechanics of walking and balance recovery in a novel pelvis exoskeleton during zero-torque control. *Robotica*, *32*(08), 1317–1330. doi:10.1017/S0263574714001568

Mooney, L. M., Rouse, E. J., & Herr, H. M. (2014). Autonomous exoskeleton reduces metabolic cost of walking. In *Engineering in Medicine and Biology Society (EMBC), 2014 36th Annual International Conference of the IEEE*, (pp. 3065-3068). IEEE. 10.1109/EMBC.2014.6944270

O'Sullivan, Nugent, & van der Vorm. (2015). *Standards for the Safety of Exoskeletons Used by Industrial Workers Performing Manual Handling Activities*. A Contribution from the Robo-Mate Project to their Future Development.

Pons, J. L. (2008). *Wearable Robots: Biomechatronic Exoskeletons*. West Sussex, UK: John Wiley and Sons Ltd. doi:10.1002/9780470987667

Sellaouti, R., Stasse, O., Kajita, S., Yokoi, K., & Kheddar, A. (2006, October). Faster and smoother walking of humanoid HRP-2 with passive toe joints. In *Intelligent Robots and Systems, 2006 IEEE/RSJ International Conference on* (pp. 4909-4914). IEEE.

Shamaei, K., Cenciarini, M., Adams, A., Gregorczyk, K. N., Schiffman, J. M., & Dollar, A. M. (2014). Design and evaluation of a quasi-passive knee exoskeleton for investigation of motor adaptation in lower extremity joints. *Biomedical Engineering, IEEE Transactions on*, *61*(6), 1809–1821. doi:10.1109/TBME.2014.2307698 PMID:24845291

Verrelst, B., Stasse, O., Yokoi, K., & Vanderborght, B. (2006, December). Dynamically stepping over obstacles by the humanoid robot HRP-2. In *Humanoid Robots, 2006 6th IEEE-RAS International Conference on* (pp. 117-123). IEEE.

Viteckova, S., Kutilek, P., & Jirina, M. (2013). Wearable lower limb robotics: A review. *Biocybernetics and Biomedical Engineering*, *33*(2), 96–105. doi:10.1016/j.bbe.2013.03.005

Wang, Y., Yang, Y., Liu, W., & Wang, R. (2015). Structure Design and Dynamic Model Analysis of Multi-degree-of-freedom Exoskeleton. In *2015 International Conference on Mechatronics, Electronic, Industrial and Control Engineering (MEIC-15)*. Atlantis Press. 10.2991/meic-15.2015.342

Yoshida, E., Esteves, C., Sakaguchi, T., Laumond, J. P., & Yokoi, K. (2006, October). Smooth collision avoidance: Practical issues in dynamic humanoid motion. In *Intelligent Robots and Systems, 2006 IEEE/RSJ International Conference on* (pp. 827-832). IEEE.

Yu, S., Han, C., & Cho, I. (2014). Design considerations of a lower limb exoskeleton system to assist walking and load-carrying of infantry soldiers. *Applied Bionics and Biomechanics*, *11*(3), 119–134. doi:10.1155/2014/585837

Zhao, Y., Zhang, W., Ge, W., & Li, S. (2013). Finite Element Simulation of Soldier Lower Extremity Exoskeleton. *Journal of Multimedia*, *8*(6), 705–711. doi:10.4304/jmm.8.6.705-711

Chapter 3
Cable-Driven Robots in Physical Rehabilitation:
From Theory to Practice

Rogério Sales Gonçalves
Federal University of Uberlândia, Brazil

Thiago Alves
https://orcid.org/0000-0001-5666-0831
Federal University of Uberlândia, Brazil

Giuseppe Carbone
DIMEG, Università della Calabria, Italy

Marco Ceccarelli
https://orcid.org/0000-0001-9388-4391
University of Rome Tor Vergata, Italy

ABSTRACT

This chapter deals with cable-driven robots when applied in physical rehabilitation. In general, neurorehabilitation is limited to physical therapy that is delivered by clinicians and potentially augmented by robotic tools to facilitate neurorehabilitation and to reduce the consequences of central nervous system injury. Among the robotic tools for rehabilitation can be considered the cable-driven manipulators. First, this chapter presents the upper and lower human limbs movements. The main rehabilitation robots are presented as exoskeletons and cable-driven manipulators. After, the cable-driven manipulators theory is introduced focusing on considerations for robot design in rehabilitation and control with safe human-machine interaction. Experimental examples with different cable-driven robot's structures are presented so that this chapter suggests that these structures can be used as a complement to conventional therapies and not as a substitute. Finally, this chapter presents the clinical evidence in cable-driven robots when applied in physical rehabilitation.

DOI: 10.4018/978-1-7998-1382-8.ch003

INTRODUCTION

Stroke is the leading cause of disability in the world and leaves a significant number of individuals with motor and cognitive deficits (Tucan et al., 2019; Gonçalves & Krebs, 2017). It can degrade upper and lower limbs functions and rehabilitation training is the most effective way to reduce these motor impairments in stroke patients. Robots can be suitable for this purpose since they can train patients for long durations with precision (Hatem et al., 2016). The robotic therapy assists the patient during training and can offer several advantages over conventional therapy as pointed out in Pennycott et al. (2012).

After a stroke the recovery of the movements is related to neural plasticity, which involves developing new neuronal interconnections, acquiring new functions and compensating for impairment. To promote neural plasticity and consequently movement's recovery, the use of robots on rehabilitation can provide an intensive, repetitive and task-oriented training which has proven effective for the movements learning (Duret et al., 2019; Takahashi et al., 2016; Takeuchi & Izumi, 2013; Colombo et al., 2012; Kuznetsov et al., 2013; Maciejasz et al., 2014, Dzahir & Yamamoto, 2014).

Repeated movements of human limbs can help the patient to regain function of the injured limb. This process involves repetitive movements that must be performed several times at various speeds. The robots are, generally, more effective and efficient in performing these exercises than humans offering longer and more frequent therapy sessions for a patient and, consequently, they are attracting significant research interest. Additionally, a self-treatment with remote supervision of several patients by a single therapist professional can be implemented using these devices (Stein, 2012; Reinkensmeyer and Boninger, 2011).

Moreover, these systems can be coupled with virtual reality simulators, enhancing conventional physical therapy by recording patient information like position, force, time and velocity. The rehabilitation process involves other activities like diagnosis and prescription of the treatment to stimulate the recovery processes.

The recorded data by the robot can be achieved and compared through patient sessions to follow the progress (or even between different patients to aid in diagnosis), providing an objective measure of the progress and outcomes of therapy (Matin & Cazorla, 2019; Ceccarelli et al., 2010, Gonçalves et al., 2015).

These measures can offer a sensitive, accurate and time-efficient approach for the assessment of sensorimotor function after neurological impairment compared to standard clinical assessments. Besides that, motor functions of stroke patients are frequently related and evaluated using drawing/tracing tasks which are simpler with robotic systems (Takebayashi et al., 2018; Dipietro et al., 2007; Hussain et al., 2017). Additionally, the use of robotic solutions has strong potential for reducing treatment costs, which are becoming less affordable due to the aging of the population (Gassert & Dietz, 2018; Stein, 2012).

Different types of robotic architectures have been developed and applied in the rehabilitation of both upper and lower human limbs like industrial robots, exoskeletons, and cable-driven robots. In general, robotic structures used in rehabilitation are industrial robots or a new structure specifically designed or adapted to reproduce rehabilitation movements (Gonçalves et al., 2015; Gonçalves et al., 2019).

Industrial robots, usually have rigid structures, are fast, but several safety issues must be considered when applying them to rehabilitation (Mao et al., 2015). Exoskeletons for rehabilitation deal with a huge variability of human limbs shape and dimensions. In this way, matching the human and robot joints rotation centers is a very difficult task (Beyl et al., 2009).

The cable-driven robots make it easier to adapt to the different human limb sizes, allowing a much wider array of patients use. Their links have very low inertial forces and cables can be designed to break at the desired tension force. In this way, the cable-driven robots can be a good solution for an inherently safe device meeting the requirements for safety, comfort, ease of manipulation, transportability, and large workspace (Ceccarelli and Romdhane, 2010; Gonçalves et al., 2015; Carbone and Ceccarelli, 2016; Alves et al., 2018).

Thus, cable-driven robots can be an efficient way of increasing the intensity of stroke rehabilitation (longer and more frequent training) in rehabilitation clinics, hospital bed or even patient's home (Schmidt, 2018).

Thus, this chapter focuses on cable-driven parallel manipulators to be applied in stroke rehabilitation to recover the movement of the human limbs. These systems can help in assisting and guiding physiotherapy motions in recovering healthy conditions of limbs.

Due to the human limb complexity, the development of these mechanical systems for rehabilitation can be a very difficult task. Thus, in the chapter, first the upper and lower human limb movements are introduced in order to understand this complexity. Next, a review of devices that are developed to be applied in rehabilitation is presented by indicating peculiarities and open issues. Then, the attention is focused on structures that are based on cable-driven parallel manipulators and their mathematical models are also presented to be used in design developments, also with experimental evaluations.

MOVEMENTS OF THE UPPER AND LOWER HUMAN LIMB

In the present section, the main upper and lower limb joints and movements are presented. The design of a rehabilitation system for a particular movement requires a full understanding of the characteristics of the joints and the movements covered by this system. The joints can be classified as (Kapandji, 2007; Moore & Dalley, 2009):

- **Slightly Movable Joints (or Semi-Mobile Joints):** They have very restricted flexibility, such as the vertebra and ribs.
- **Rigid Joints (or Immovable Joints):** Do not allow flexibility, such as the skull.
- **Hinged Joints**: Allow flexibility in only one direction, such as the knee and the elbow.
- **"Ball and Socket" Joints**: Allow radial movement in almost any direction, such as the shoulder.
- **Saddle joints**: allow movement back/forth and up/down, but do not allow rotation. The only saddle joint in the human body is the thumb, but they have limited rotation.
- **Plane Joints (or Gliding Joints)**: Occur between two flat bones surfaces which are held together by ligaments, such as some of the wrists and ankles bones in human.
- **Ellipsoidal Joints**: Like a "ball and socket" joint, allowing the same type of movement but with a lesser magnitude and limited rotation, such as the wrist.
- **Pivot Joints:** Allow rotation around an axis, such as the neck and forearms.

The upper limb (arm) and the lower limb (leg) movements are presented in the following in order to understand the rehabilitation movements.

Upper Limb

The upper limb is characterized by its mobility, ability to handle, strike, and perform fine motor activities like objects manipulation. These characteristics are more noticeable while performing manual tasks with hands. The hand can touch, press, hold and handle, and is able to perform a variety of motor tasks, as well as transmit sensory information about temperature, shape, and texture of objects (Moore & Dalley, 2009).

The upper limb main movements, except for the finger movements, are shown in Fig. 1.

Is it common, to describe the motions based on the anatomic planes of the body defined by the medical community (Vaida et al, 2017; Nadas et al., 2017). These planes are shown in Fig. 2 as:

- **Coronal Plane (Frontal Plane):** Divides the body into anterior and posterior sides.

Figure 1. Main movements of the upper limb as function of a kinematic model

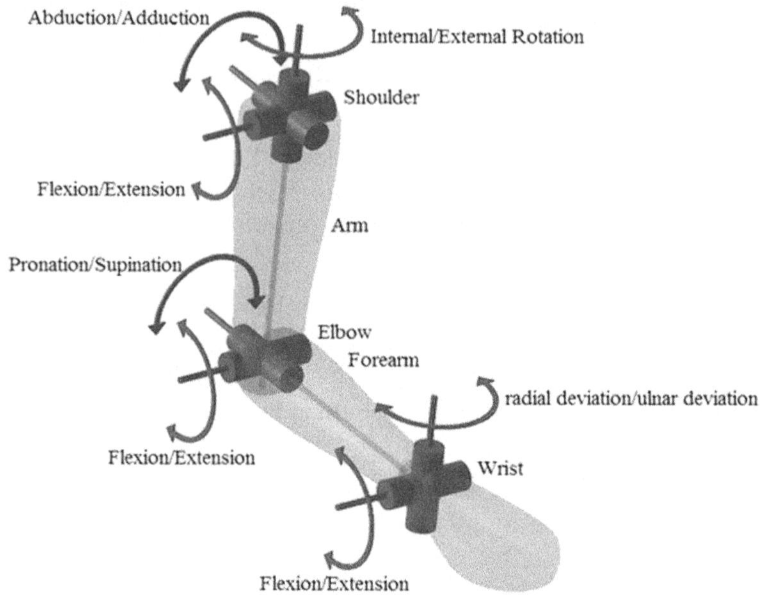

- **Sagittal Plane:** Divides the body into left and right sides.
- **Transverse Plane (Cross-Sectional):** Divides the body into superior and inferior sides.

Based on the body planes, shown in Fig. 2(a), a corresponding Cartesian system is defined for the robotic system as is shown in Fig. 2(b).

The Shoulder

The shoulder is the proximal joint of the upper limb and is the most mobile joint of the human body, promoting the movements of the arms in the three references planes. In a simplified model, some authors claim that the shoulder has three degrees of freedom (DOF), despite, in reality, it allows more different movements (Kapandji, 2007).

The shoulder, actually, consists of five joints: glenohumeral, subdeltoid, scapulothoracic, acromioclavicular and sternocostoclavicular where the first two (first group) is the true main joint (shoulder joint) linked to a ‘false’ joint, the scapulothoracic joint (the last three). The movement of the shoulder girdle in the horizontal plane (retraction and protraction of the shoulder girdle) involve this ‘false’ joint.

Figure 2. Schemes for human body referencing: (a) anatomical body planes; (b) cartesian coordinate system

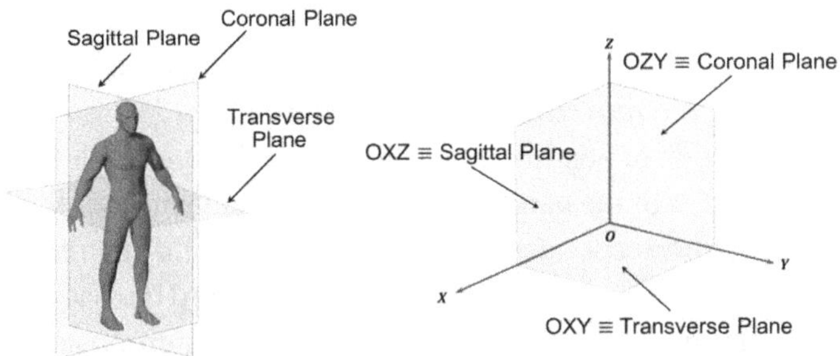

The flexion is a movement of great range, up to 180º and is performed in the sagittal plane as well as extension, but this one is a small range movement, up to 50º (Kapandji, 2007).

The adduction is a movement that is performed in the coronal plane and is only possible when it is combined with a movement of extension (which adduction is minimal) or a movement of flexion which adduction can reach up to 45º. The abduction also performed in the coronal plane and it is the movement of the upper limp away from the trunk, its range is 180º when the arm comes to lie vertically above the trunk.

The adjunct rotation occurs when the arm rotates along its long axis.

The Elbow

The elbow is the intermediate joint of the upper limb. Anatomically, the elbow consists of a single joint with only one joint cavity and, physiologically, it has two distinct functions: flexion-extension and supination/pronation. Flexion-extension involves the humeroulnar and the humeroradial joints; pronation-supination involvers the superior radio-ulnar. Flexion occurs when the hand is displaced toward the shoulder and extended in the opposite direction. Supination-pronation is the rotation about the longitudinal axis of the elbow. Supination-pronation is the forearm rotation movement about the longitudinal axis of the elbow and involves the superior and inferior radio-ulnar joints. This rotation also introduces a degree of freedom into the wrist (Kapandji, 2007).

The Wrist

The wrist is the distal joint of the upper limb and it consists of many bones and joints which make it extremely complex (and mobile as well). The articular complex of the wrist consists of radio e mid carpal joints and it has two degrees of freedom, that combined with the forearm rotation mentioned above (pronation-supination) add a third degree of freedom.

The flexion-extension of wrist occurs around a transverse axis (lying in a coronal plane) and controlling the movements in the sagittal plane. The flexion of the wrist corresponds to the motion of bending the joint (decreasing the angle) and moving the anterior (palmar) surface of the hand towards the anterior aspect of the forearm. The extension of the wrist corresponds in straightening the joint (increasing the angle) and moving the posterior (dorsal) surface of the hand toward the back of the forearm.

The adduction-abduction of the wrist occurs in an anteroposterior axis (lying in a sagittal plane) and controlling movements in the coronal plane. The adduction (or ulnar deviation) corresponds to the hand movement toward the midline of the body, moving the little finger side of the hand toward the medial side of the forearm and forming an obtuse angle with the medial border of the forearm. The abduction corresponds to the hand movement away from the midline of the body moving thumb side of the hand toward the lateral of the forearm. The range of adduction is about 45° and abduction is not superior to 15°. The active flexion or extension range are both 85°, but it can exceed 90° in pronation.

The movement of circumduction is, in fact, the combination of the flexion-extension with adduction-abduction movements, in which the hand traces a conical surface (not regular) in space.

The Hand

The hand is part of the upper limb distal of the forearm, consisting of the carpal, metacarpal, and phalanges. It is formed by the wrist, palm, dorsum of the hand and fingers, including the thumb. The hand is richly supplied by sensory endings for touch, pain, and temperature.

Except for the thumb, which has only two phalanges (proximal and distal phalange), each finger has three phalanges; the fingers joints follow the arrangement of their phalanges. Thus, the thumb has two joints: metacarpophalangeal and interphalangeal, and other fingers have three joints: metacarpophalangeal, connecting the metacarpal bones to the proximal phalange, proximal interphalangeal, connecting the proximal to the intermediate phalange and, distal interphalangeal which join the intermediate and the distal phalanges; each one has a DOF for flexion/extension. The metacarpophalangeal joint has an extra DOF to accomplish adduction/abduction. The movement in the metacarpophalangeal joint of the thumb is limited to flexion-extension.

The flexion of the fingers indicates an angular range in the direction of closing the hand and extension in the opposite direction. The abduction is the lateral movement, away from the axial line i.e., the movement of the fingers away from the middle finger and adduction toward the middle finger. The opposition is the diagonal movement of the thumb contacting the fingers.

In Carbone et al. (2017), the upper limb has the amplitude for its joint movements determined with a sampling procedure conducted by medical specialists on a group of patients between 43 and 83 years old. The movements measured are Flexion/Extension, Abduction/Adduction, and Pronation/Supination. It was used a goniometer to measure all joint angles with respect to the initial position arm segments. The goniometer was for medical use and allow an accurate measurement of these amplitudes.

In the measurement of the shoulder flexion-extension, Fig. 3, the test-participants aligned the moving arm with the midline of the humerus. Then the participant is requested to raise his arm forward until the maximum angle is reached. The extension was represented if moving the arm in the opposite direction (Carbone, 2017).

For the measurement of the shoulder abduction-adduction, Fig. 4, the patient stands with the upper limb alongside the body (initial position) and then extends the arm upwards, for adduction. For abduction, the arm moves in the opposite direction.

For the measurement of the elbow flexion, Fig. 5, the shoulder is held at flexion of 90° and then the forearm is brought as closely as possible to the arm.

For the measurement of forearm pronation-supination, Fig. 6, the elbow is held at flexion of 90°, the forearm in a half-supine position, and an object is held by the patient's hand. In the supination, the forearm rotates the object until the palm facing down. The pronation represents the same movement in the opposite direction (palm facing upward).

For the measurement of wrist extension-flexion, Fig. 7, the elbow is held at flexion of90°, the forearm in the prone position (palm facing down), the patient then extends the wrist upwards and flexes it downwards.

Figure 3. A CAD model for shoulder flexion-extension measurements: (a) upper limit; (b) lower limit
(Carbone et a., 2017)

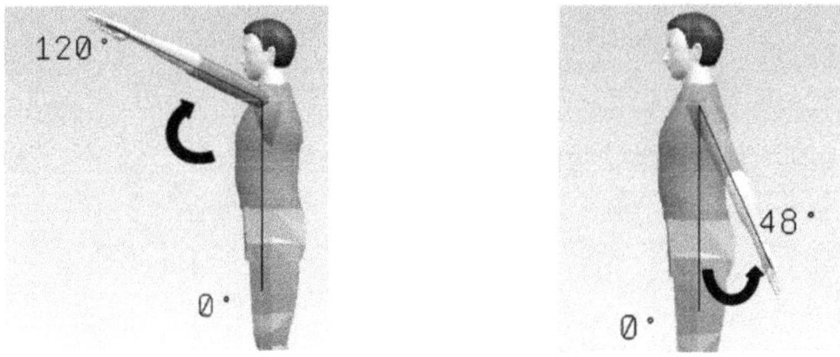

Figure 4. A CAD model for shoulder abduction-adduction measurements: (a) upper limit; (b) lower limit
(Carbone et a., 2017)

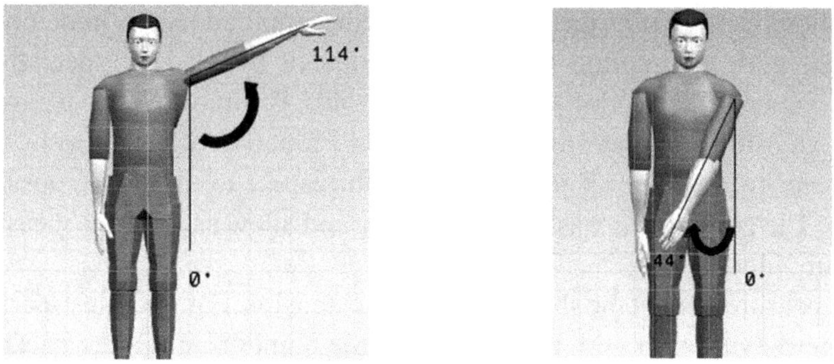

Figure 5. A CAD model for elbow flexion measurements: upper and lower limits
(Carbone et a., 2017)

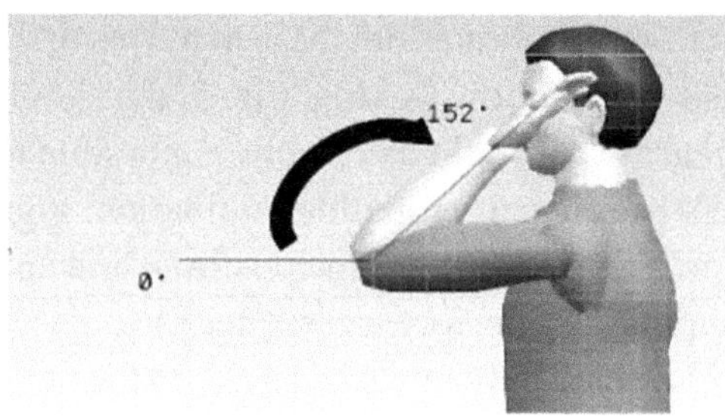

Figure 6. A CAD model for forearm pronation-supination measurements: (a) initial position; (b) upper limit; (c) lower limit

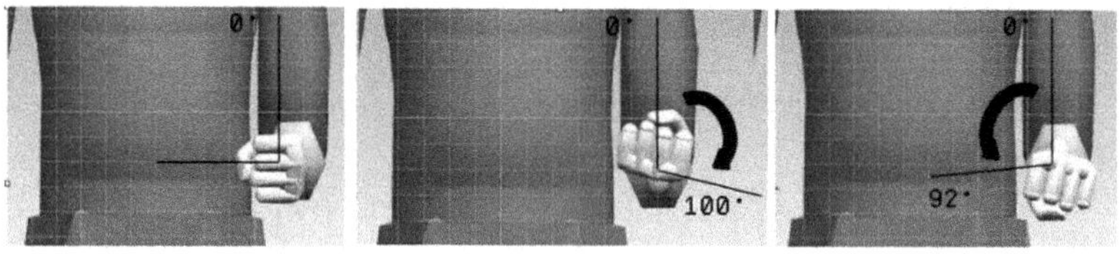

Figure 7. A CAD model for wrist flexion-extension measurements
(Carbone et a., 2017)

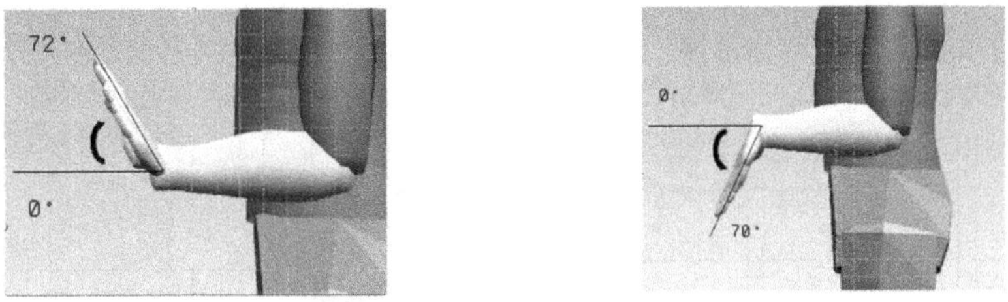

A summary of the experimental measures is presented in Table 1.

Lower Limb

The lower limb has two basic and important functions, of supporting the body weight and also to allow locomotion. The lower limb main movements, except for the fingers foot movements, are shown in Fig. 8.

The Hip

The hip is the proximal joint of the lower limb that allows the limb to assume any position in space. The hip has three axes and three DOF. The movements of the hip occur at a ball-and-socket joint, the hip joint.

Table 1. Summary of the experimental measures in terms of operation ranges (mean angles), the standard error of the mean (S.E.) and standard deviation (S.D.)

Joint Movement	Right Side			Left Side		
	Mean	S.E.	S.D.	Mean	S.E.	S.D.
Shouder Flexion	93.75°	2.32°	10.38°	98.3°	3.28°	14.68°
Shouder Extension	48.67°	3°	12.73°	48.78°	2.62°	11.1°
Shouder Abduction	89.15°	1.69°	7.57°	90.9°	2.03°	9.07°
Shouder Addction	90.9°	2.03°	9.07°	24.55°	2.08°	9.3°
Ebow Flexion	25.3°	1.78°	7.95°	137.3°	1.61°	7.2°
Ebow Pronation	131.1°	4.39°	19.65°	81.1°	3.33°	14.89°

(Carbone et a., 2017)

Figure 8. Main movements of the lower limb as function of a kinematic model

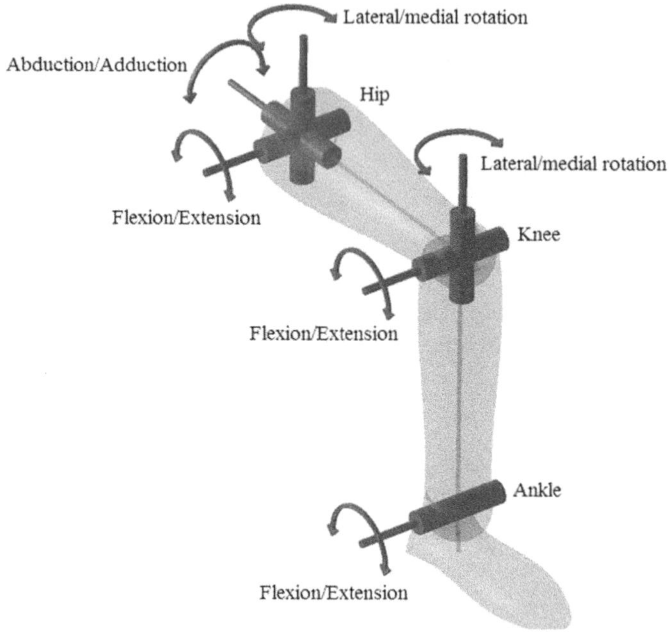

The hip joint has a more limited range of movement when compared to the shoulder, also a ball-and-socket joint, but on the other hand, it is more stable. These features meet the requirement of supporting body weight and locomotion.

The flexion-extension motions occur about the transverse axis and are the forward and backward movements of the hip. The adduction-abduction occur about the anteroposterior axis and are the lateral movements of the hip. The movements of medial and lateral rotation occur about the vertical axis and coincide to the longitudinal axis of the limb when the hip joint is in the straight position. The range of hip joint movements allowed depends on the knee position (Kapandji, 2010).

The Knee

The knee is the intermediate joint of the lower limb. It allows the distance between the trunk and the ground to be varied and works essentially by axial compression under gravity.

The knee main joint has one degree of freedom related to the transverse axis, around which occur movements of flexion and extension along the sagittal plane. It also has an additional DOF related to rotation around the horizontal axis of the leg. This second degree of freedom occurs with the knee flexed and is important to running and to the optimal orientation of the foot relative to the irregularities of the ground. Due to the knee structure, it is impossible to rotate the knee when it is fully extended.

The Ankle

The ankle is the distal joint of the lower limb. It is a hinge joint and has only one degree of freedom that controls the movements of the leg relative to the foot, which occur along the sagittal plane. These movements are essential for walking on flat or rough surfaces. During single limb support, this joint is exposed to extreme mechanical conditions, when it supports the entire body weight during walking, running or jumping.

This joint, when associated with the hip and knee motions, allows the foot to make a three-dimensional movement to adapt to irregularities on the ground.

The Foot

The joints of the foot are many and complex. The foot can move about the vertical axis of the leg and about its own horizontal and longitudinal axis, in addition to flexion and extension, which occur at the ankle. The movement of adduction occurs in the vertical axis when the tips of the toes move towards the plane of symmetry of the body and face inwards. The movement of abduction occurs when the tips of the toes move away from the plane of symmetry and point outwards.

The plantar vault acts as a shock absorber, which is essential for flexibility of the gait and helps support the body on the ground for running, walking and the maintenance of the erect posture.

Amplitudes of the Upper Limb to Patients

To correctly evaluate the disability, classical physiotherapy uses goniometric measurements in order to quantify the limitations of motion. This helps to design appropriate therapeutic interventions and assess effectiveness (Hirschhorn et al., 2015).

To design rehabilitation devices the same pre-evaluation may be required because a disabled limb must be moved only within their physiological limits and a robot should be programmed accordingly (Beebe & Lang, 2008).

An analysis of human motions is helpful to design a novel biomechanical model. To analyze the effects of stroke in the human motion ranges, in Major et al. (2008) was performed a study on a group of 21 patients (11 females and 10 males,) aiming at the design of robotic rehabilitation solutions. The age limits were from 40 to 85 years old and beyond, which represent a variable due to higher stiffness of joints in the elderly.

Although rehabilitation ideally might target all segments, it is unpractical to address more than 45 joints of the human upper limb. Thus, a goniometer was used to measure the range of motion of the shoulder flexion-extension and abduction-adduction; the elbow flexion and pronation-supination; the wrist flexion-extension, metacarpal-phalangeal, proximal and distal inter-phalangeal flexion. This simplified model may possibly maintain the upper-limb range of motion for several activities of daily living (ADL) (Gates et al. 2015).

It was compared the right and left side since the laterality of a subject influences slightly the muscle size and could impact the bone structure development. No significant side-dependent differences were identified. The mean angles and their standard error and standard deviation are shown in Table 2.

Gender is experienced as a source of difference in anatomy and limb motion characteristics (Kapandji, 2007). This can be evaluated as reported in the data of Table 3.

The study also found that age impacts the motion amplitude and the obtained angles are significantly lower for the elderly (elderly having stiffer joints, lower motion ranges) as shown in Table 4. The threshold for significance was 0.05 for both tests (Major et al., 2008).

Table 2. Mean angles of the measured joints, the standard error of the mean (S.E.) and standard deviation (S.D.)

	N	Mean	S.E.	S.D.
Shoulder Flexion	40	96.03	2.02	12.76
Shoulder Extension	36	48.72	1.96	11.77
Shoulder Abduction	40	90.03	1.31	8.29
Shoulder Adduction	40	24.93	1.35	8.55
Elbow Flexion	40	134.2	2.36	14.94
Elbow Pronation	40	79.43	2.21	14
Elbow Supination	40	83.05	2.2	13.91
Radio Carpal Flexion	36	59.75	2.52	15.12
Radio Carpal Extension	36	45.14	2.28	13.69
Metacarpophalangeal Flexion	42	73.29	2.79	18.07
First Interphalangeal Flexion	42	82.86	2.14	13.87
Second Interphalangeal Flexion	42	54.36	2.95	19.14

(Major et al., 2008)

Table 3. Mean angles of the investigated joints, in case of male and female subjects

	Male	**Female**
Shoulder Flexion	95.45	96.6
Shoulder Extension	46.67	50.77
Shoulder Abduction	89.11	90.77
Shoulder Adduction	25.85	24
Elbow Flexion	133.35	135.05
Elbow Pronation	73.67	84.14
Elbow Supination	83.22	82.91
Radio Carpal Flexion	55.32	64.71
Radio Carpal Extension	39.95	50.94
Metacarpophalangeal Flexion	71.6	74.82
First Interphalangeal Flexion	79.95	85.5
Second Interphalangeal Flexion	50.2	58.14

(Major et al., 2008)

Table 4. Mean angle differences according to age groups

	Mean Angle	
	< Mean (N) >	**Mean (N)**
Shoulder Flexion	100.86 (22)	90.11 (18)
Shoulder Extension	47.60 (20)	50.13 (16)
Shoulder Abduction	93.27 (22)	86.06 (18)
Shoulder Adduction	26.18 (22)	23.39 (18)
Elbow Flexion	135.45 (22)	132.67 (18)
Elbow Pronation	79.73 (22)	79.06 (18)
Elbow Supination	85.64 (22)	79.89 (18)
Radio Carpal Flexion	64.38 (21)	53.27 (15)
Radio Carpal Extension	49.24 (21)	39.40 (15)
Metacarpophalangeal Flexion	72.13 (24)	74.83 (18)
First Interphalangeal Flexion	79.67 (24)	87.11 (18)
Second Interphalangeal Flexion	51.29 (24)	58.44 (18)

(Major et al., 2008)

These findings represent a possible guideline for the robot (for amplitudes being safe): to be programmed in an individualized manner for gender, age, size/weight, health conditions, etc. Also, as women have a higher range of motion for some of the joints, the patients might use data safe for both categories (obtained from males) or individualization for the genders.

MECHANICAL SYSTEMS FOR REHABILITATION

Mechanical and mechatronics systems are, generally, more efficient in performing the necessary rehabilitation movement than human. Moreover, these systems can record information like position, force, and velocity, and if necessary, they can assist the patient in guiding the movement of its limb. These data can also be archived and accessed to follow the patient's progress in therapy. Thus, the rehabilitation robots allow the therapist to assist several patients at the time, while increasing the repeatability of therapies and the measurability of outcome (Gonçalves et al., 2015; Rosati et al., 2017). The mechanical and mechatronics systems developed and applied for rehabilitation can be divided into robots, exoskeletons, and cable-driven manipulators.

Rehabilitation Robots

Several mechanical systems have been developed for the rehabilitation of the human limbs. An industrial robot used for rehabilitation can also be considered a rehabilitation robot (Ionescu, 2003); it has advantages like a three-dimensional workspace with many degrees of freedom, which enables flexibility to produce different trajectories (useful for motion rehabilitation). They are attached to the human limb in order to execute the programmed motions. The MIT-Manus, Fig. 9 (a), is probably the most known example of a robot designed for neurorehabilitation (Krebs et al., 2004). This robot has two degrees of freedom that interact with the patient's arm over a working plane, moving, guiding, or perturbing the upper limb movement. It can also record information such as position, velocity, and force (Gonçalves et al., 2015).

The InMotion ARM™, Fig. 9 (b), is a clinical/commercial version of MIT-MANUS. It is a robotic arm with two degrees of freedom and is the upper limb rehabilitation most researched device. This robotic device was tested in over 1000 patients, with clinical evidence of improvements in Functional Independence Measure (FIM) scores (InMotion, 2018; InMotion, 2019).

An exoskeleton is a "mechanism with joints that correspond to those of the human body and which move with the body to which it is attached". This same definition can be applied to exoskeletons used for rehabilitation of human limbs (Ionescu,

Figure 9. Rehabilitation robots: (a) MIT-MANUS; (b) InMotion ARM ™.
(a) (Krebs et al., 2003); (b) (InMotion, 2018)

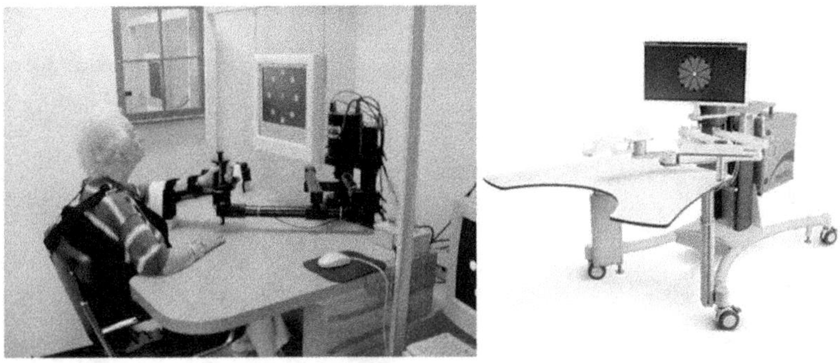

2003). There are exoskeletons for different applications like to assist in locomotion. An exoskeleton advantage is that it can perform movements in a three-dimensional workspace or for a specific motion when fixed around the patient's limb to provide the physiotherapy movement. They also can be used to increase the strength and endurance of the user. An exoskeleton disadvantage is the complexity of adjusting it for different patients, as well as the construction complexity due to the several mechanical components. The exoskeleton can extend, complement, substitute or enhance some human functions and capabilities or empowers a part of the human limb where it is worn (Carpino et al., 2013).

The Cyberdyne exoskeleton, named HAL (Hybrid Assistant Limb), Fig. 10 (a), is one of the commercial most successful product of this type. It can increase the muscle strength of the wearer, or to be used in rehabilitation. The HAL has sensors that analyze and calculate how much force is attributed to the body of a person, and to the HAL itself (Gonçalves et Al., 2015; Cyberdyne, 2019).

The Hocoma's LOKOMAT®, Fig. 10 (b), is probably the commercial robotic system most relevant in gait rehabilitation. It is an adjustable bilateral orthosis that is used over a treadmill attached to the patient's lower limbs acting on knee and hip flexion-extension of both legs while supporting part of the patient weight (Jezernik et al., 2003). The Lokomat improves the therapy outcome by providing highly intensive, individualized training in a motivational environment of constant feedback (Hocoma, 2019b).

In Bang et al. (2016) gait training using LOKOMAT was compared to treadmill training in 18 post-stroke patients. The gait velocity, cadence, and step length were significantly higher in the LOKOMAT group.

Some other exoskeleton examples applied to rehabilitation are the ARM (Assisted Rehabilitation and Measurement), Fig. 11 (a), which allows three-dimensional movements and the ARMIN, Fig. 11 (b), an exoskeleton with 6 DOF that can be

Figure 10. Robotic system for gait rehabilitation: (a) Cyberdyne HAL schematic; (b) LOKOMAT
(a) (Cyberdyne, 2019) ; (b) (Jezernik et al., 2004)

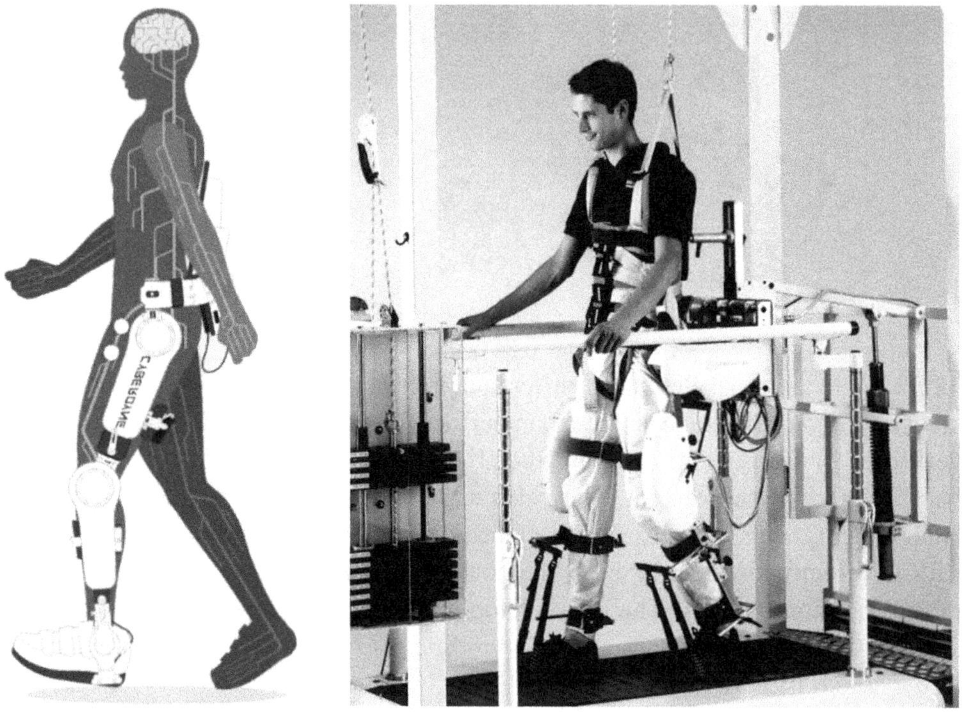

Figure 11. Robotic system for arm rehabilitation: (a) ARM; (b) ARMIN
(a) (Kahn et al., 2001); (b) (Novak et al., 2015)

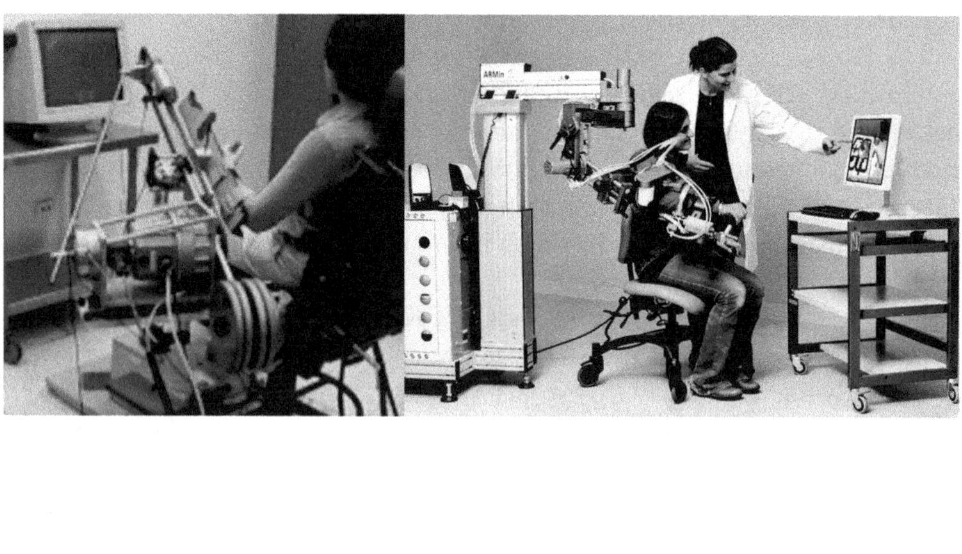

fixed around the patient's arm and provide some physiotherapy movements (Kahn et al., 2006; Nef & Riener, 2005).

Cable-Driven Parallel Manipulators

Another alternative that has been studied over the past few years are the cable-based parallel manipulators.

Cable-driven robots are a special class of parallel manipulators in which the end-effector is directly actuated by cables and they have several advantages like lightweight structure with small moving parts, a large workspace, and intrinsically safe features due to cable flexibility (allowing safe manipulation near humans). Disadvantages include redundancy (as cables are unilateral actuators), interference between cables and the environment (Barbosa et al., 2018; Gonçalves et al., 2016; Gonçalves et al., 2014), elasticity (Khosravi & Taghirad, 2011) and backlash (Mahler et al., 2014). The end-effector is linked by one or more cables to a fixed base, Fig. 12. The cable lengths are variable (can extend or retract) in order to control the end effector's position and orientation, while preventing cables to became slack (Rosati et. Al, 2017; Alves et al., 2018).

The use of the technology of cable-driven robots in rehabilitation lies in the need for building simple and portable robotic devices which can assist the patients to execute the natural movements of their limb. Compliance is fundamental in rehabilitation robotic devices because it allows for safe human-robot interaction and preserves the causal relationship between patient effort and resulting arm movement, even when robotic assistance is provided (Wolbrecht et Al., 2008).

Moreover, robot compliance allows patients to make movement errors (that would not be permitted by a stiff controller), what can help to stimulate motor learning, which is an error-driven process (Rosati et. al, 2017).

From this perspective, simple, lightweight structures are preferable with respect to rigid robotic links, as the robot-mediated exercise must be as close as possible to the free motion of the arm.

The "fully constrained manipulator", Fig. 12 (b), is a type of cable-driven parallel manipulators in which the pose of the end effector can be completely determined by the cable's configuration. In the under-constrained manipulator, Fig. 12 (a), the position and orientation of the end effector cannot be completely defined by the configuration of cables, and gravity may affect it. The large workspace of these structures enables them to be adapted to different patients and training. Additionally, the mechanical structure is easy to assemble/disassemble, can be easily reconfigured (in order to perform different therapies) and transported. As, generally, the actuators are located on the fixed base, the structure can be reconfigured only by changing the actuators positions and/or the attachment points of cables. The structures are

Figure 12. Cable-manipulator structural diagrams: (a) under-constrained; (b) fully constrained

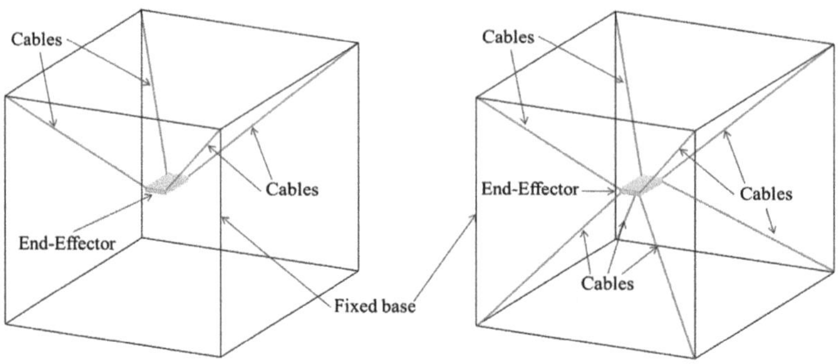

modular and have good inertial behavior since these systems have small moving masses, consisting basically of cables and end-effectors, as the actuators are mounted at the base and the cable-driven systems have the capability to support the weight of the injured limb (Gonçalves et al., 2015).

The cable-driven robots have low cost and simple maintenance, which are an important characteristic of self-administrated treatment by the patient at home. In the clinical aspect, the use of cables instead of rigid links, present in industrial robots and exoskeletons, makes the patient feel less constrained. All these characteristics mentioned above made the cable-driven structures ideal for rehabilitation application (Homma et al., 2002).

The cables can pull the end effector, but do not push it, therefore feasible tasks are limited due to static/dynamic characteristics of the cables, thus the workspace evaluation becomes forces dependent and can have a complex and irregular shape (Laribi et al., 2019; Hiller et al., 2009; Tavolieri, 2008).

Cable-driven manipulators can be classified as an *i-j* structure where *i* represents the number of connections points at the base and *j* the number of connection points of cables at the end effector. Several cable-driven robots have been designed specifically for stroke rehabilitation.

The CALOWI (Cassino wire low-cost robot), Fig. 13 (a), has a 4-4 architecture and is composed by a mechanical cubic-frame structure, a controller, a PC for programming and monitoring and an end-effector (suitable for rehabilitation). The cables are driven by four DC motors and transmission systems using pulleys (Tavolieri, 2008; Ceccarelli et al., 2010).

The Multi-Axis Cartesian-based Arm Rehabilitation Machine (MACARM), Fig. 13 (b), is a cable-based rehabilitation robot for rehabilitation of human upper limbs, which have eight motors and can provide six degree-of-freedom to the end-effector

Figure 13. Cable-based robots: (a) CALOWI; (b) MACARM
(a) (Ceccarelli et al., 2010); (b) (Mayhew et al., 2005)

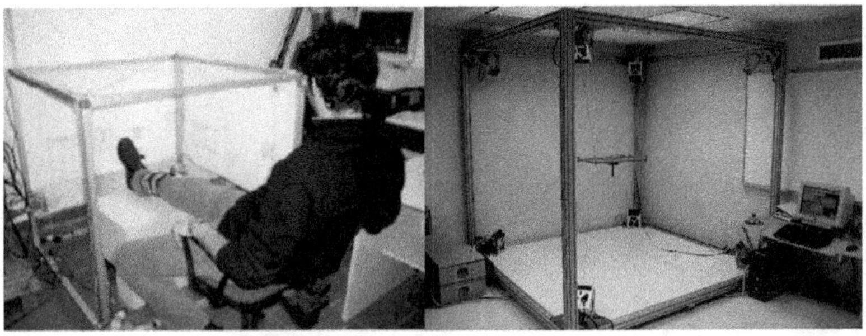

(connected to the fixed frame by eight cables). The hand is supported in the end-effector, making the arm rehabilitation possible (Mayhew et al., 2005).

The STRING-MAN, Fig. 14 (a), is a robotic system for supporting gait rehabilitation and motor function restoration. This system is designed to support gait restoration for several kinds of injury (Surdilovic & Bernhardt, 2007).

The CaMaReS (Cable Manipulator for Rehabilitation of Shoulder), Fig. 14 (b), was developed at the Laboratory of Robotics and Automation at Federal University of Uberlândia, Brazil. This device is a 4-2 cable-driven parallel manipulator which allows three-dimensional motion through the desired trajectory. To fit any patient, the position of the cables on the base and in the splint can be adjusted (Nunes et al., 2011a-c; Gonçalves and Carvalho, 2010; Gonçalves et al., 2013).

Freebal, Fig. 15 (a), is a dedicated weight-support system for upper-limb rehabilitation that uses a sling system with ideal spring mechanisms which provide compensation forces. The device has a large range of motion and low movement impedance and, additionally, can be easily moved, serviced and used in arm rehabilitation with either sitting or standing subjects. It does not need external power, force sensors or active controllers, however, it is indicated for passive therapy (to apply to patients with some functional movement control) (Stienen et al., 2007; Stienen, 2009).

The ArmeoBoom®, Fig. 15 (b), is an arm support device designed for home environments and provides an adjustable amount of weight support and allow patients to perform self-directed, free movement exercises of the impaired arm in a large tridimensional workspace. Interactive exercises for rehabilitation (games) are performed by moving the affected arm and using a webcam plus a potentiometer. The workspace can be adjusted to the patient maximum and inside the game environment. The amount of support is reduced progressively when there is an improvement. An assessment software provides the patient's progress to determine

Figure 14. Cable-based robots: (a) STRING-MAN; (b) CaMaReS
(a) (Surdilovic & Bernhardt, 2007); (b) (Gonçalves et al., 2013)

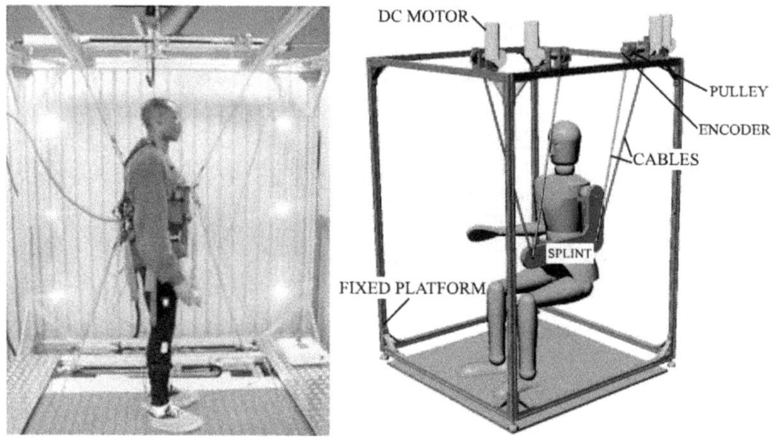

the next appropriate challenge and to promote the optimum therapy and best possible outcomes (Hocoma, 2019a).

The DIEGO®, Fig. 15 (c), developed by Tyromotion (2019) is a versatile, unilateral and bilateral rehabilitation device for shoulders and arms. This device adapts to the patient individual needs (assist-as-needed control with intelligent gravity compensation). DIEGO® also has a virtual reality system that stimulates and facilitates training with a three-dimensional therapeutic area, enabling the transfer of learning into ADL (Tyromotion, 2019).

The NeReBot ("NEuro REhabilitation roBOT"), Fig. 16 (a), is a 3 DOF robot designed to treat patients during the acute phase of the stroke. In NeReBot the patient upper limb is supported and manipulated by three cables, operated by electric motors independently. The compliance of this device is given by the under-actuated structure and unilateral actuation (cable). In this way, the cables assist the patient's limb motion along a pre-planned trajectory, while allowing out-of-path voluntary movements. The voluntary movements are allowed even when robotic assistance is provided. The NeReBot was used in stroke rehabilitation and improves the activities of daily living of stroke survivors when used as an additional treatment and in partial substitution of conventional rehabilitation therapy (Rosati et al., 2005; Stefano et al., 2014; Rosati et al., 2017).

MariBot (Marisa robot), Fig. 16 (b), is an evolution of NeReBot with five degrees of freedom. It has a hybrid structure and implements the same principle of NeReBot (three cables to suspend the patient upper limb), and has, additionally, two DOF (actuated by two motors) which are used to move the cables about the plane. In this way, it adapts the position of cable entry points according to the limb motion. This

Figure 15. Arm supports with gravity compensation: (a) Freebal; (b) ArmeoBoom®; (c) DIEGO®
(a) (Stienen et al., 2007); (b) (Elsa, 2019); (c) (Tyromotion, 2019)

architecture yields greatly improved capabilities in controlling human upper-limb motion at the joint level and also in the horizontal plane (Rosati et al., 2005; Rosati et al., 2017).

DESIGN CONSIDERATIONS AND SAFETY ISSUES

Unlike the robots used in industries or medical procedures (which are controlled by a qualified person), in a rehabilitation device, the patient is also the operator, so in this situation, the risks and safety issues are more critical (Carbone et al., 2017; Boschetti et al., 2019).

Most robots for rehabilitation are end-effector-based and exoskeleton-based and differ in how the movement is transferred from the robot to the patient. The end-effector robots have the advantage of their structure simplicity, consequently, the control algorithm is generally less complicated. However, most movements produced are complex and it is difficult to isolate specific joint, what determine injuries risks and complicated algorithms. The exoskeleton robot system mirrors the skeletal structure of the limb, associated with the joint motions. Exoskeletons allow

Figure 16. Cable-driven robots: (a) NeReBot; (b) MariBot
(Rosati et al., 2017)

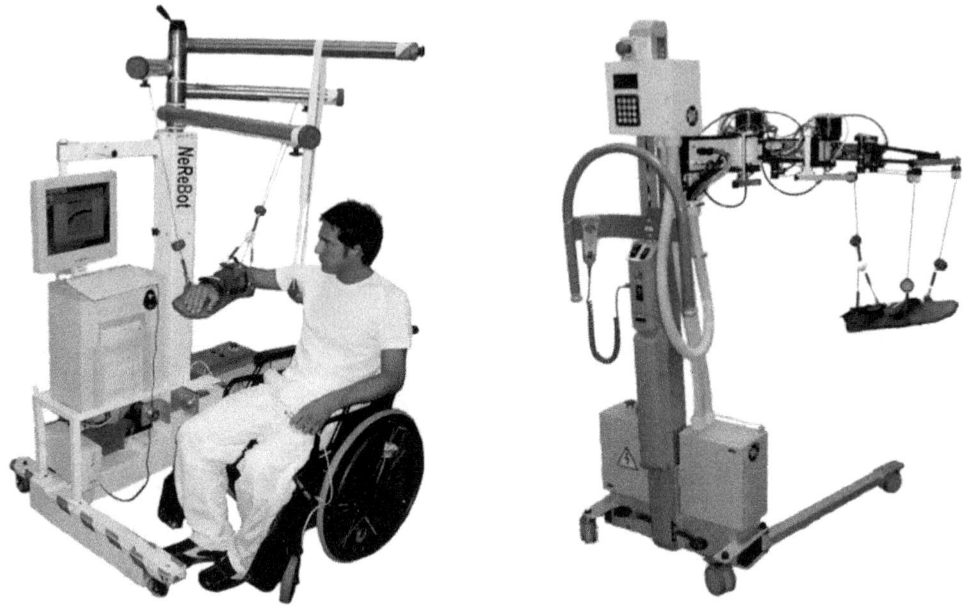

independent control of patient limbs motion in many joints. Manipulator segments lengths need to be adapted to the patient segments in order to avoid injuries. Also, the rotation center position of human body joints may change during the movement. Thus, setting these devices for different patients ensuring safety and comfort, even when assistance is present, is not a simple task and require special mechanisms. In this way, compared to end-effector-based systems, the complexity of the mechanical and control is greater in exoskeleton-based systems (Kiguchi et al., 2003; Nadas et al., 2017).

The human-robot interaction (HRI) is a field that studies the challenges associated with people and robots placed in shared settings. The assistive robotics aims the provision and administration of motivation, encouragement, and rehabilitation for humans that suffer from motor impairments and cognitive deficits. In the context of human-robot interaction, the rehabilitation devices can measure and apply forces and torques to the human's limb in order to encourage and assess the training of motor functions (Nadas et al., 2017).

Safety plays an important role in the design of robots for rehabilitation and any risk should be avoided for both the patient and the therapist.

Although there are no official standards set for rehabilitation robots, some key aspects can be extracted by considering other medical standards like Medical devices (2015) and human-robot interaction standards in the ISO Standards (2016).

Vaida et al. (2017) and Major et al. (2016) classified the human-robot interaction modalities in robot-assisted rehabilitation as shown in Fig. 17.

The modalities are briefly described below:

- **Assistive**: The robot assists by partially supporting the patient limb weight or leading the limb to perform the task.
- **Active**: The patient moves his limb voluntarily and the robot only monitors the limb without providing any forces on it.
- **Passive**: The robot makes the whole movement without any patient contribution;
- **Passive-Mirrored**: It is a bimanual configuration in which the healthy limb guides the affected one.
- **Corrective**: the patient's movement stops when the task can't be performed as indicated and the robot operates in an active mode.
- **Path-Guidance**: The robot assists the movement in a pre-determined trajectory and corrects any errors.
- **Resistive**: Robot opposes the movement, providing a certain resistive force.

Some aspects of the robotic rehabilitation devices can affect safety, for instance, these devices can move the patient limb outside its operation ranges, which could cause unwanted damage to the already affected limb. To solve this issue the desired operation workspace must be determined previously. During rehabilitation therapy, the joint is moved very slowly by the therapist or by an external-powered device. In

Figure 17. Diagram of HRI modalities in robot-assisted rehabilitation

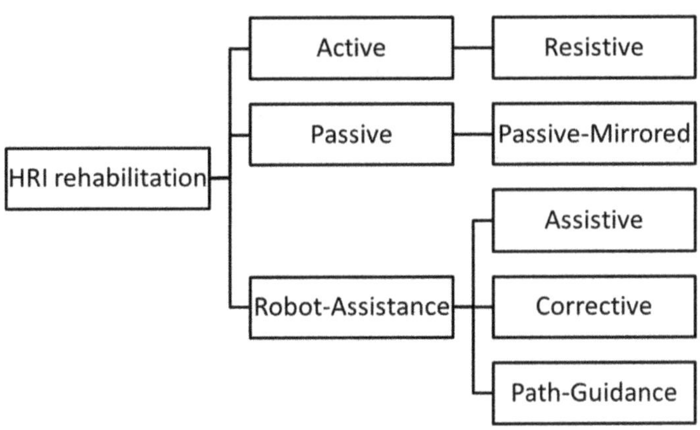

robotic rehabilitation, the robots can move the patient limb at a speed that is not safe, which also can incur additional injury. Finally, the robots can apply forces/torques that are not safe for the human limb and, therefore, the desired operation forces/torques need to be previously determined. In end-effector-based robots, a splint must be securely attached to the human limb, allowing the limb to perform the movement imposed by the robot. This type of attachment is not always appropriate because the limb is already injured and can compromise patient comfort. A safe procedure must be implemented to avoid any of the mentioned risks. It is also important the forces/torques not only be limited through software but also by hardware constraints (Gonçalves et al., 2015; Carbone et al., 2017).

Therefore, compliant devices which preserve the relationship between the patient's effort and the resulting movement of the limb may be utilized. They are easier to design, cheaper and safer compared to exoskeletons, but they do not allow control of single human joints and cannot measure, for instance, displacements, speeds, and torques. In this way, the cable-driven robots are a compliant device and can fill these requirements (Rosati et al., 2009).

MATHEMATICAL MODELING OF A CABLE-DRIVEN ROBOT

As mentioned above, there are many design difficulties to develop a suitable mechanical system for moving an injured human limb, either by using industrial robots, exoskeleton, or a cable-driven device. It is not easy to locate the position of the rotational center of the human joint, in general, the joint is modeled as a spherical joint or as a hinge joint and located inside the body (Gopura & Kiguchi, 2009).

The kinematic modeling of a cable-driven robot is very similar to the modeling of a parallel structure (Côté, 2003; Tsai, 1999). The inverse kinematic problem is to find the cable lengths, ρ_i, depending on the end-effector pose, and the forward kinematic problem consists of finding the end-effector poses for a given set of cable lengths ρ_i.

The parameters used for the kinematic model are shown in Fig. 18, where the reference *OXYZ* is connected to the base and *Cxyz* in the mass center of the mobile platform (Gonçalves et al., 2015).

The inverse kinematic model of the proposed parallel structure can be expressed as:

$$\rho_i = \left\| c + Qv_i - p_i \right\| \tag{1}$$

Figure 18. Kinematic parameters

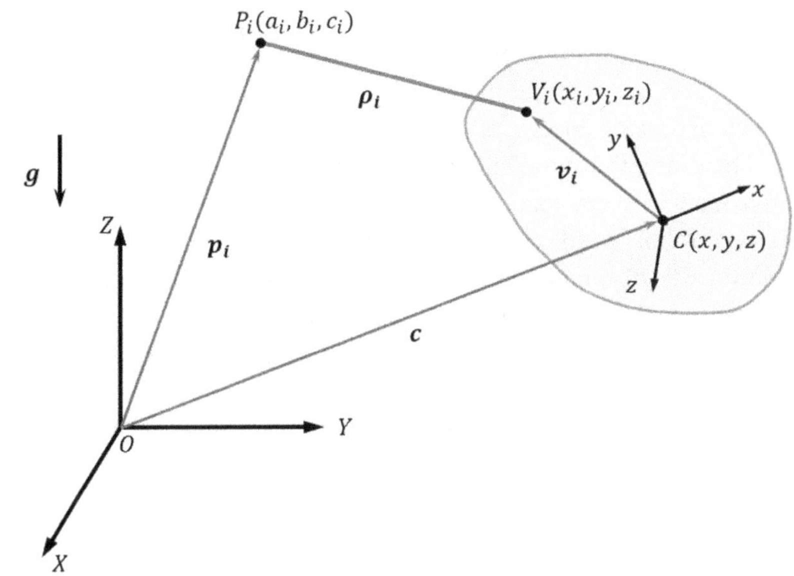

$$Q = \begin{bmatrix} \cos\beta\cos\gamma & -\cos\beta\sin\gamma & \sin\beta \\ \sin\theta\,\sin\beta\cos\gamma + \cos\theta\,\sin\gamma & -\sin\theta\sin\beta\sin\gamma + \cos\theta\cos\gamma & -\sin\theta\cos\beta \\ -\cos\theta\sin\beta\cos\gamma + \sin\theta\sin\gamma & \cos\theta\sin\beta\sin\gamma + \sin\theta\cos\gamma & \cos\theta\cos\beta \end{bmatrix} \quad (2)$$

With i varying from 1 to n (number of cables), where (Nunes et al., 2011a-c):

- p_i is the position vector of point P_i with components a_i, b_i and c_i in relation to fixed reference frame.
- v_i is the position vector of point V_i with components x_i, y_i and z_i related to the moving frame.
- C (c_x, c_y, c_z) is the position vector of the center of gravity of the moving platform.
- Q is the rotation matrix between fixed and moving frames obtained by a rotation of θ about x-axis followed by a second rotation β about the new y-axis and a third rotation γ about the new z-axis.
- ρ_i is the distance between points P_i and V_i (length of cable i).

When the cable-based manipulator performs a certain task, the end effector exerts force and moment on the external environment, and the forces are transmitted by the extension and retraction of the cables and ensure the condition of pulling the cables. Static force analysis is important in determining the quality of force transmission, which is a key aspect of energy efficiency and is necessary to obtain a feasible workspace. Therefore, the static analysis is made, considering that all cables must remain in tension under any load (Cafolla et al., 2019).

Cable-driven robots used in rehabilitation movements are, generally, slow and therefore inertia effects can be ignored. However, Newton-Euler formalism can be applied to obtain the dynamic analysis (Tsai, 1999), where the cable inertia can be ignored, the cables are considered rigid bodies and the inertia of the end-effector (splint/hand strap) should be considered.

The equilibrium equations for forces and moments acting on each cable can be expressed as:

$$\sum_{i=1}^{n} F_i = \sum_{i=1}^{n} F_i \hat{\rho}_i = P \text{ and } \sum_{i=1}^{n} t_i = \sum_{i=1}^{n} \hat{\rho}_i \times Qv_i = M \tag{3}$$

That can be written in matrix form:

$$[J]^T [F] = [W] \tag{4}$$

Where vector F represents the cable tension, which are forces that must be done by actuators, W is the vector of external forces and moments applied to the system, which are the limb and the end-effector weight and, J is the Jacobian matrix of the structure, $\hat{\rho}$ is the unitary vector defining the cable direction to the actuator.

For the structure with i cables, the Jacobian can be written as Eq. (5).

Equations (3) and (5) are used to evaluate the cable tension for a given trajectory in relation to the cable-based parallel architecture kinematics.

$$J = \begin{bmatrix} \hat{\rho}_1 & \hat{\rho}_2 & \cdots & \hat{\rho}_i \\ \hat{\rho}_1 \times Qv_1 & \hat{\rho}_2 \times Qv_2 & \cdots & \hat{\rho}_i \times Qv_i \end{bmatrix} \tag{5}$$

The workspace is one of the most important characteristics of the manipulators. The workspace is the set of position and orientation configurations in which the end effector is controllable, with tensions in cables that are positive, whose force values lie between a minimum and maximum in order to maintain cables in tension and to avoid breaking of cables, the end effector is far from singularities, and the wrapping of cables is avoided (Hiller et al., 2009; Merlet, 2004; Barrette and Gosselin, 2005).

EXAMPLES OF DESIGNS FOR CABLE-DRIVEN ROBOTS

In cable-driven robots, cable lengths are adjustable in order to control the end effector's position and orientation. This makes it easier to adapt to the different human limb sizes, allowing a much wider array of patients use. Based on exercises type and complexity, cables can be either added or removed. However, the more cables are introduced into the structure, the more complex and expansive it becomes. In the present section, three cable-driven robots design are presented as examples such a: LAWEX (Carbone et al., 2017), CUBE (Cafolla et al., 2019), and CAR (Alves et al. 2018; 2019).

LAWEX, Limb Wire Driven Exercising device

A cable-driven robot solution named LAWEX is presented in this topic. This proposed device, Fig 19, has four cables, allowing three degrees of freedom (Carbone et al., 2017). Figure 19 (a) shows the LAWEX 3D CAD model and Fig. 19 (b) shows the built prototype.

Figure 20 (a) shows the workspace of the robot LAWEX in the XZ plane (the sagittal plane) as coming from design procedure with simulations for operation characterization. In the same way, Fig. 20 (b) presents the robot and human workspace in the XY plane (top-view). The human forearm motion ranges in the 0-90 degrees in the XZ plane and 0-45 degrees in the XY plane, with 0 degrees in XZ plane. Figure 20 (c) superposes both workspaces, in 3D. The human forehand range of

Figure 19. A cable driven robot solution: (a) 3D CAD model; (b) built prototype LAWEX in 2016
(Carbone et al., 2017)

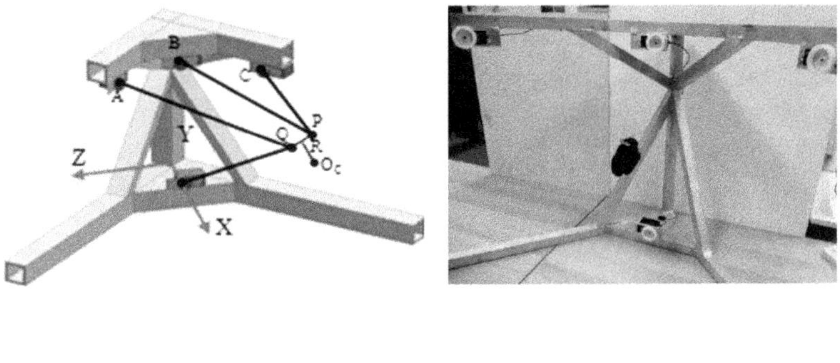

Figure 20. Simulation design results for the cable-driven robot in Fig. 19: (a) robot workspace in the sagittal plane (XZ), (b) robot workspace in the XY plane and (c) isometric view of the robot workspace
(Carbone et al., 2017)

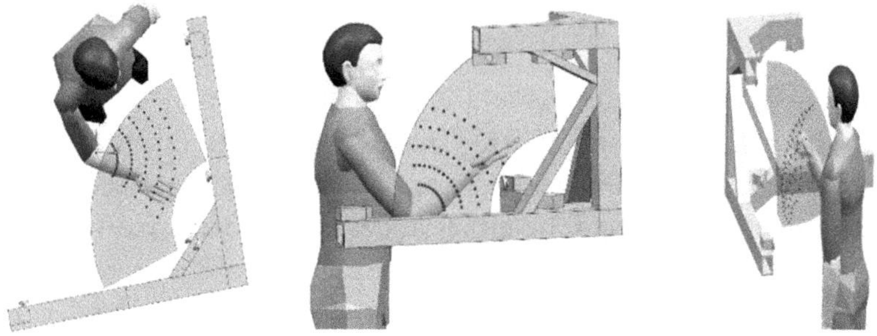

motion is covered by the robot workspace, proving that the robot is capable of the forearm rehabilitation task, Fig. 21 (Carbone et al., 2017).

Four 24 V DC motors with an encoder are used in the prototype. These actuators have a 1:150 reduction ratio, achieving a torque of 1.5 N·m. Each motor receives a speed input that depends on the current (acquired by the encoder) and the desired position difference. While the actuator is moving, the speed decreases until the desired position is reached. A dual H-Bridge motor driver module, L298N, has been chosen to control both the rotation sense and the speed. This motor driver allows speed and direction control of two DC motors. The control is performed by an Arduino MEGA that sends the sense of rotation and the desired speed using PWM to the motor drivers to reach the desired position. The synchronization of the

Figure 21. Numerical Evaluation of LAWEX Robot workspace in Fig. 20: (a) XY plane; (b) XZ plane; (c) 3D workspace
(Carbone et al., 2017)

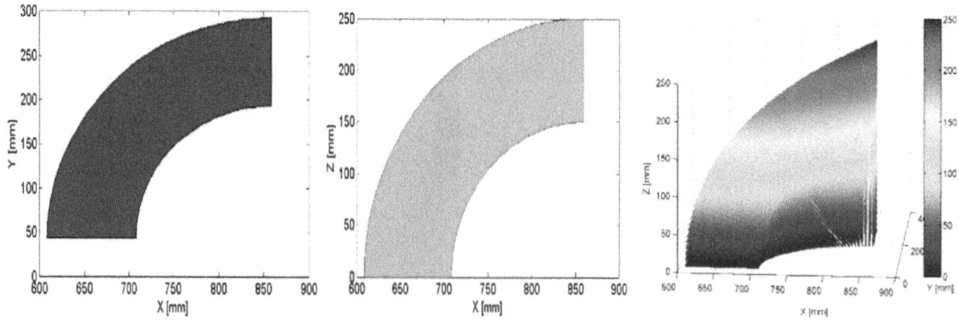

four motors is fundamental to maintain the cables in tension and allow the system to work properly.

CUBE, Cable-Driven Device for Upper and Lower Limb Exercising

The CUBE, Fig. 2 (a), is a 5 DOF parallel manipulator with a cable-driven architecture based on six cables for the rehabilitation of both upper and lower limbs. The kinematic of the device is characterized by a fixed frame with adaptable geometry, defined by the variable lengths h and d, Fig. 22 (b). The end-effector is ring-shaped and is worn by the user as a wristband. Its position is controlled by the six cables (end-effector with 3 DOF in position) while its orientation is assumed to be constrained by fixed support for the elbow on the Z-axis and fixed on the X and Y-axis. Three cables are attached to the upper part of the structure in points A_1, A_2 and A_3, and they converge into point A_H of the end-effector, Fig. 22 (b). The remaining three cables are attached to the lower part of the structure in points B_1, B_2 and B_3, and they converge into point B_H of the end-effector, Fig. 22 (b), (Cafolla et al., 2019).

Six 24V DC motors with an encoder are used. These actuators have a 1:150 reduction ratio, achieving a torque of 1.5 N·m. Each motor receives a speed input that depends on the current (acquired by the encoder) and the desired position difference. While the actuator is moving, the speed decreases until the desired position is reached. A dual H-Bridge motor driver module, L298N, has been chosen to control both the verse of rotation and the speed. This motor driver allows speed and direction control of two DC motors. The control is performed by an Arduino MEGA that sends the sense of rotation and the desired speed using PWM to the motor drivers to reach the desired position. The synchronization of the six motors is fundamental to maintain the cables in tension and allow the system to work properly.

Figure 22. CUBE design: (a) 3D CAD design; (b) a kinematic scheme (Cafolla et al., 2019)

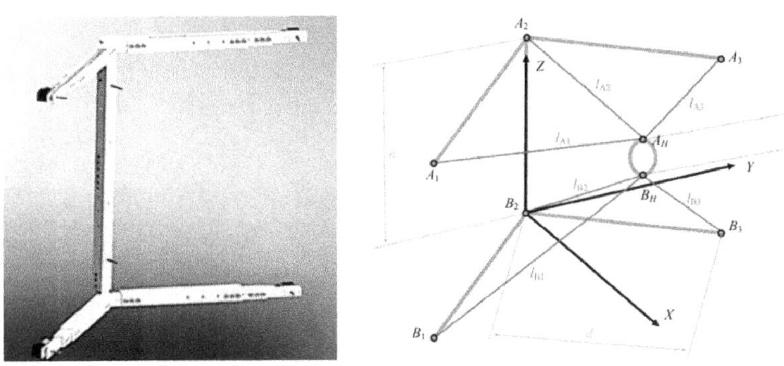

The reachable workspace of the CUBE design with constraints on cable lengths and tensions (which must be always positive) is shown in Fig. 23, it is adequate for the motion range of rehabilitation exercises. A control box, including microcontroller, motor drivers, power supply and voltage regulators, was designed and placed on the prototype making it fully portable (Cafolla et al., 2019).

CAR, Cable-Actuated Robot

The Cable-Actuated Robot (CAR), Fig. 24, was developed at the Laboratory of Robotics and Automation at Federal University of Uberlandia, Brazil. It is a 2 degrees of freedom cable-driven robotic structure with an end-effector (handle/strap).

The structure consists of an aluminum fixed frame, DC motors, load cells, and rotary encoders. The motors are powered by 24 V DC voltage, the nominal torque and power are respectively 10 N.m (48 N.m peak) and 46 W; rotational speed up to

Figure 23. A numerical evaluation of CUBE design in Fig. 22: (a) Isometric view; (b) front view; (c) lateral view; (d) top view
(Cafolla et al., 2019)

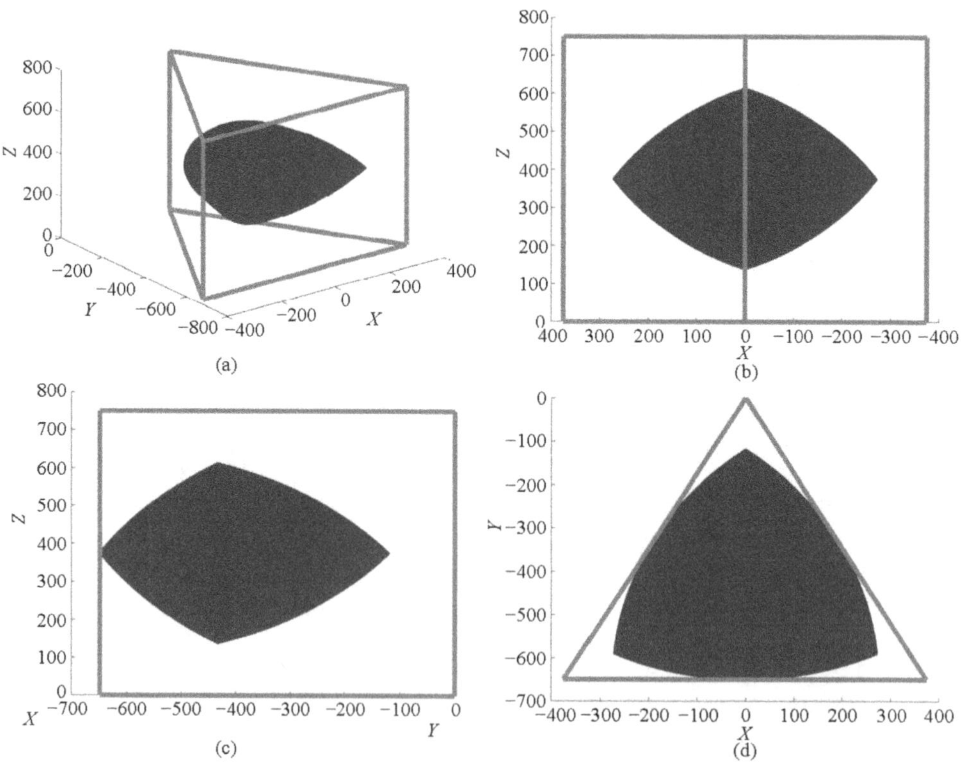

Figure 24. Cable-driven device CAR: (a) 3D CAD design; (b) built prototype

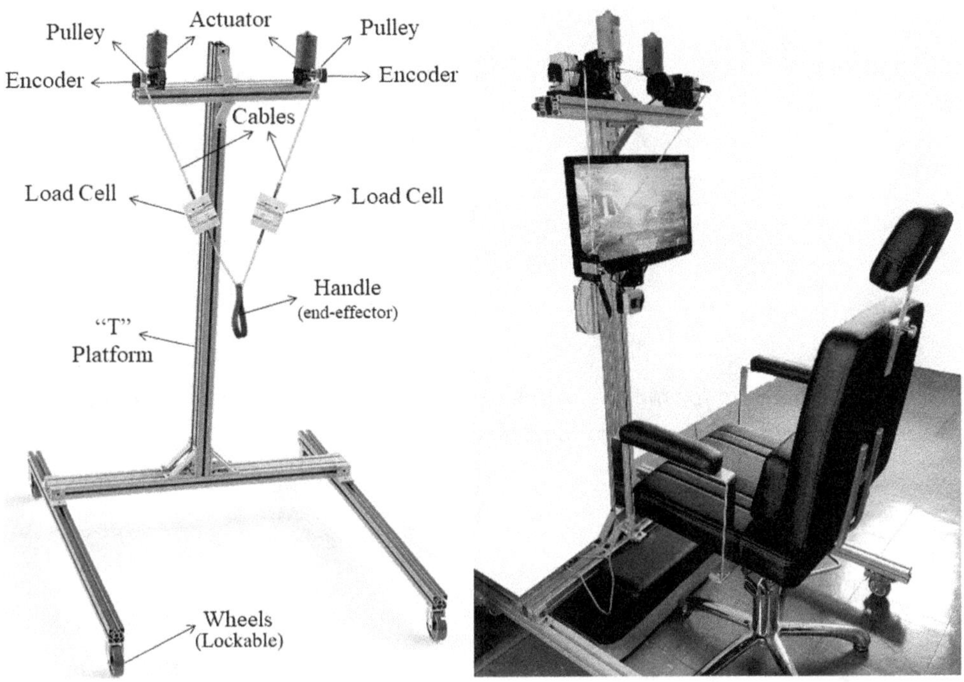

45 RPM. Each load cell has a 20 kgf capacity and the resolution of the measuring system is 0.025 kgf. The encoders are incremental and produce 500 ppr (pulses per revolution). Theses components are detailed on the CAD design in Fig. 24 (a). Figure 24 (b) shows the built prototype. The control is a position proportional and it is performed by an Arduino MEGA with MATLAB (through Arduino Hardware Support Package to enable communication) and a full bridge motor driver VNH2SP30 (up to 2 actuators and 30 A). The actuators speed input depends on the current and the desired position difference and, while moving, the speed decreases until the desired position is reached. The Arduino MEGA sends the sense of rotation and the desired speed using PWM to the motor driver to reach the desired position. The results of a validation performed with this device and a goniometer indicated high accuracy and repeatability levels, which the maximum absolute error was not greater than 1.55° and mean relative error was 2 ± 1% (Alves et al., 2019). This device can be used for both upper and lower limb. Simulations for the upper and lower limb movement using the CAR device is shown in Fig. 25 and Fig. 26 respectively. A numerical evaluation of CAR workspace is shown in Fig. 27.

Figure 25. Simulation for the upper limb using the CAR device: (a) initial position; (b) intermediate position and (c) final position

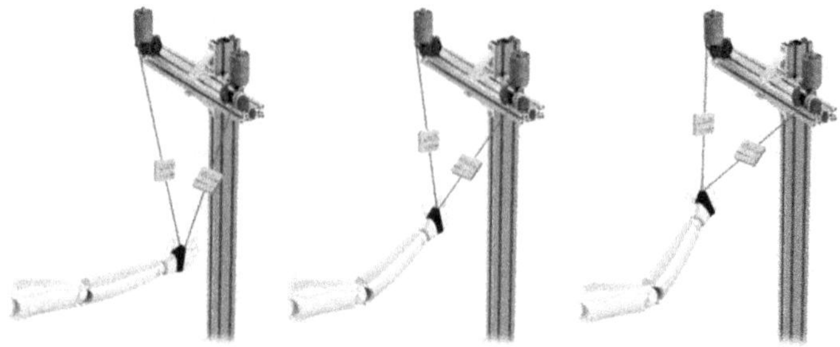

Figure 26. Simulation for the lower limb using the CAR device: (a) initial position; (b) intermediate position and (c) final position

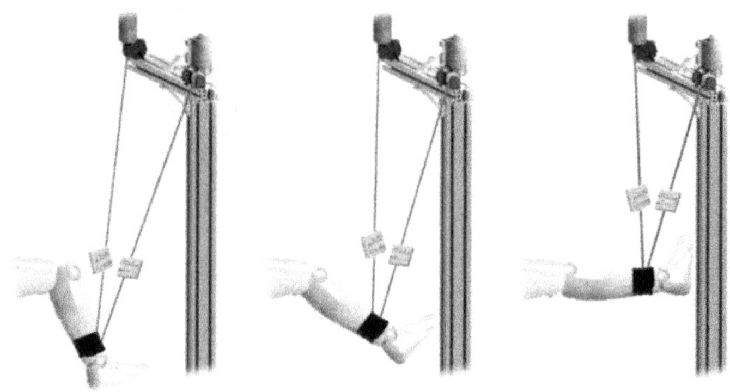

Figure 27. A numerical evaluation of CAR workspace

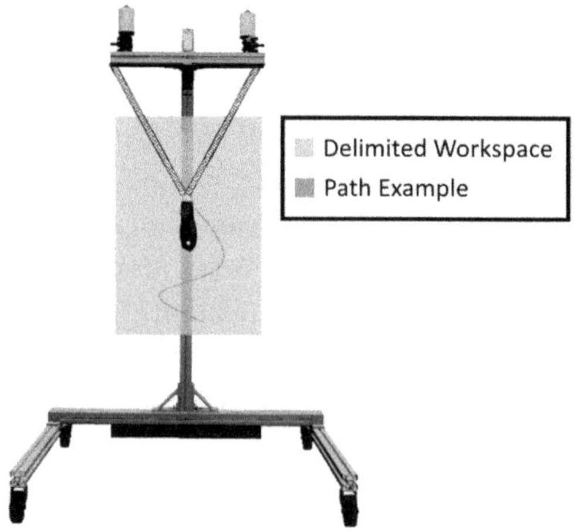

Clinical Evidence

Despite the potential benefits of robotic rehabilitation after stroke, there is no evidence that robotic training, using cable-driven robots, brings functional improvements compared to standard therapy: although motor benefits have been observed, robotic approaches have disparate responses depending on the rehabilitation problem, the device used and the characteristics of the patient (Rosati et al., 2010; Mehrholz et al., 2012; Masiero et al. 2014b; Rosati et al., 2017).

The successful stroke rehabilitation must be early intensive and task-specific therapy. A Cochrane review based on randomized controlled trials (RCTs), showed that early robotic training, during the acute or subacute phase, can enhance motor learning and improve functional abilities more than chronic-phase training (Masiero et al., 2011).

Arm function can be limited by a reduction in the selectivity of movements after a stroke. This is due to the involuntary coupling of shoulder abduction and elbow flexion, which limits the ability to reach. The cable-based system Freebal can help to reduce active shoulder abduction torques, resulting in a greater range of motion. Chronic stroke patients received reach training using the Freebal system and a rehabilitation game. After the training, reach performance (distance and direction) had improved in most of the patients, which indicates that this system can be a valuable training modality for stroke rehabilitation (Prange et al., 2009).

In Krabben et al. (2012) a study examined whether training using the Freebal improves unsupported arm function. Chronic stroke patients received gravity compensated reach training together with a motivating computer game. The work area of the hemiparetic arm increased significantly and also the Fugl-Meyer scores.

A multicenter RCT was performed in Prange et al. (2013) to compare the training effect using the ArmeoBoom® device and conventional reach training (with dose-matched approach) on the recovery of hand-arm function in 70 subacute stroke patients. The Fugl-Meyer scores improved by 10 points on the conventional group and 8 points in the ArmeoBoom® group, which indicates similar gains compared to conventional reach training, and that it is suitable for application in clinical practice. Additionally, therapists and patients reported positive experiences (Prange et al., 2013).

In a similar way, in Prange et al. (2015), a multicenter RCT was performed with 70 subacute stroke patients which received training with either the ArmeoBoom® (combined with computerized exercises) or dose-matched conventional training. Fugl-Meyer scores and reach distance improved significantly within both groups and, additionally, the ArmeoBoom® group reported higher interest/enjoyment during training than the conventional one.

The NeReBot was tested in two different randomized controlled trials (Masiero et al., 2007; Masiero et al., 2011; Masiero et al., 2014a). These studies involved stroke survivors in the acute/subacute phase, in which the brain has an increased plasticity capacity and so the benefits brought by rehabilitation are generally greater. To compare with standard rehabilitation, the RCTs tested two different robotic protocols: the first using robotic therapy in addition to traditional treatment (Masiero et al., 2007) and another replacing partially the standard rehabilitation program with a dose-matched approach (Masiero et al., 2011, Masiero et al., 2014a). Clinical results showed that the NeReBot substitutive therapy protocol could be considered comparable to the traditional one. Patients treated with the additional protocol developed a greater recovery of motor function and coordination than patients treated with the standard protocol alone (at the end of treatment and at follow-up).

These results raise the hypothesis that a mixed robotic protocol, in which additional robotic training is performed in the very early stage post-stroke may lead to anticipation of patient discharge (Rosati et al., 2017).

Thanks to the low-cost of cable-driven robots, the cost of such a protocol may be comparable to that of conventional therapy (Masiero et al., 2014c).

Stefano et al. (2014), compared several treatment protocols using NeReBot, with different combinations of robotic and non-robotic exercises. This study showed that robotic technology can be a valuable and economically sustainable aid in management post-stroke rehabilitation.

However, it is unclear whether the differences in treatment costs of each patient would favor robotic treatment compared to the conventional ones and a cost analysis study would be required to verify this. Moreover, the inconsistencies in the baseline characteristics of patients included in comparative studies make it difficult to objectively compare the groups.

CONCLUSION

The robotic rehabilitation design must consider all aspects, primarily the safety, ergonomics, and the motion control to not exceed the limits of joint movements, forces, and torques. This chapter presents the main upper and lower limb joints and movements as the design of a rehabilitation system requires a full understanding of the characteristics of the joints and the movements covered by it. The upper limb should be addressed by considering the amplitude for its main joint movements for healthy and stroke patients to analyze the effects of stroke in the human motion ranges.

Three types of mechanical systems for rehabilitation are introduced: rehabilitation robots, exoskeletons, and cable-driven parallel manipulators. Each of them has advantages and disadvantages related to performance, reconfigurability/ transportability, safety, comfort, and costs.

In a rehabilitation device for home use, the patient is also the operator, so in this situation, the risks and safety issues are more critical. The robotic rehabilitation devices need to provide a large workspace at the same time they must not move the patient limb outside its operation ranges, causing damage to the already affected arm. The speed and forces/torque need to be adjustable and controllable for the rehabilitation process. Related to the forces/torques, it is important not only to be limited through software but also by hardware constraints, in this way, the cables on the cable-driven robots can be designed to break at the desired tension force filling this requirement.

The great variability of the human limbs complicates the adjustment of the equipment (like exoskeletons) and one must calibrate the system for each patient. Cable-driven robots make it easier to adapt to the different human limb sizes, allowing a much wider array of patients use and meeting the requirements for safety, comfort and large workspace. Thus, cable-driven robots can be a good solution for an inherently safe device.

Some patients may need intelligent active assistance, which should be provided by the robotic rehabilitation system (by applying forces and torques to the human's limb) in order to assist and encourage them. Serious games for rehabilitation (computational exercises for rehabilitation purposes) can also help with patient motivation during therapy.

Finally, this chapter presented the design of three cable-driven devices to be applied for upper and/or lower limbs rehabilitation. The development of these devices is justified by a large number of people with upper and lower limb problems due to stroke. Simulations to determine the workspace had been carried through proving it can reproduce upper and lower limb movements and are feasible to be applied in rehabilitation therapy at clinics, hospitals or even patient's home, additionally, providing information about patient performance and progress. The authors of this chapter are working in an improved version of these devices with more sensors and better capability to correlate with the human limb's motions. After, experimental tests with patients will be performed. These devices will be a tool to assist health professionals like physiotherapists, doctors and the patient as self users.

The use of cable-based devices in the post-stroke treatment (in addition to traditional therapy) has shown encouraging clinical results, despite the limited capabilities of such systems in controlling precisely the motion of the patient's limb. These findings indicate that a low-cost device can achieve similar gains as conventional therapy and is suitable for application in clinical practice. However, more research and RCTs are needed to study this kind of robotic device involved in rehabilitation training.

REFERENCES

Alves, T., D'Carvalho, M. C., & Gonçalves, R. S. (2018). Controle "Assist-As-Needed" em Estruturas Robóticas Atuadas por Cabos para reabilitação das Articulações do Corpo Humano. ENEBI 2018, Águas de Lindóia, SP, Brazil. (in Portuguese)

Alves, T., D'Carvalho, M.C., & Gonçalves, R.S. (2019). Assist-As-Needed Control in A Cable-Actuated Robot for Human Joints Rehabilitation. *Journal of Mechanical Engineering and Biomechanics*.

Bang, D.H., & Shin, W.S. (2016). Effects of robot-assisted gait training on spatiotemporal gait parameters and balance in patients with chronic stroke: A randomized controlled pilot trial. *NeuroRehabilitation*, *38*(4), 343–349.

Barbosa, A. M., Carvalho, J. C. M., & Gonçalves, R. S. (2018). Cable-driven lower limb rehabilitation robot. *Journal of the Brazilian Society of Mechanical Sciences and Engineering*, *40*(5), 245. doi:10.100740430-018-1172-y

Beebe, J. A., & Lang, C. E. (2009). Active Range of Motion predicts Upper Extremity Function Three months post-stroke. *Stroke*, *40*(5), 1772–1779. doi:10.1161/STROKEAHA.108.536763 PMID:19265051

Beyl, P. (2009). Safe and compliant guidance in robot-assisted gait rehabilitation using proxy-based sliding mode control. *IEEE 11th International Conference on Rehabilitation Robotics*. 10.1109/ICORR.2009.5209505

Boschetti, G., Carbone, G., & Passarini, C. (2019). Cable Failure Operation Strategy for a Rehabilitation Cable-Driven Robot. *Robotics*, *8*(1), 17. doi:10.3390/robotics8010017

Cafolla, D., Russo, M., & Carbone, G. (2019). CUBE, a Cable-driven Device for Limb Rehabilitation. *Journal of Bionic Engineering*, *16*(2).

Laribi, M. A., Carbone, G., & Zeghloul, S. (2019). On the Optimal Design of Cable Driven Parallel Robot with a Prescribed Workspace for Upper Limb Rehabilitation Tasks. *Journal of Bionics Engineering*, *16*, 503. doi:10.100742235-019-0041-4

Carbone, G., & Ceccarelli, M. (2016). *Sistema a cavi per assistenza motoria* [Cable driven system for motion assistance]. Patent Application n.102016000038975.

Carbone, G., Gherman, B., Ulinici, J., Vaida, C., & Pisla, D. (2017). Design Issues for an Inherently Safe Robotic Rehabilitation Device. In *Mechanisms and Machine Science 49: Advances in Service and Industrial Robotics* (Vol. 49, pp. 1025–1032). Springer. doi:10.1007/978-3-319-61276-8_110

Ceccarelli, M., Pîslă, D., & Graur, F. (2010). Design and operation issues for parallel robotic devices in the rehabilitation of stroke patients. *International Journal of Mechanics and Control, 11*(1), 51–57.

Ceccarelli, M., & Romdhane, L. (2010). Design Issues for Human-Machine Platform Interface in Cable Based Parallel Manipulators for Physiotherapy Applications. *Journal of Zhejiang University. Science A, 11*(4), 231–239. doi:10.1631/jzus.A1000027

Colombo, R., Sterpi, I., Mazzone, A., Delconte, C., & Pisano, F. (2012). Taking a Lesson from Patients' Recovery Strategies to Optimize Training During Robot-Aided Rehabilitation. *IEEE Transactions on Neural Systems and Rehabilitation Engineering, 20*(3), 276–285. doi:10.1109/TNSRE.2012.2195679 PMID:22623406

Côté, G. (2003). *Analyse et conception de mécanismes parallèles actionnés par cables* (M.Sc. dissertation). Université Laval, Quebec. (in French)

Cyberdyne. (2019). *Cyberdyne HAL*. Retrieved from https://www.cyberdyne.jp/english/products/HAL/

Dipietro, L., Krebs, H. I., Fasoli, S. E., Volpe, B. T., Stein, J., Bever, C., & Hogan, N. (2007). Changing Motor Synergies in Chronic Stroke. *Journal of Neurophysiology, 98*(2), 757–768. doi:10.1152/jn.01295.2006 PMID:17553941

Duret, C., Grosmaire, A.-G., & Krebs, H. I. (2019). Robot-Assisted Therapy in Upper Extremity Hemiparesis: Overview fo an Evidence-Based Approach. *Frontiers in Neurology, 10*, 412. doi:10.3389/fneur.2019.00412 PMID:31068898

Dzahir, M. A. M., & Yamamoto, S.-I. (2014). Recent Trends in Lower-Limb Robotic Rehabilitation Orthosis: Control Scheme and Strategy for Pneumatic Muscle Actuated Gait Trainers. *Robotics, 3*(2), 120–148. doi:10.3390/robotics3020120

Elsa, Hocoma Armeo Boom - Elsa. (2019). Disponível em: http://www.elsa.web.tr/tr/urun/robotik-rehabilitasyon/hocoma-armeo-boom

Gassert, R., & Dietz, V. (2018). Rehabilitation robots for the treatment of sensorimotor deficits: A neurophysiological perspective. *Journal of Neuroengineering and Rehabilitation, 15*(1), 46. doi:10.118612984-018-0383-x PMID:29866106

Gates, D.H., Walters, L.S., Cowley, J., Wilken, J.M., & Resnik, L. (2016). Range of Motion Requirements for Upper-Limb Activities of Daily Living. *Am J Occup Ther, 70*(1). doi: . doi:10.5014/ajot.2016.015487

Gonçalves, R. S., & Carvalho, J. C. M. (2010). Desenvolvimento de uma estrutura robótica paralela atuada por cabos para reabilitação dos movimentos do ombro. *VI Congresso Nacional de Engenharia Mecânica - CONEM 2010*. (in Portuguese)

Gonçalves, R. S., & Carvalho, J. C. M. (2014). Robot modeling for physical rehabilitation. In *Robotics: Concepts, Methodologies, Tools, and Applications*. IGI Global. . doi:10.4018/978-1-4666-4607-0.ch058

Gonçalves, R. S., Carvalho, J. C. M., & Lobato, F. S. (2016). Design of a robotic device actuated by cables for human lower limb rehabilitation using self-adaptive differential evolution and robust optimization. *Bioscience Journal*, *32*(6). doi:10.14393/BJ-v32n1a2016-32436

Gonçalves, R. S., Carvalho, J. C. M., Ribeiro, J. F., & Salim, V. V. (2015). *Cable-Driven Robot for Upper and Lower Limbs Rehabilitation. In Handbook of Research on Advancements in Robotics and Mechatronics* (pp. 284–315). IGI Global. doi:10.4018/978-1-4666-7387-8.ch011

Gonçalves, R. S., Carvalho, J. C. M., Rodrigues, A. A. O., & Barbosa, A. M. (2013). Cable-Driven Parallel Manipulator for Lower Limb Rehabilitation. *Applied Mechanics and Materials*, *459*, 535–542. doi:10.4028/www.scientific.net/AMM.459.535

Gonçalves, R. S., & Krebs, H. I. (2017). MIT-Skywalker: Considerations on the Desing of a Body Weight Support System. *Journal of Neuroengineering and Rehabilitation*, *14*(1), 88. doi:10.118612984-017-0302-6 PMID:28877750

Gonçalves, R. S., Soares, G., & Carvalho, J. C. M. (2019). Conceptual design of a rehabilitation device based on cam-follower and crank-rocker mechanisms hand actioned. *Journal of the Brazilian Society of Mechanical Sciences and Engineering*, *41*(7), 277. doi:10.100740430-019-1772-1

Gopura, R. A. R. C., & Kiguchi, K. (2009). Mechanical designs of active upper-limb exoskeleton robots. *11th Int. Conf. on Rehabilitation Robotics*, 178-187.

Hatem, S. M., Saussez, G., Della Faille, M., & .... (2016). Rehabilitation of Motor Function after Stroke: A Multiple Systematic Review Focused on Techniques to Stimulate Upper Extremity Recovery. *Frontiers in Human Neuroscience*, *10*(September), 1–22. PMID:27679565

Hiller, M., Hirsch, K., Bruckmann, T., Brandt, T., & Schramm, D. (2009). Common aspects in methods for the design of mechatronic systems - Applications in automotive and robotic systems. *XII International Symposium on Dynamic Problems of Mechanics – DINAME 2009*.

Hirschhorn, A. D., Lockhart, J. W., & Breckenridge, J. D. (2015). Can a physical activity monitor provide a valid measure of arm elevation angle? A study to assess agreement between the SenseWear Mini Armband and the universal goniometer. *BMC Musculoskeletal Disorders*, 16–46. PMID:25886361

Hocoma. (2019a). *Armeo® Therapy Concept*. Retrieved from https://products.iisartonline.org/products/24/marketing/BRO_Armeo_Therapy_Concept_130225_en.pdf

Hocoma. (2019b). Retrieved from http://www.hocoma.com/

Hogan, N. (1985). Impedance control: an approach to manipulation: Part I- Theory, Part II- Implementation, Part III - Applications. *ASME Journal of Dynamic Systems. Measurement and Control*, *107*(1), 1–24. doi:10.1115/1.3140702

Hussain, A., Budhota, A., & Contu, S. (2017). Quantitative assessment of motor functions post-stroke: Responsiveness of upper-extremity robotic measures and its task dependence. *IEEE International Conference on Rehabilitation Robotics*, 1037–1042. 10.1109/ICORR.2017.8009386

InMotion. (2018). *InMotion ARM™ Interactive Therapy System - Bionik Labs*. Retrieved from http://bionikusa.com/healthcarereform/upper-extremity-rehabilitiation/inmotion2-arm/

InMotion. (2019). *Helps Traumatic Brain Injury Patient's - Bionik Labs*. Accessed Feb 13, 2019, <https://www.bioniklabs.com/products/inmotion-arm>

Ionescu, T.G. (2003). Terminology for the Mechanism and Machine Science. *Mech. and Mach. Theory*, (38), 819-825.

Jezernik, S. (2003). Robotic orthosis lokomat: A rehabilitation and research tool. Neuromodulation: Technology at the Neural Interface. *Wiley Online Library*, *6*(2), 108–115.

Jezernik, S., Colombo, G., & Morari, M. (2004). Automatic Gait-Pattern Adaptation Algorithms for Rehabilitation with A 4-Dof Robotic Orthosis. IEEE Transactions on Robotics and Automation, 20(3), 574-582.

Kahn, L. E., Averbuch, M., & Rymer, W. Z. (2001). *Comparison of Robot-Assisted Reaching to Free Reaching in Promoting Recovery from Chronic Stroke*. National Institute on Disability and Rehabilitation Research.

Kahn, L. E., Zygman, M. L., Rymer, W. Z., & Reinkensmeyer, D. J. (2006). Robot-assisted reaching exercise promotes arm movement recovery in chronic hemiparetic stroke: A randomized controlled pilot study. *Journal of Neuroengineering and Rehabilitation*, 1–13. PMID:16790067

Kapandji, A. I. (2007). The Physiology of the Joints: Vol. 1. *The Upper Limb* (6th ed.). Churchill Livingstone.

Kapandji, A. I. (2010). The Physiology of the Joints: Vol. 2. *Lower Limb* (6th ed.). Churchill Livingstone.

Khosravi, M., & Taghirad, H. D. (2011). Dynamic analysis and control of cable driven robots with elastic cables. Transactions – Canadian Society for Mechanical Engineering 35(4), 543-557. doi:10.1139/tcsme-2011-0033

Kiguchi, K., Iwami, K., Yasuda, M., Watanabe, K., & Fukuda, T. (2003). An exoskeletal robot for human shoulder joint motion assist. *IEEE/ASME Transactions on Mechatronics*, *8*(1), 125–135. doi:10.1109/TMECH.2003.809168

Krabben, T., & Prange, G. B. (2012). Influcnce of gravity compensation training on synergistic movement patterns of the upper extremity after stroke, a pilot study. *Journal of Neuroengineering and Rehabilitation*, 9–44. PMID:22824488

Krebs, H.I., Bruce V., and Hogan N. (2009). A working model of stroke recovery from rehabilitation robotics practitioners. *Journal of NeuroEngineering and Rehabilitation, 6*(1).

Krebs, H.I., Finley, M.A., Dipietro, L., Ohlhof F, J., Whitall, J. & Bever, C.T. (2004). *Does MIT-MANUS upper extremity robot testing create a learning effect in healthy adults?* Academic Press.

Krebs, H. I., Palazzolo, J. J., Dipietro, L., Ferraro, M., Krol, J., Rannekleiv, K., ... Hogan, N. (2003). Rehabilitation Robotics: Performance-Based Progressive Robot-Assisted Therapy. *Autonomous Robots, Springer*, *15*(1), 7–20. doi:10.1023/A:1024494031121

Kuznetsov, A. N., Rybalko, N. V., Daminov, V. D., & Luft, A. R. (2013). Early Poststroke Rehabilitation Using a Robotic Tilt-Table Stepper and Functional Electrical Stimulation. Stroke Research and Treatment. doi:10.1155/2013/946056

Maciejasz, P., Eschweiler, J., Gerlach-Hahn, K., Jansen-Troy, A., & Leonhardt, S. (2014). A survey on robotic devices for upper limb rehabilitation. *Journal of Neuroengineering and Rehabilitation, 11*(3), 1–29. PMID:24401110

Mahler, J., Krishnan, S., Laskey, M., & Sen, S. (2014). Learning accurate kinematic control of cable-driven surgical robots using data cleaning and Gaussian process regression. *IEEE International Conference on Automation Science and Engineering*. 10.1109/CoASE.2014.6899377

Major, K. A., Major, Z. Z., & Carbone, G. (2008). Ranges of motion as basis for robot-assisted post- stroke rehabilitation. *Human & Veterinary Medicine - International Journal of the Bioflux Society*, *8*(4), 192–196.

Major, K. A., Major, Z. Z., Carbone, G., Pîslă, A., Vaida, C., Gherman, B., & Pîslă, D. (2016). Ranges of motion as basis for robot-assisted post-stroke rehabilitation. *J Bioflux Soc Hum Vet Med*, *8*(4), 192–196.

Mao, Y., Jin, X., Gera Dutta, G., Scholz, J. P., & Agrawal, S. K. (2015). Human Movement Training with a Cable Driven ARm EXoskeleton (CAREX). *IEEE Transactions on Neural Systems and Rehabilitation Engineering*, *23*(1), 84–92. doi:10.1109/TNSRE.2014.2329018 PMID:24919202

Martin, E. M., & Cazorla, M. (2019). *Rehabilitation Technology: Assistance from Hospital to Home*. Hindawi Computational Intelligence and Neuroscience. doi:10.1155/2019/1431509

Masiero, S., Armani, M., Ferlini, G., Rosati, G., & Rossi, A. (2014a). Randomized trial of a robotic assistive device for the upper extremity during early inpatient stroke rehabilitation. *Neurorehabilitation and Neural Repair*, *28*(4), 377–386. doi:10.1177/1545968313513073 PMID:24316679

Masiero, S., Armani, M., & Rosati, G. (2011). Upper-limb robot-assisted therapy in rehabilitation of acute stroke patients: Focused review and results of new randomized controlled trial. *Journal of Rehabilitation Research and Development*, *48*(4), 355–366. doi:10.1682/JRRD.2010.04.0063 PMID:21674388

Masiero, S., Celia, A., Rosati, G., & Armani, M. (2007). Robotic-assisted rehabilitation of the upper limb after acute stroke. *Archives of Physical Medicine and Rehabilitation*, *88*(2), 142–149. doi:10.1016/j.apmr.2006.10.032 PMID:17270510

Masiero, S., Poli, P., Armani, M., Ferlini, G., Rizziello, R., & Rosati, G. (2014c). Robotic upper limb rehabilitation after acute stroke by nerebot: Evaluation of treatment costs. *BioMed Research International*. PMID:24967345

Masiero, S., Poli, P., Rosati, G., Zanotto, D., Iosa, M., Paolucci, S., & Morone, G. (2014b). The value of robotic systems in stroke rehabilitation. *Expert Review of Medical Devices*, *11*(2), 187–198. doi:10.1586/17434440.2014.882766 PMID:24479445

Mayhew, D., Bachrach, B., Rymer, W. Z., & Beer, R. F. (2005). Development of the MACARM-a novel cable robot for upper limb neurorehabilitation. *Proc. of ICORR 9th International Conference on Rehabilitation Robotics*, 299-302. 10.1109/ICORR.2005.1501106

Medical Devices. (2015). *Recognized essential principles of safety and performance of medical devices*. BS ISO 16142-2.

Mehrholz, J. (2012). Electromechanical and robot-assisted arm training for improving generic activities of daily living, arm function, and arm muscle strength after stroke. *Cochrane Database of Systematic Reviews*, *6*, CD006876. PMID:22696362

Moore, K. L., & Dalley, A. F. (2009). *Clinically oriented anatomy* (6th ed.). Philadelphia: Lippincott Williams & Wilkins.

Nadas, I., Vaida, C., Gherman, B., Pisla, D., & Carbone, G. (2017). Considerations for designing robotic upper limb rehabilitation devices. *AIP Conference Proceedings*, *1917*, 030005. doi:10.1063/1.5018278

Nef, T., & Riener, R. (2005). ARMin: Design of a novel arm rehabilitation robot. *IEEE 9th Int. Conf. Rehabilitation Robotics - ICORR2005*, 57–60.

Novak, D. E., & Riener, R. (2015). Control Strategies and Artificial Intelligence in Rehabilitation Robotics. *AI Magazine*, *36*(4), 23. doi:10.1609/aimag.v36i4.2614

Nunes, W. M., Ribeiro, J. F., Carvalho, J. C. M., & Gonçalves, R. S. (2011a). Kinematics modeling and workspace analysis of a cable-based parallel manipulator for shoulder rehabilitation. *21th Brasilian Congress on Mechanical Engineering – COBEM 2011*.

Nunes, W. M., Ribeiro, J. F., Carvalho, J. C. M., & Gonçalves, R. S. (2011b). Workspace Analysis of a Cable-Base Parallel Manipulator for Rehabilitation of Shoulder Movements. *Proceedings of the XIV International Symposium on Dynamic Problems of Mechanics*.

Nunes, W. M., Rodrigues, L. A. O., Oliveira, L. P., Ribeiro, J. F., Carvalho, J. C. M., & Gonçalves, R. S. (2011c). Cable-Based Parallel Manipulator for Rehabilitation of Shoulder and Elbow Movements. *International Conference on Rehabilitation Robotics*. 10.1109/ICORR.2011.5975503

Cafolla, D., Russo, M., & Carbone, G. (2019). CUBE, a Cable-driven Device for Limb Rehabilitation. *Journal of Bionics Engineering*, *16*, 492. https://doi.org/10.1007/s42235-019-0040-5

Pennycott, A., Wyss, D., Vallery, H. K., Marganka, V., & Riener, R. (2012). Towards more effective robotic gait training for stroke rehabilitation: a review. Journal of NeuroEngineering and Rehabilitation, 9(65).

Prange, G. B. (2009) An explorative Study into Changes in Reach Performance After Gravity Compensation Training in Chronic Stroke Patients. *IEEE 11th International Conference on Rehabilitation Robotics.*

Prange, G. B., Kottink, A. I. R., Buurke, J. H., Eckhardt, M. M. E. M., van Keulen-Rouweler, B. J., Ribbers, G. M., & Rietman, J. S. (2015). The effect of Arm Support combined with rehabilitation games on upper-extremity function in subacute stroke: A randomized controlled trial. *Neurorehabilitation and Neural Repair*, *29*(2), 174–182. doi:10.1177/1545968314535985 PMID:24878589

Prange, G. B., Kottink, A. I. R., Buurke, J. H., & Rietman, J. S. (2013). Application of arm support training in sub-acute stroke rehabilitation: first results on effectiveness and user experiences. *IEEE International Conference on Rehabilitation Robotics.* 10.1109/ICORR.2013.6650470

Reinkensmeyer, D. J., & Boninger, M. L. (2011). *Technologies and combination therapies for enhancing movement training for people with a disability. J Neuroeng Rehab.*

Rosati, G. (2010). The place of robotics in post-stroke rehabilitation. *Expert Review of Medical Devices*, *7*(6), 753–758. doi:10.1586/erd.10.49 PMID:21050086

Rosati, G., Masiero, S., & Rossi, A. (2017). "On the Use of Cable-Driven Robots in Early Inpatient Stroke Rehabilitation." Advances in Italian mechanism science. *International Journal of Mechanics and Control*, *18*(2), 551–558.

Schmidt, R. A. (2018). Motor control and learning: a behavioral emphasis (6th ed.). Human Kinetics Publishers.

ISO Standards. (2016). Robots and robotic devices. *Collaborative robots*, PD ISO/TS 15066:2016.

Stefano, M., Patrizia, P., Mario, A., Ferlini, G., Rizzelli, R., & Rosati, G. (2014). Robotic Upper Limb Rehabilitation after Acute Stroke by NeReBot: Evaluation of Treatment Costs. *BioMed Research International*, *2014*, 1–5. doi:10.1155/2014/265634 PMID:24967345

Stein, J. (2012). Robotics in rehabilitation: technology as destiny. *Am J Phys Med Rehabil, 91*(11), S199–S203.

Stienen, A. H. A. (2007). Freebal: dedicated gravity compensation for the upper extremities. *Proceedings of the 2007 IEEE 10th International Conference on Rehabilitation Robotics*. 10.1109/ICORR.2007.4428517

Stienen, A. H. A. (2009). *Development of novel devices for upper-extremity rehabilitation* (Ph.D. dissertation). University of Twente.

Surdilovic, D., Zhang, J., & Bernhardt, R. R. (2007). STRING-MAN: Wire-robot technology for safe, flexible and human-friendly gait rehabilitation. *IEEE 10th Int. Conf. on Rehabilitation Robotics*, 446-453.

Takahashi, K., Domen, K., Sakamoto, T., Toshima, M., Otaka, Y., Seto, M., ... Hachisuka, K. (2016). Efficacy of Upper Extremity Robotic Therapy in Subacute Poststroke Hemiplegia. *Stroke*, *45*(5), 1385–1388. doi:10.1161/STROKEAHA.115.012520 PMID:27006452

Takebayashi, T., Takahashi, K., Amano, S., Uchiyama, Y., Gosho, M., Domen, K., & Hachisuka, K. (2018). Assessment of the Efficacy of ReoGo-J Robotic Training Against Other Rehabilitation Therapies for Upper-Limb Hemiplegia After Stroke: Protocol for a Randomized Controlled Trial. *Frontiers in Neurology*, *9*, 730. doi:10.3389/fneur.2018.00730 PMID:30210446

Takeuchi, N., & Izumi, S.-I. (2013). Rehabilitation with Poststroke Motor Recovery: A Review with a Focus on Neural Plasticity. Stroke Research and Treatment. doi:10.1155/2013/128641

Tucan, P., Vaida, C., Plitea, N., Pisla, A., Carbone, G., & Pisla, D. (2019). Risk-Based Assessment Engineering of a Parallel Robot Used in Post-Stroke Upper Limb Rehabilitation. *Sustainability*, *11*, 2893.

Tavolieri, C. (2008). *Design of a cable-based parallel manipulator for rehabilitation applications* (Ph.D. dissertation). University of Cassino, Italy, and INRIA, France.

Tyromotion. (n.d.). *Diego*. Retrieved from https://tyromotion.com/en/produkte/diego/

Vaida, C., Carbone, G., Major, K., Major, Z., Plitea, N., & Pisla, D. (2017). ACTA TECHNICA NAPOCENSIS. *Applied Mathematics. Mechanical Engineering*, *60*, 91–102.

Wolbrecht, E. T., Chan, V., Reinkensmeyer, D. J., & Bobrow, J. E. (2008). Optimizing compliant, modelbased robotic assistance to promote neurorehabilitation. *IEEE Transactions on Neural Systems and Rehabilitation Engineering*, *16*(3), 286–297. doi:10.1109/TNSRE.2008.918389 PMID:18586608

Chapter 4
Intelligent Processes in Automated Production Involving Industry 4.0 Technologies and Artificial Intelligence

Alessandro Massaro
https://orcid.org/0000-0003-1744-783X
Dyrecta Lab srl, Italy

Nicola Contuzzi
Dyrecta Lab srl, Italy

Angelo Galiano
Dyrecta Lab srl, Italy

ABSTRACT

The chapter presents different case studies involving technology upgrading involving Industry 4.0 technologies and artificial intelligence. The work analyzes four cases of study of industry projects related to manufacturing process of kitchen, tank production, pasta production, and electronic welding check. All the cases of study concern the analysis of engineered processes and the inline implementation of image vision techniques. The chapter discusses other topics involved in the production process such as augmented reality, quality prediction and predictive maintenance. The classic methodologies to map production processes are matched with innovative technologies of image segmentation and data mining predicting defects, machine failures, and product quality. The goal of the chapter is to prove how the combination of image processing techniques, data mining approaches, process simulation, chart process modeling, and process reengineering can constitute a scientific research project in industry research.

DOI: 10.4018/978-1-7998-1382-8.ch004

INTRODUCTION

Industry 4.0 facilities are important about the upgrade of manufacturing processes (Rubmann et al., 2015). Automatism and image vision techniques can represent a good solution in order to improve quality control procedures (Massaro et al., 2018) according with ISO 9001:2015 standard, thus implementing Industry 4.0 control systems. Also traceability represent an important topic about process digitalization and process mapping. In this direction different technologies such as barcode, QR-code and RadioFrequency IDentification –RFID- can be adopted to trace and map production (Lotlikar et al., 2013). Traceability models should be integrated into management tools (Khabbazi et al., 2010) thus improving the knowedge base –KB-. Augmented reality – AR- can represent a further facility for monitoring manufacturing processes (Caudell et al., 1992), aided manufacturing (Novak-Marcincin et al., 2013), and quality check (Segovia et al., 2015). Specifically, by adopting marker-based approach in AR systems (Siltanen, 2012), it is possible to improve computer vision, image processing and computer graphics techniques. Image vision and image processing are suitable for automatic processes involving real time quality control of products. In this direction a study has been performed for welding monitoring of electronic boards by executing 2D and 3D image processing algorithms (Massaro & Vitti et al., 2018). In particular the 3D image processing could enhance and show hidden information (Cicala et al., 2014) thus contributing to view better possible defects and other construction details. Other sensors can be applied on production lines in order to achieve predictive maintenance (Massaro & Galiano et al., 2018). According with predictive calculus, data mining algorithms can be implemented into an information system in order to predict failures of single machines or of the whole production lines (Massaro & Maritati et al., 2018). In order to control the production during the time, can be adopted different charts mapping processes such as 4M charts (Favi et al., 2017), Plan Do Check Act -PDCA- (Chakraborty, 2016), Xm-R (Fouad et al., 2010), and p control charts able to check defect rate (Wang, 2009) and process quality (Acosta-Mejia, 1999). These charts can be joined with data mining and artificial intelligence –AI- algorithms to formulate a predictive maintenance scheduling model. In literature data mining has been implemented for multilevel alerting systems enabling predictive maintenance (Bastos et al., 2014), thus suggesting different industry research implementations and data flow architectures also in other application fields (Massaro, Maritati, Savino et al., 2018). Still remaining in the context of predictive maintenance, some authors applied artificial neural networks –ANNs- for the prediction of mechanical component failures and degradation prediction (Zhang, 2015), formulating innovative processes inherent in optimizing prediction and defining optimal training dataset (Krenek et al., 2016). These studies prove that different technologies can be applied in order to optimize production processes. According with the state of the art, the goal of the proposed chapter is to discuss preliminary research industry projects involving automated processes embedding artificial intelligence, data mining algorithms, and image processing tools together with Industry 4.0 enabling technologies. Specifically in this chapter are discussed some requirements of industry projects, following research and development –R&D- guidelines (Frascati, 2015): different cases of studies proves how it is possible to provide knowledge gain in production processes, supporting at the same time production traceability and quality assessment. The knowledge gain can be performed by innovative algorithms which can be structured in different ways depending on the specific requirements of the case of study. Knowledge gain can be achieved also by formulating association rules between the production processes or by applying scientific methodologies for data processing. In any case the first step to increase the knowledge in industry is to digitize the information as for Industry 4.0 logic. Some important topics of Industry 4.0 are Internet of Things -IoT-, Industrial Internet, Smart Manufacturing, and Cloud based Manufacturing (Vaidya et al. 2018). These topics (Oztemel et al., 2018) are implemented in different models enabling Industry 4.0 facilities (Basl et al., 2019). In this scenario artificial intelligence –AI- (Skobelev at al., 2017) could improve adaptive processes by predictive maintenance, collaborative robotics and rapid prototyping (OECD 2017). An upgraded Industry 4.0 architecture oriented on Industry 5.0 logic includes artificial intelligence, big data systems and IoT facilities (Özdemir et al., 2018). Following the state of the art the proposed chapter shows some cases of study involving Industry 4.0 facilities thus enhancing how the topics of Industry 4.0 can be applied in practical cases.

The chapter is structured by discussing the following main specifications of four industry research projects describing how it is possible to increase the information thus optimizing production processes and quality procedures:

- **Case of Study 1**: Upgraded production lines for kitchen manufacturing implementing Industry 4.0 facilities.
- **Case of Study 2**: Engineered processes optimized for tank production.
- **Case of Study 3**: Quality process and monitoring in pasta industry.
- **Case of Study 4**: Electronic board welding check and classification.

For each project will be discussed the main requirements focusing the attention on research topics defined in the 'Frascati' manual (Frascati, 2015).The chapter provides an overview about enabling technologies suitable for the design and the optimization of industry production processes.

CASE OF STUDY 1: INTELLIGENCE IN MANUFACTURING PROCESS OF KITCHEN

Main Specifications

In this section are discussed the facilities of an industry project based on an upgrade of the technology related the production of kitchens. Before to design the new production line has been mapped the "AS IS"

production processes. This process mapping allowed to map the new processes (processes "TO BE") taking into account the necessary found facilities. In the table 1 are indicated the main specifications and the main functions necessary to improve production processes and related to the new production processes.

Where the listed facilities are necessary:

- To intervene in real time in case of potentially dangerous anomalies that can cause a stopped production, allowing a more rapid resolution of problems (rapid knowledge of the type of error).
- To update the machine software in real time (network connection).
- To check the operating status of parts of the production line deemed "essential" for production.
- To reduce production downtime.
- To manage the kitchen production remotely.
- To interface the computer numerical control –CNC- machines directly with the technical office and with the design drawings to be imported in input of the numerical control interfaces.
- To speed up production.
- To drastically reduce maintenance.
- To produce new models of kitchens.

The project therefore involves the acquisition of:

- Necessary equipment and works for the development of prototypes and for the enhancement of automation systems.
- Related technical services of a technical-specialist type to support the direct implementation of the design of new products (through new Computer-Aided Design –CAD- projects to be sent to the CNC boards of the various machines enabled for the production), to the installation of a network that combines different technologies (optical reading, technology PLC, ethernet connectivity, programmable logic controller –PLC- network, augmented reality –AR- technology, CNC machine interfaces, management interface

Table 1. Production upgrade facilities (production of kitchen top)

Machine/Function	Short Description of the Facilities Considered Important for the Project Purposes
Increased warehouse capacity	-increase in warehouse capacity (primary point for the planning of production increase as the warehouse usually represents the "bottleneck" for the increase in production volumes); - Tools modification improving knowledge base (for the integration of information associated with the warehouse, the creation of lists with connection to the management system will be re-analyzed).
Transport automation (automatic connection of the batteries formed in the warehouse up to the loader)	- Roller and chain conveyors; - Automatic pantograph loader (with inverter control); - Photocell barriers; - Control by inverter of the transport load. The automation serves to increase the production volume in safety and through transport control.
Connection automation (panel warehouse transport and edge banding)	- Roller and / or belt conveyor (speed control by inverter).
Kitchen Top recirculation automation (beading machine unloading)	- Conveyor controlled by inverter; - Pick and Place manipulator.
Pantograph-beading machine automation and banding exit	- Conveyor with speed controlled by inverter; - Sectioning of the transport with independent motorization (greater control).
Packing automation	- Gearmotor controlled by inverter; - Structure and transport; - Tool for panel centering.
Kitchen top processing machine (electronic panel numerically controlled)	- axis control driven by inverter through brushless motors (drastic maintenance reduction); - automatic lubrication; - construction of dimensional drawings of pantograph input; - application tool for work procedures oriented on material optimization
Double-sided squaring machine for longitudinal processing (first pass) of rectangular panels	- high speed working groups; - possibility to vary the type of edge thickness to be machined; - possibility to vary different setting fields; -machine control tool (positioning axes); -electric adjustment of the track advancement speed, thickness height according to the panel thickness, movable upright opening according to the piece size; - automatic management of edge exchange; - separate management of processing and diagnostics; -ability to work at a maximum speed of 50 m/min on the first pass in the case of thin edge application.
Double-sided squaring machine for transversal machining (second pass) of rectangular panels	- software tool interface for operator; - statistics tool; - PC numerical control (management of single machines or of complete squaring lines); - positioning axis control interface; - electric adjustment of the track advancement speed, thickness height according to the panel thickness, movable upright opening according to the piece size;
Loader, unloader and turntable (to be integrated with the squaring machines)	Increase of the speed up of the handling bringing until 15-17 cycles / 1'.
Edge squaring machine for the processing of the kitchen doors	- software tool interface for operator; - statistics software; - PC numerical control (management of single machines or complete squaring lines); - positioning axis control interface; - electric adjustment of the track advancement speed, thickness height according to the panel thickness, movable upright opening according to the piece size.

modules) for a complete exchange of data and digitized information according with Industry 4.0 logic.

For the production line upgrade can be implemented the following enabling technologies (integrating information along the production chain from the raw material supplier to the customer, and integrating information between machines of the production line):

- First enabling technology allowing integration of maintenance systems for the machine diagnosis by remote control.
- Second enabling technology allowing continuous monitoring of process parameters with vertical integration (pertaining to the integration of the different phases of the production chain).
- Third enabling technology implemented by augmented reality (augmented reality supporting the assembly production process).
- Fourth enabling technology allowing simulation of interconnected machines to optimize production processes.

The general functional scheme that contains the enabling technologies listed above is shown in Fig. 1. In this scheme are also indicated the optical reading stations enabling production traceability thus gaining the production knowledge. Preliminary have been identified the following control point for the optical traceability system:

1. Cut and input section (traceability of warehouse materials).
2. Cut-off and sectioning exit and squaring-in entrance.
3. Edge-banding exit and drilling line entry.
4. Inserter, drilling, inlet, and outlet section.
5. Assembly exit and packaging entry.
6. Packaging exit and transport line entry (logistic traceability).

Other facilities are:

- PLC/ethernet network for the management of machinery diagnostics, for production control and for interfacing of CNC processing.
- Design software for passing information from project cad files to CNC production machine interfaces adopting:
    - Static models related to standard kitchen designs.
    - Dynamic models (product upgrades) related to the 'ad hoc' design of particular furniture (dynamic modification in cutting, edging, finishing, etc.).

- Software module for the integration of the design software, compatible with the optical traceability system.
- Integration of optical readers with dedicated processors and database system with app service linking and with the management system.
- Integration of the entire information data flow with the management system.
- Augmented reality workstation with 1 viewer as prototype suitable for training contents and for viewer assembly procedures.
- Neural network algorithm (data mining algorithms implemented by Rapid Miner, or Orange Canvas, or KNIME, Weka or other similar tools) predicting machine malfunctions and consecutively defects and formulating a maintenance planning).
- Production mapping (by adopting PDCA cycle charts or other cause-effect diagrams);
- Simulation of production activities.
- Formulation of a business plan template.
- Production indicators indicating the following performance parameters.
  - Production speed per line.
  - Suitability of goods shipped.
  - Packaging withdrawals.
  - Sales volume.
  - Operational quality (for example% of incorrect forms / forms sent, or % of deliveries made / number of scheduled deliveries, or even % number of requests for compensation / number of shipments).

Figure 1. Functional scheme of project "UpgradedKitchens"

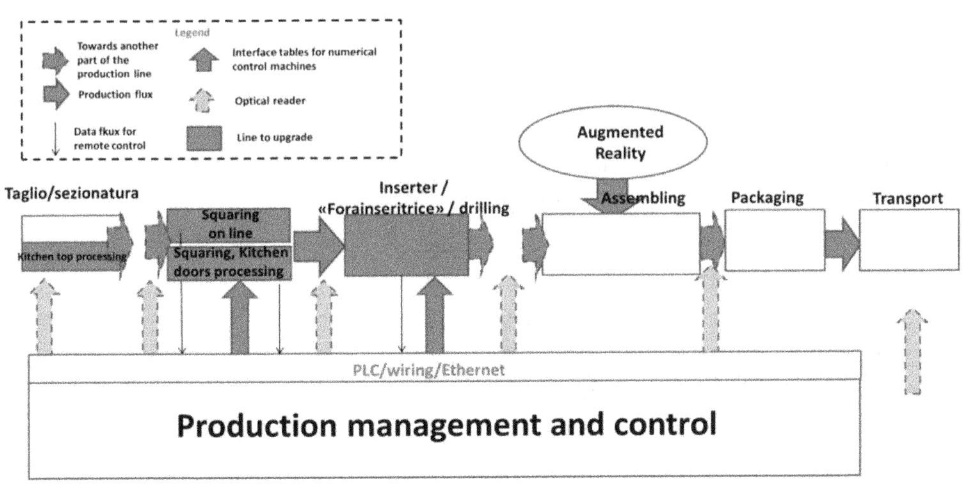

- Technology efficiency of new production machinery in terms of failure rate reduction.
    - Easiness of intervention on the production processes.
    - Efficiency of logistics synchronized with traceability.
    - Management efficiency of warehouse and material stocks.
    - Percentage machine failures over time.
- Other important elements to quantify are:
    - Comparison of productivity with respect to the historical production data.
    - Calculation of the incidence of costs on sales.
    - Quantification and control of anomalies.
    - Comparison of production data with respect to the budget.
    - Competition performance of a specified kitchen.
    - Comparison of production data with respect to the standard.
    - Comparison of production with the set objectives.
    - Customer expectations.
    - Sales prediction (useful for the business intelligence).
    - Warehouse supply prediction (business intelligence).

Project Results

We illustrated now some development of the previous specifications addressing the research in scientific topics.

Concerning augmented reality, in Fig. 2 is illustrated the system implemented in the project enabling by a remote access the interactive process to intervene in specific production processes. In this case the marker-based AR approach is fundamental in order to improve inline control processes.

In Fig. 3 is illustrated the optical traceability layout of the new production line where are indicated the following process monitoring positions:

- **Position 1**: Bench and exit.
- **Position 2**: Transfer to central belts.
- **Position 3**: Roll table frame and exit belts.
- **Position 4**: Roller transfer 850x4000x900.
- **Position 5**: Roller transfer and wheels 850x2500x900.
- **Position 6**: Roller transfer 850x4450x900.
- **Position 7**: Double lift with lung.
- **Position 8**: Roller transfer 850x4450x900.

Figure 2. Augmented reality in "UpgradedKitchens" project (enabling technology of Industry 4.0)

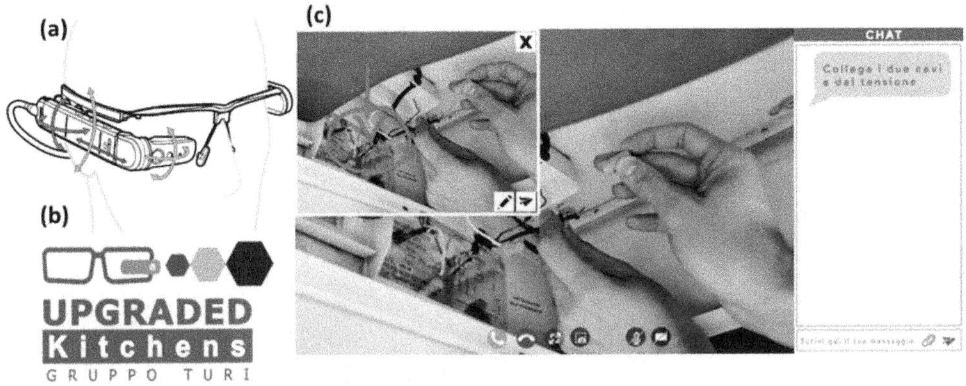

In order to improve the knowledge gain by innovative algorithms, has been applied the ANN algorithm by considering the defects data of the production line. In Fig. 4 is illustrated an example of ANN data processing performed by a KNIME workflow (Probabilistic Neural Network –PNN- based on the Dynamic Decay Adjustment –DDA- method). In this normalized graph, the predicted defect can be estimated also by the average value which is low (the production line will work properly). For the calculation of Fig. 4 has been considered a relative percentage of 80% as first partition (learning partition) and a dataset of 24 records (24 rows). This last example proves how the mapping of the upgraded production line can provide important information about product quality and production line predictive maintenance. The knowledge gain can be adopted also to improve quality processes according with ISO 9001:2015 standard.

Figure 3. Production optical traceability in "UpgradedKitchens" project

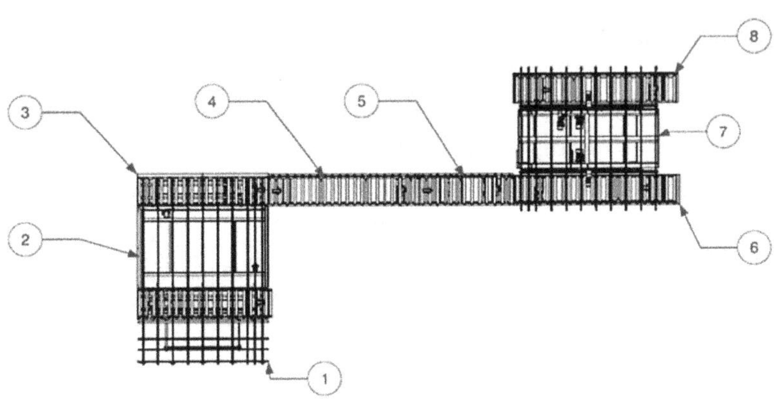

Figure 4. Defect prediction of the new production line. Inset: KNIME experimental ANN workflow

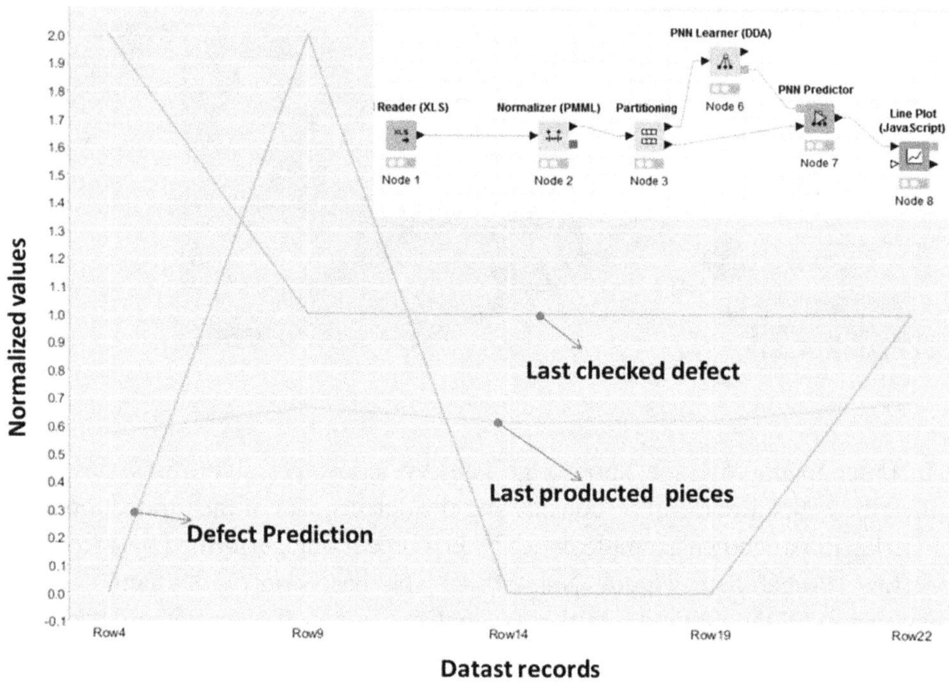

All the listed specifications, technologies and algorithms represent a prototype platform gaining the knowledge base of the company thus optimizing production and business intelligence.

CASE OF STUDY 2: INTELLIGENCE IN TANK PRODUCTION

Main Specifications

In this section are discussed the main requirements of a second research project involving enabling technologies and artificial intelligence for knowledge gain in a specific production process. Also for this case of study, an upgrade of the production line is necessary in order to optimize the production processes. In Table 2 are listed the facilities of the new upgraded system able to produce tanks.

In this case, the design idea consists of creating a prototype production line that also incorporates technologies of:

Table 2. Prototypes of tank production line

Prototype/Machine Function	Short Description of the Facilities Considered Important for the Project Purposes
Load system	- cutting Lines; - unwinding; - cutting group systems; - control panel; - PLC integration; - prototype data extraction interface .
Roller stapler	- welding setting system; - PLC integration; - prototype data extraction interface.
Mechanical components prototypes to adapt to the production line	- mechanical component prototype.

1. Artificial Intelligence for the Analysis of Production KPIs and Predictive Maintenance

KPIs are variables that companies use to measure, track and analyze production performance. This data is usually used to evaluate a company's success and to determine if and how the established operational and strategic objectives are achieved. Measuring performance is fundamental in every business area to be able to implement improvement processes. In a manufacturing company it is therefore very important to be able to identify the strategic Key Performance Indicators –KPIs- to evaluate the production trend in terms of efficiency, service level and process quality.

However, KPIs must meet the following requirements:

- KPIs must be numbers.
- **Practicality:** KPIs must be integrated with current business processes.
- **Directionality:** KPIs must indicate if a company is improving.
- Operations can be related to the practical context to measure an actual change.

Predictive maintenance requires advanced analysis of operational data with the aim of determining asset conditions and intervening where and when needed. It requires analysis directly derived from data science, understood as a multidisciplinary field in which quantitative methodologies converge, such as statistics, data mining, operational research, machine learning, with the aim of transforming data into knowledge.

Specifically, the artificial intelligence engine will process the machine data (logs, operating status variables, set speeds, machine downtime records, etc.) that will travel through the PLC network and will be stored in an appropriate database. The predictive maintenance, as suggested by the state of the art, will develop on three levels of alarm:

- **Level 1:** Analysis of exceeding threshold values (real time alerting).
- **Level 2:** Short / medium term malfunction forecast.
- **Level 3:** Long-term malfunction forecast.

2. Image Vision

qualitative analysis of welds or of certain control points with particular reference to the verification of welding. For this module we hypothesize to use the image segmentation methodology.

3. Mapping of Production Processes:

"AS IS", "TO BE" maps, Xm-R maps, PDCA modules, 4M graphs will be drawn up.

Project Goal

The above specifications (facilities) are necessary to allow:

- To intervene in real time in case of potentially dangerous anomalies that can cause a stopped production, allowing a more rapid resolution of problems (rapid knowledge of the type of error).
- To update the machine setting in real time (network connection).
- To check the operating status of parts of the production line deemed "essential" for production.
- To reduce production downtime.
- To manage production remotely.
- To speed up production
- To drastically reduce maintenance
- To produce new product models with higher quality standards.

These improvements may be implemented considering different subgroups inserted in different levels (Fig. 5). First of all, is necessary to have a dynamic control of each operating unit (Operational Level) that ensure that all machines operate at the maximum efficiency, that the quality of the product responds to the identified KPIs and that the energy consumption is minimized. These devices can be transmitters, actuators or solenoid valves, with the purpose of detecting and manipulating the process and recording all events. This subgroup communicates (in both directions) with the elements present in the Control Level and supply data to Artificial Intelligence.

A supervision and coordination system determine the Control Level. This subgroup, using the analysis provided by the Artificial Intelligence, determines and sets the production of each unit in order to optimize their operation, manages the control of the execution of the operations, coordinates the system and reports the operational data, collecting and maintaining the queues of production data, raw materials, and semi-finished products for the units under its control. This system, also, responds to emergencies of any unit present in the previous level, in collaboration with the dynamic control system and Artificial Intelligence, in order to stop or reduce the activities of the units to compensate for emergencies.

The Artificial Intelligence subgroup defines the activities of verification and optimization of processes in real time. It uses the information taken by the sensors of the Operational Level and, independently and in self-learning mode, communicates the improvement solutions to the Control Level, that has, also, the final goal to implement them in all phases of the production process.

The management of this subgroup is entrusted to AI algorithms that use external supports (image vision, comparative thermography, pressure switches, etc.) that optimize the system.

The subgroup Area Level defines the activities to coordinate the production resources necessary to achieve the desired final products. It is capable of performing plant programming functions in order to produce the products required for these orders in the optimal service combination. This subgroup coordinates the system and reports the operational data to the highest levels and manages the personnel functions and the internal training and qualification of personnel.

In Fig. 5 is illustrated a block diagram of the proposed model improving tank production processing by a knowledge gain adopting the production upgrade and artificial intelligence.

The improvements therefore involve the acquisition of:

- Necessary prototype components and works for the development of prototypes and for the enhancement of production systems compatible with the interconnection of an artificial intelligence engine.
- Related technical services of a technical and scientific type to support the direct implementation of the design of new processes/products and a suitable system for a complete exchange of data and information.

Specifically, as regards innovation with respect to what was found in the state of the art, the following points are identified:

Figure 5. Functional design scheme of the innovative model supporting tank production processes

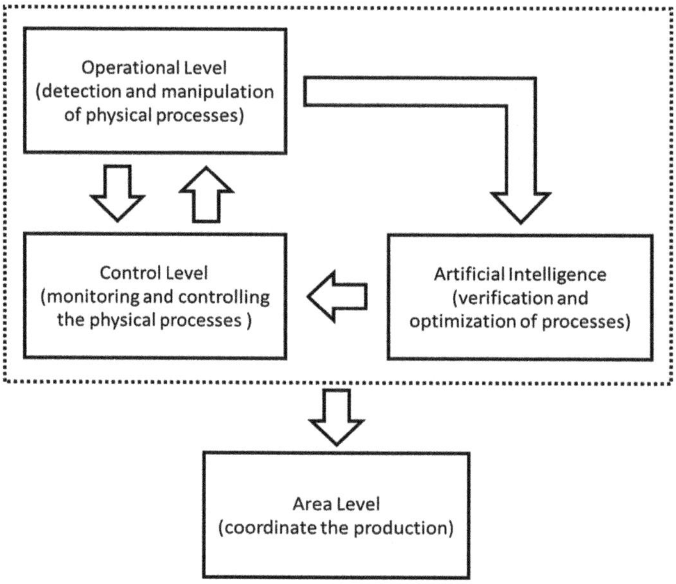

- The scientific methodologies of mapping of the processes will be combined with the realization of mechanical prototype components, which, integrated in an information system, will provide data for the elaboration of KPIs and predictive maintenance.
- The KPIs and the predictive maintenance will be applied to a specific case study, trying to overcome the critical issues related to the creation of a training model (training) as appropriate as possible to the case study in order to reduce the prediction error.
- Several data mining / artificial intelligence algorithms will be combined, correlating different attributes (multi-attribute data processing), formulating combined algorithms.
- New quality analysis processes will be formulated by applying image vision techniques based on image segmentation (weld analysis).

In Fig. 6 is illustrated the block diagram of the new production line involving artificial intelligence data processing and image vision techniques thus improving production process control and tank quality.

All the listed facilities can increase the knowledge associated with the production processes thus optimizing the production quality. Other technologies such image vision can be applied to further improve quality check as for welding check.

Figure 6. Functional scheme of the project "TankProductionAI"

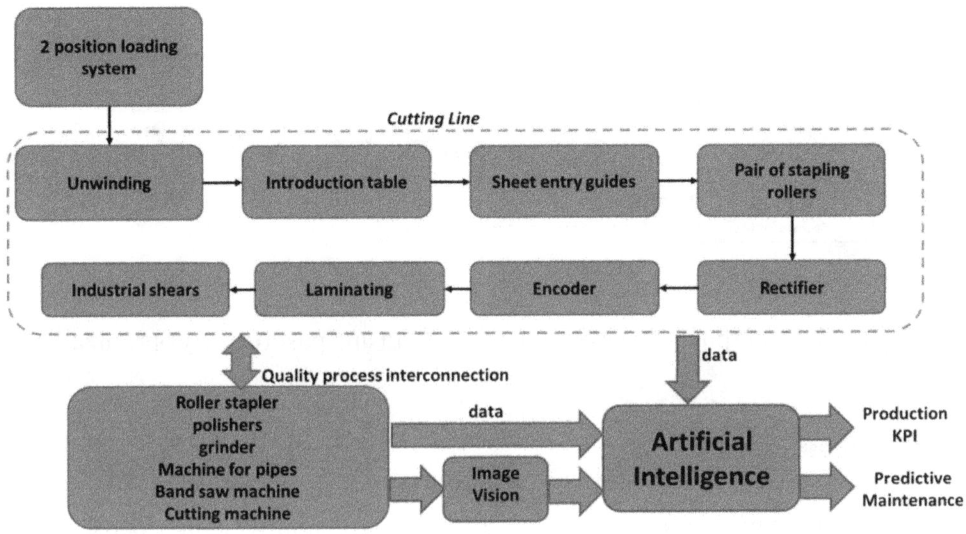

CASE OF STUDY 3: PROCESS ENGINEERING APPROACH AND IMAGE VISION TECHNIQUES IN PASTA INDUSTRY

Main Specifications

In this section is discussed a research industry project of pasta production process optimization. The research project covers the preliminary study of communication technologies and sensors applicable to pasta processing. The study is focused on Industry 4.0 enabling technologies applicable to the entire supply chain of pasta. The research starts from a first mapping of production processes useful for understanding how the Internet of Things –IoT- and sensing technologies in general can provide added value to the pasta industry processes. Of particular importance is the identification of potential communication and data transfer channels related to process upgrades and machinery monitoring. In this context, process simulations represents an important topic of the research which is oriented production monitoring methods. In Fig. 7 is illustrated a conceptual scheme of the research to be carried out. In this scheme are identified the main following phases:

- Identification of software and hardware enabling technologies suitable for the current production process of the company.
- Integration of these technologies in a prototype system relevant to the entire production chain and useful for process management.

Figure 7. Functional scheme of the project "Pasta IoT 4.0"

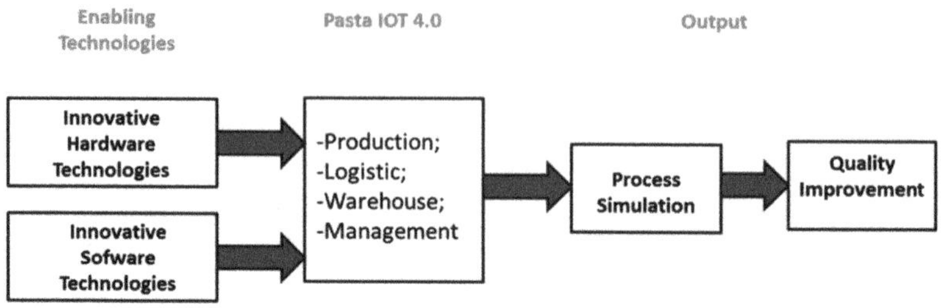

- Technology output (process simulation output and pasta quality optimization).

Project Results

The illustrated main phases are related to the whole supply chain. Focusing the attention on pasta quality monitoring, it is possible to check and classify defects during the production process by image vision techniques. Specifically, in the example of Fig. 8 is shown the image processed of a piece of pasta. The image segmentation provides the highlighting of the pasta defects viewed as white spots. By applying the "Analyze particles" function of ImageJ tool it is possible to extract the geometrical features of the pasta defects represented as ellipses enclosing white spots. Figure 8 (a) shows the defects spots enhanced by the red color. Figure 8 (b) represents the extracted ellipses characterizing the defects which are listed in the table of Fig. 8 (c) indicating intensity statistics (mean, modal, median, min. & max. gray value, standard deviation, integrated density, and major and minor axis lengths). The preliminary image pre-processing setting is an important first stage of the algorithm to apply. This step can be improved by setting RGB color thresholds as illustrated in Fig. 8 (d). An example of defect classification results is plotted in Fig. 8 (e), reporting the areas of the classified defects expressed in pixels^2 (the same solution can be expressed in mm^2 by using the 'set scale' function). This tool can be adopted for the inline quality processing, and can be improved by an artificial neural network predicting during the time defects classes by processing historical data of the pasta production line. By setting threshold limits it is also possible to define automatic alerting regions.

The pasta defects which the image processing algorithm can classify are:

- White spots.
- Dark spots.
- Misshaped pasta.

Figure 8. Image vision technique: (a) clustered region enhancing pasta defects (white spots) of a zoomed pasta region; (b) calculated ellipses enclosing defects; (c) data table output indicating the calculated geometrical parameters of the defects; (d) histogram color setting for the image processing; (e) reporting of the classified defects

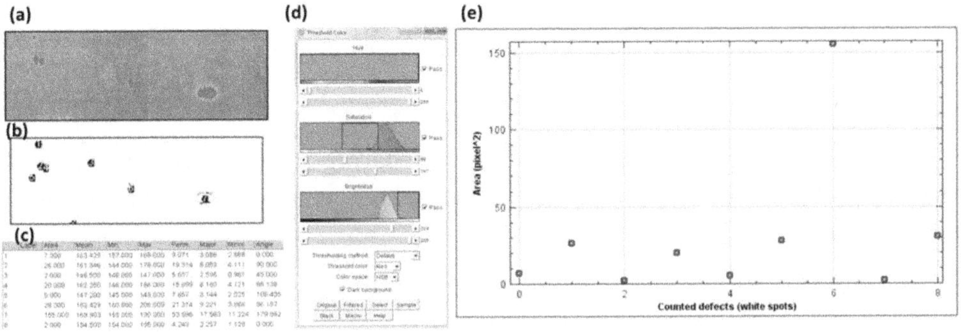

- Stickiness.
- Cracking.

Concerning white spots, possible causes are:

- Presence of whitish particles in the flour that are difficult to hydrate.
- Irregular granulometry in the flour with large diameter particles present alongside fine particles.
- Uneven kneading, too dry or not enough kneaded, or unevenly moistened.

The general quality improvement of pasta product can be achieved by optimizing all the production processes concerning also logistic, traceability, energy consumption, warehouse management, resources and material management. The process illustrated in Fig. 7 can be applied for other typologies of products different from food.

CASE OF STUDY 4: ELECTRONIC WELD QUALITY MONITORING BY ADVANCED IMAGE VISION TECHNIQUE

Main Specifications

In this section is discussed the main specifications of a project idea mainly focused on the quality check of Pin Through Hole –PTH- and Surface-Mount Technology –SMT- electronic boards. As for the previous case of study, has been used an image processing segmentation technique able to identify and to classify the welding regions.

The goal of the project is to define the correct geometrical features of circuit welding, thus identifying automatically defects which differ from the standard. Specifically the image processing tool verifies:

- The geometric profiles of the welds.
- The correct assembly of the electronic components mounted on a PTH /SMT board.

The executed methodology consists in comparing the processed image of a perfectly welded board with a defective one (where welding is smaller or bigger than expected).

Project Results

In Fig. 9 is shown an example of a segmented image targeting the geometrical features of welding of an electronic board by using ImageJ "*Analyze Particles*" function. The image processing can provide further information by predicting for example defects by means of artificial intelligence algorithms: by analysing the historical data and defect conditions it is possible to predict if defects will increase or not. The image segmentation of Fig. 9 has been executed by adopting the Watershed Method which is a transformation defined on a grayscale image using mathematical morphology (Rivest et al., 1993). In order to improve automatism in image acquisition time, has been implemented OpenCV libraries realizing multithreading task execution of the implemented script.

A block diagram representing AI prediction is illustrated in Fig. 10. This diagram describes how it is possible to improve automatically the quality check of electronic boards by means of two types of quality processes: the first one is based on the automatic extraction of the geometrical features, and the second one is related to a post-processing check. In particular the steps that should be implemented for the automatism are the following ones:

1. Webcam or camera setting (shooting area, lighting conditions, noise threshold setting, etc.).
2. Recognition of markers or components useful for the image alignment.
3. Acquisition of electronic board layout image.
4. 2D and 3D image segmentation.
5. Application of the welding verification algorithm based on weld classification and comparison.
6. Optimization of the detection error.
7. Definition of measurement procedures suitable for the new quality processes.

Figure 9. Example of image processing: extraction of welding areas (project: "Image Processing Board Quality- IPBQ")

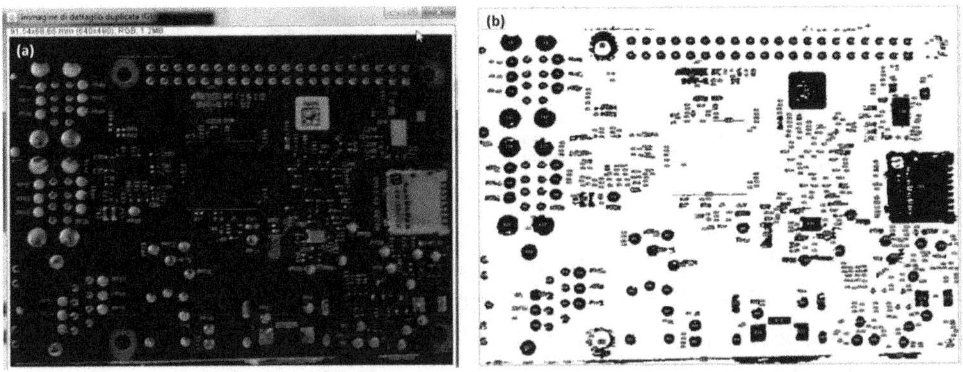

The 3D image processing is joined to the 2D image processing in order to highlight in three dimension the geometrical features of the welds thus supporting the defect identification. The 3D image processing is performed in post-processing modality. In Fig. 11 is shown an example of a 3D processing reconstructing the 3D profile of the weldings (3D Surface Plot function of ImageJ tool). By comparing the profile distribution of the same weldings will be possible to check if there are irregular weldings on the same analysed area.

The classified irregularities performed by the algorithm are:

- Irregular welding geometries.
- Big/small welds.
- Non solder.
- Irregular patterns.
- Agglomerates (coupled welds).
- No welds made (detection by welds counting).
- Dirty welds.
- Excess/less deposited welding materials.
- Component missing.
- Component shift.
- Component lift.
- Component damage.

Defects can be archived automatically into a database useful for quality check and predictive maintenance.

The proposed image processing algorithm and the model of Fig. 10 can be adopted for the quality check of other kind of products different from electronic boards. The

Figure 10. Automatic quality process and AI facilities for welding defect prediction

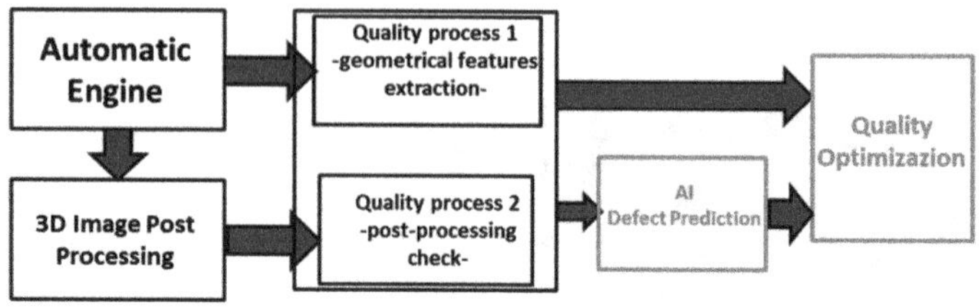

hardware technology together with the algorithm efficiency define the precision of the quality check: a good image resolution, for example acquired by means of a good microscope, can provide details also in micro-scale dimension as for precision products (Cicala et al., 2014).

CONCLUSION

The topics analysed in this chapter are important in order to design intelligent production processes and to formulate new quality monitoring procedures following the basic principle of the 'Frascati' R&D guideline about knowledge gain. Starting from the logic of Industry 4.0 have been presented four industry research projects oriented on the design of intelligent production process supporting production efficiency and quality management. The goal of the chapter is to provide an overview about technologies which can improve production processes. The attention is focused on artificial intelligence algorithms and image vision techniques showing how it is

Figure 11. ImageJ: 3D Surface Plot of electronic weldings. (a) 2D photo of an electronic board; (b) extraction of a 2D region; (c) 3D profile welding reconstruction

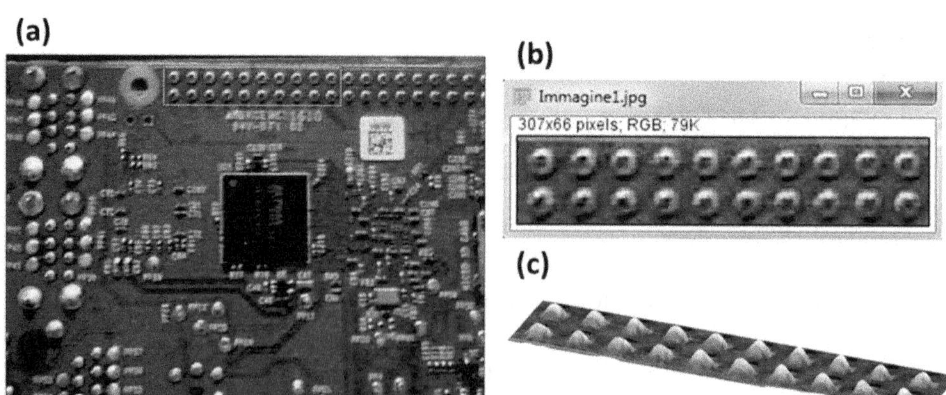

Table 3. Comparison of the industry 4.0 facilities proposed in the cases of study

Case of Study	Industry 4.0 Facilities
1	- AR for the optimization of product assembly processes; - Artificial intelligence (defect prediction); - Production data storage and mapping; - CNC production process; - Optical reader (digital production traceability).
2	- Production process levels improved by digitized information; - Image vision for production process monitoring; - Artificial intelligence for the improvement of the verification and optimization of the production processes.
3	- Production processes simulation; - Image vision (pasta defect monitoring).
4	- Image vision (2D and 3D image segmentation techniques for electronic weld monitoring); - Artificial Intelligence (defect prediction).

possible to gain the knowledge base through a digitized data flow. Concerning quality check, has been discussed how image segmentation approaches can contribute to classify defects such as not perfect welds of electronic boards or spot pasta defects or defect of kitchen top parts. An important aspect is to apply the charts for mapping process and the related historical data to the predictive algorithm thus enabling predictive maintenance.

The combined use of image vision techniques and of artificial intelligence represent a new issue in industry research thus reaching important results in automatic and adaptive manufacturing processes. The self learning automatism of artificial intelligence can contribute to look in a new era of Industry 5.0. In the table below are summarised the Industry 4.0 facilities adopted in the proposed cases of study.

All the listed facilities are important for the optimization of quality processes according with ISO 9001:2015 standard.

All the proposed cases of study indicate an overview of technologies suitable for the application of Industry 4.0 facilities. By properly combining artificial intelligence algorithms with different monitoring methodologies will be possible to improve auto-adaptive production processes by designing innovative Industry 5.0 facilities. The auto-adaptive Industry 5.0 processes are characterized by the following implementation difficulties (Table 4).

Table 4. Difficulties for the upgrade in Industry 5.0 facilities

Difficulty	Description
Interconnection between software and hardware facilities	Mechatronic board are necessary in order to improve control and actuation. The AI engine should be correctly interconnected in order to adapt in real time the machine working condition. The best synchronization between software and sensing/actuator hardware will provide the best efficiency of the production machine according with working time.
Real time adaptation of the production velocity based on sensor detection	The production velocity could change in function of the sensor reading. The AI algorithm should predict correctly possible failures in order to ensure the optimum production thus eliminating possible defects.
Production process adaptation based on historical data processing	The historical data could provide in a long term possible failures of the production lines. A correct predictive maintenance approach based on AI processing historical data could eliminate this risk.
Construction of a correct training data set for the mechatronic board model	The actuation and the feedback control of a mechatronic board could be performed by means of a training dataset constructed by considering correct machine operations. The parameter setting should be planned by considering different working conditions in order to formulate a correct training dataset. The self learning AI algorithms will optimize the control and the actuation mechanisms. An error in the training dataset is crucial for the production process efficiency.
Self-adaptive process	A good self-adaptation of the production machine can be performed by analyzing simultaneously multiple attributes (temperature, machine energy consumption, vibration, ecc.). For an optimum analysis, the AI multi attribute analysis should be applied on not correlated attributes (no errors due to redundancies). A low computational cost is required in order to provide a real time self-adaptive process. In this direction could be considered cloud computing facilities, but they are characterized by risks of internet network failures.
Correct choose of hardware and mechanical components	The electronic boards should control and move mechanical components according with the designed spaces and specific functionalities. The compatibility between hardware, software and mechanical parts will guarantee the production quality.

ACKNOWLEDGMENT

The work has been carried out during the development of the following different industry projects: "*Progetto di tecnologia di comunicazione macchina / processi orientato alla produzione della pasta 'Pasta IoT 4.0'*" [Project of machine / process communication technology oriented to the production of Pasta 'Pasta IoT 4.0']"; "*Linea di produzione prototipale di serbatoi con digitalizzazione dell' informazione ed upgrade della produzione mediante intelligenza artificiale: 'TankProductionAI'* "[Prototype production line of tanks with digitalization of the production and upgrade of production by artificial intelligence]; "*Modello di innovazione di processo e di upgrade di prodotto di cucine componibili 'UpgradedKitchens'*" [Innovation process modeling and product upgrade of modular kitchens 'UpgradedKitchens']; "*Sistemi di*

Image processing 2D/3D idonei per la qualità ed il montaggio di schede elettroniche con tecnologia PTH e SMT 'Image Processing Board Quality- IPBQ' "[2D / 3D Image Processing Systems suitable for the quality and assembly of electronic boards developed by PTH and SMT technology 'Image Processing Board Quality – IPBQ'].

REFERENCES

Acosta-Mejia, C. A. (1999). Improved *p* charts to monitor process quality. *IIE Transactions*, *31*(6), 509–516. doi:10.1080/07408179908969854

Basl, J., & Doucek, P. (2019). A Metamodel for Evaluating Enterprise Readiness in the Context of Industry 4.0. *Information*, *10*(89), 1–13.

Bastos, P., Lopes, P. I., & Pires, L. (2014). Application of data mining in a maintenance system for failure prediction. In Safety, Reliability and Risk Analysis: Beyond the Horizon. Taylor & Francis Group.

Caudell, T. P., & Mizell, D. W. (1992). Augmented reality: an application of heads-up display technology to manual manufacturing processes. *Proceedings of the 25th Hawaii International Conference on System Sciences*, 2(1), 659. 10.1109/HICSS.1992.183317

Chakraborty, A. (2016). Importance of PDCA cycle for SMEs. *SSRG International Journal of Mechanical Engineering, 3*(5), 30-34.

Cicala, C., Massaro, A., Velardi, L., Senesi, G. S., & Valentini, A. (2014). Self-assembled pillar-like structures in nanodiamond layers by pulsed spray technique. *ACS Applied Materials & Interfaces*, *6*(23), 21101–21109. doi:10.1021/am505974d PMID:25402729

Favi, C., Germani, M., & Marconi, M. (2017). 4M approach for a comprehensive analysis and improvement of manual assembly lines. *Procedia Manufacturing*, *11*(1), 1510–1518. doi:10.1016/j.promfg.2017.07.283

Fouad, R. H. & Mukattash, A. (2010). *Statistical process control tools: a practical guide for Jordanian industrial organizations*. Academic Press.

Frascati Manual. (2015). *The Measurement of Scientific, Technological and Innovation Activities- Guidelines for Collecting and Reporting Data on Research and Experimental Development*. OECD.

Khabbazi, M. R., Yusof Ismail, M. D., Ismail, N., & Mousavi, S. A. (2010). Modeling of traceability information system for material flow control data. *Australian Journal of Basic and Applied Sciences*, *4*(2), 208–216.

Krenek, J., Kuca, K., Krejcar, O., Blazek, P., & Jun, D. (2016). Application of artificial neural networks in condition based predictive maintenance. In D. Król, L. Madeyski, & N. Nguyen (Eds.), *Recent Developments in Intelligent Information and Database Systems. Studies in Computational Intelligence* (Vol. 642). Cham: Springer. doi:10.1007/978-3-319-31277-4_7

Lotlikar, T., Kankapurkar, R., Parekar, A., & Mohite, A. (2013). Comparative study of barcode, QR-code and RFID system. *International Journal of Computer Technology & Applications*, *4*(5), 817–821.

Massaro, A., Galiano, A., Meuli, G., & Massari, S. F. (2018). Overview and application of enabling technologies oriented on energy routing monitoring, on network installation and on predictive maintenance. *International Journal of Artificial Intelligence and Applications*, *9*(2), 1–20. doi:10.5121/ijaia.2018.9201

Massaro, A., Maritati, V., Galiano, A., Birardi, V., & Pellicani, L. (2018). ESB platform integrating KNIME data mining tool oriented on Industry 4.0 based on artificial neural network predictive maintenance. *International Journal of Artificial Intelligence and Applications*, *9*(3), 1–17. doi:10.5121/ijaia.2018.9301

Massaro, A., Maritati, V., Savino, N., Galiano, A., Convertini, D., De Fonte, E., & Di Muro, M. (2018). A study of a health resources management platform integrating neural networks and DSS telemedicine for homecare assistance. *Information*, *9*(176), 1–20.

Massaro, A., Meuli, G., Savino, N., & Galiano, A. (2018). A precision agriculture DSS based on sensor threshold management for irrigation field. Signal & Image Processing *International Journal*, *9*(6), 39–58.

Massaro, A., Vitti, V., & Galiano, A. (2018). Automatic image processing engine oriented on quality control of electronic boards. *Signal and Image Processing: an International Journal*, *9*(2), 1–14. doi:10.5121ipij.2018.9201

Novak-Marcincin, J., Barna, J., Janak, M., & Novakova-Marcincinova, L. (2013). Augmented reality aided manufacturing. *Procedia Computer Science*, *25*(1), 23–31. doi:10.1016/j.procs.2013.11.004

OECD. (2017). *The Next Production Revolution: Implications for Governments and Business*. Paris: OECD Publishing.

Özdemir, V., & Hekim, N. (2018). Birth of Industry 5.0: Making Sense of Big Data with Artificial Intelligence, 'The Internet of Things" and Next-Generation Technology Policy'. *OMICS: A Journal of Integrative Biology*, *22*(1), 65–76. doi:10.1089/omi.2017.0194 PMID:29293405

Oztemel, E., & Gursev, S. (2018). Literature Review of Industry 4.0 and related Technologies. *Journal of Intelligent Manufacturing*, *30*(176), 1–56.

Rivest, J. F., Soille, P., & Beucher, S. (1993). Morphological gradients. *Journal of Electronic Imaging*, *2*(1), 326–336.

Rubmann, M., Lorenz, M., Gerbert, P., Waldner, M., Justus, J. & Harnish, M. (2015). *Industry 4.0: the future of productivity and growth in manufacturing industries*. The Boston Consulting Group Report.

Segovia, D., Mendoza, M., Mendoza, E., & González, E. (2015). Augmented reality as a tool for production and quality monitoring. *Procedia Computer Science*, *75*(1), 291–300. doi:10.1016/j.procs.2015.12.250

Siltanen, S. (2012). *Theory and applications of marker-based augmented reality*. JULKAISIJA – UTGIVARE – Publisher.

Skobelev, P. O., & Borovik, S. Y. (2017). On the Way from Industry 4.0 to Industry 5.0: from Digital Manufacturing to Digital Society. *International Scientific Journal Industry 4.0*, *2*(6), 207-311.

Vaidya, S., Ambad, P., & Bhosle, S. (2018). Industry 4.0 –A Glimpse. *Procedia Manufacturing*, *20*(1), 233–238. doi:10.1016/j.promfg.2018.02.034

Wang, H. (2009). Comparison of *p* control charts for low defective rate. *Computational Statistics & Data Analysis*, *53*(12), 4210–4220. doi:10.1016/j.csda.2009.05.024 PMID:20161085

Zhang, X., Xiao, L., & Kan, J. (2015). Degradation prediction model based on a neural network with dynamic windows. *Sensors (Basel)*, *15*(1), 6996–7015. doi:10.3390150306996 PMID:25806873

KEY TERMS AND DEFINITIONS

Artificial Intelligence: Artificial intelligence (AI) is intelligence demonstrated by machines by adopting self-learning algorithms.

Artificial Neural Network (ANN): ANNs are computational networks composed of multiple nodes named neurons interacting with each other. The nodes can take input data and perform simple operations on the data.

Augmented Reality: Is an interactive experience of a real-world environment where the objects that reside in the real-world are enhanced by computer-generated perceptual information.

Data Mining: Is the process of discovering patterns in large data sets involving methods at the intersection of machine learning, statistics, and database systems.

Image Segmentation: Is the process of partitioning a digital image into multiple segments represented by sets of pixels.

Image Vision: Image processing techniques implementing image processing algorithms.

Industry 4.0: Is a name given to the current trend of automation and data exchange in manufacturing technologies.

Chapter 5
Framework for Quick and Intuitive Programming of Robot Applications

Asad Tirmizi
Flanders Make, Belgium

Patricia Leconte
Flanders Make, Belgium

Karel Janssen
Flanders Make, Belgium

Jean Hoyos
Flanders Make, Belgium

Maarten Witters
Flanders Make, Belgium

ABSTRACT

This chapter proposes a framework to make the programming of cobots faster, user-friendly and flexible for assembly tasks. The work focusses on an industrial case of a small (10kg) air compressor and investigates the technologies that can be used to automate this task with human-robot collaboration. To this end, the framework takes a radically different approach at the motion stack level and integrates the cobot with a constraint-based robot programming paradigm that enhances the robot programming possibilities. Additionally, the framework takes inputs from the operator via speech recognition and computer vision to increase the intuitiveness of the programing process. An implementation is made with focus on industrial robustness and the results show that this framework is a promising approach for the overall goal of achieving flexible assembly in the factories by making robot programming faster and intuitive.

DOI: 10.4018/978-1-7998-1382-8.ch005

INTRODUCTION

Manufacturing is an activity that has been in flux since the industrial revolution. In its earliest days the focus was on minimizing energy losses and maximizing production numbers by focusing on the manufacturing techniques. A couple of centuries of innovation led to many breakthrough technologies like assembly lines, Programmable Logic Controllers, Manufacturing Execution Systems, Product Lifecycle Management etcetera. The modern world we inhabit is in many ways shaped by these innovations. It allows us to produce in huge numbers. However, this entire manufacturing process is very rigid. Rigid in the sense that the mass-produced items are al the same. They are generic in nature and try to appeal to a mass clientele. Any type of customization to suit the local tastes of a market is very difficult. Customization of products requires flexibility at the assembly lines. However, the entire manufacturing industry is composed of technologies, whose raison d'être is production in bulk. They are an antithesis to the very concept of customized production.

However, there is a clear market trend towards increasingly customized products. Manufacturing industry is changing at a breath-taking pace. Those companies that have given customization options on their products, instantly benefit with an increased market-share. It is in human nature to prefer a product that is customized for them. With the advent of many new regions of the world in manufacturing, the competition is very stiff and companies need to find ways to keep their market share. Customization of products is one such avenue that guarantees traction in the market. Therefore, we see a clear push from the companies to shift towards mass-customization. This trend gets the necessary push as technologies like robotics, artificial intelligence and faster processing speeds provide a theoretical avenue to make flexible assembly systems viable.

Robotics, as a reconfigurable technology, can be leveraged to extend automation to unstructured tasks that can greatly facilitate such flexible production (Pan, 2012). For the manufacturing industry, this means a shift from mass production to mass customization, thus the ability to support small batch-size production. If a flexible assembly line is realized today with the current state-of-art in robotics, it would require frequent reprogramming of the robots every time the product is even slightly changed. This reprogramming comes at a high cost, both temporally and economically. Thus, the manufacturing process is not really flexible even though the hardware possesses the potential.

A typical factory requires the services of robotic integration companies to make any changes to their production process for a new product batch. The setup time for a typical application is around 3 months or greater (Jager, 2015). This poses a major hurdle towards more frequent reprogramming, especially for Small and Medium

Figure 1. A flexible assembly work cell using a modern collaborative robot (cobot), that is safe for humans to work alongside, being taught by demonstration

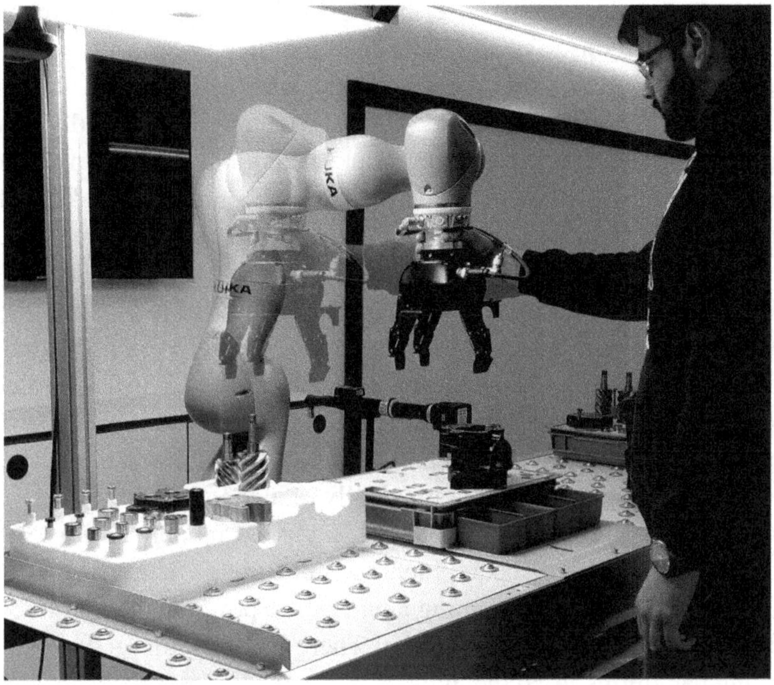

Enterprises (SME) to stay up to speed with the global demands. There is a strong interest from the industry to push down the cost of programming robots by making them fast and intuitive to program (Tirmizi, 2019). They envisage a future where integrators will have access to frameworks that allow quick programming of new robotic applications and any modifications required for a reprogramming will be done on the shop floor by interacting intuitively with the robots.

Realizing this vision requires an understanding of why programming robots is so tedious and time consuming. In current commercial systems, robot programs are made by explicitly specifying coordinate frames (either in joint or cartesian space) for all positions relevant for a task. This approach, although conceptually simple, quickly becomes tedious and error-prone for programming complex applications. This is especially the case when programming modern robots with redundant degrees of freedom, a complex sensor suite, and the ability to interact with humans and operate in uncertain environments. The traditional programming methods become overly complex and highly unsuitable to optimally take advantage of such collaborative robots, also informally referred to as cobots.

Their manufacturers do provide programming suites that attempt to make it easier to take advantage of their capabilities. However, they follow the same classical concept of generating motions as described above. Thereby, any development done on these platforms is constrained to the same robot in the same setting. Moreover, these programming frameworks are text-based built as a customized version of programming languages like PASCAL, ADA, C, etcetera. Thus an individual tasked with programming robots need to have a background in programming, mechanics, dynamics, robotic basics and sensors. These are highly marketable skills, therefore it costs integrators a significant sum of money to keep these skills in-house. These costs are transferred to the end-customers. As long as the end-customer is large scale industry (LSI) that has to re-program every once in a while, this model can be borderline sustainable but this is not feasible for SME's or even LSI's that want to produce in small batch sizes. Considering the importance of SME's to a country's economy, it is highly desirable that the benefits of flexible assembly systems via robotics is transferred to them as well. This is only possible if the programming effort for a functional work cell is dramatically reduced.

Adding further complexity to the above, a typical modern factory has robots from different OEM's (original equipment manufacturers). The big brands of robot OEM's, all have different programming languages to program their products. These programming suites are not interoperable. There is a lack of vendor-independent programming paradigms. Also the integrators are many times dedicated to a single OEM. Therefore an integrator might be an expert in programming robots from FANUC™ but unable to do so for KUKA™. Aggravating the problem further the integrators are typical specialized in certain types of applications. A good integrator for a welding application might focus on just that and might not be proficient enough for programming the same robot for a handling application. In a nutshell, industry for slightest of product changes, need to go through the expensive exercise of finding the right integrators and robots that are compatible with their current setup. To do so frequently on razor-sharp deadlines is a very difficult task. Therefore, we can conclude that the current state of production lines is a non-starter when the aim is to use the same robot setup for different applications at hand. This is only possible when:

- Frequent reprogramming is not a severe drawback.
- When programming new application is not an overly complex task.

Moreover the conventionally programmed applications can neither be scaled up nor reused. Thus every reprogramming requires a restart from scratch. The programming frameworks of today are not skill-based. As in, it is not yet a standard practice to have ready-made skill templates that can be parameterized for the task at hand. A combination of whom provides a complete application. *To sum it up, there is a profound need for vendor-independent frameworks to intuitively program robotic assembly applications from any manufacturer in an intuitive fashion.* It is also desirable that the programmed applications or robot skills can be reused and scaled to deal with customization demands.

MAIN CONTRIBUTIONS

Realizing the shortcomings in the traditional robot programming frameworks and their debilitating effect on future industrial production, a consortium of research institutes and industrial partners started an effort to upgrade the robot programming by taking a radically different approach that plugs the current short-comings of robot programming frameworks. The work presented here is part of its realization and aims to make programming complex robotic tasks easier for operators without robot programming expertise. We use intuitive Human-Robot interfaces that connect with robot skills, programmed in both conventional and non-conventional ways like **constraint based programming.** This makes automated solutions possible in workshop scenarios where they are not affordable today. The application selected to demonstrate and validate the framework's potential consists of assembling an air compressor with highly parameterized surfaces as shown in Fig. 2. This application was selected after extensive feedback from industry to find out a representative assembly task that requires sub-tasks that are considered the domain of assembly requiring humans.

Such an assembly requires programming of robot skills like:

- Following pick and place trajectories.

Figure 2. The air compressor components kitted on the assembly tray

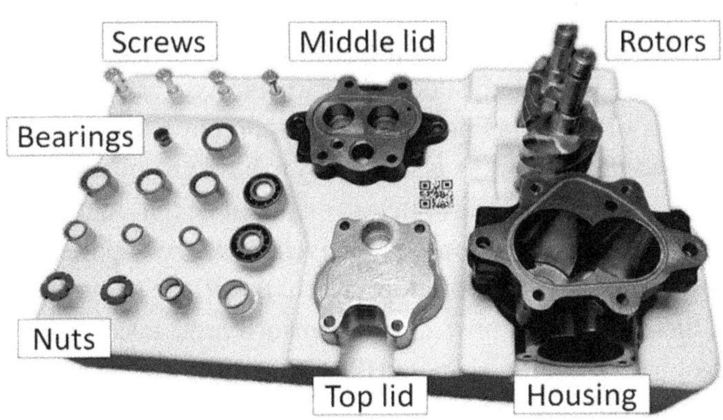

- Insertion of rotors, placing of lids, etc.
- Executing the pick and drop with the gripper.
- Identifying the correct component required in the assembly process.

Traditional methods of programming with a conventional, non-sensitive robot would not result in a cost-effective solution, especially in the case of changes to the model requiring modifications. Moreover, the solution would be highly inflexible, as it would require a very structured and controlled production environment.

In this work we present a framework that incorporates a Human-Machine Interface (HMI) that provides access to intuitive input methods for programming the robot like speech recognition software, vision-based component detection and programming by demonstration to intuitively program the robot behavior. These high level inputs connect to constraint-based programming approach (defined below) for generating robot motions, which results in robot-skills that are vendor-independent, scalable and reusable on any other robot. Therefore it is possible to deal with variability and customization requirements of a modern factory.

In the "constraint-based programming" approach, a robot task is described as a set of constraints and objective function(s) on a set of controllable variables of interest. In such a paradigm, the output variables can, e.g., consist of the position of an object in a camera image, the relative position and orientation of two workpieces that have to be assembled, or the contact force while performing contour racking. Constraints are more modular compared to traditional procedural programming of point-based trajectories. Constraints also allow to specify only what is actually needed for the task, leaving open as many degrees of freedom as possible for task execution optimization.

In this work, we use the expression graph-based Task Specification Language (eTaSL) (Aertbeliën 2014), which is a domain specific language based on Lua that allows to specify constraints by means of expression graphs. Automatic differentiation techniques are implemented for these expression graphs, facilitating the computation of jacobians and feedforward control terms. Executing an eTaSL specification corresponds to solving a constrained optimization problem, by translating it into a numerical form and using a dedicated solver to resolve it. In (Aertbeliën 2014), a detailed description of eTaSL is presented. To appreciate fully the strength of the framework proposed in this work, it is necessary to have an understanding of eTaSL.

The contribution of this work is to iterate on top of eTaSL to take it to a maturity level that robot skills can be programmed for a real industrial application. Moreover, eTaSL is good at complex tasks that involve relative motions between several entities in different frames. It is very powerful in sensor-based tasks that might require reactive abilities. However, for simple tasks like Point-to-Point motions the native software can have a more straight-forward implementation. Therefore the aim is to create a framework that wraps eTaSL along with other key hardware and software components into a middleware that connects it to high-level interfaces thus providing a mean to take advantage of both conventional and non-conventional (eTaSL based) robot programming. The section on framework architecture shall deep dive into this complex interplay. Eventually our developments allow us to present a Human-Robot Interaction framework that provides, vendor independent, reusable, composable, and scalable solutions.

BACKGROUND

The difficulty of programming robots is well-known. For several years now, research efforts have been ongoing to make this practice easier and faster. The holy grail has always been to enable a novice to communicate with the robot to adjust its behaviour safely with minimal downtime. There are two main ways of programming robots that are the current state of practice. Online programming where the robot is programmed by a teach-pendant. The worker guides the robot to points of interest and stores them. Subsequent points get assigned a motion profile to traverse between these points as required by the task at hand. This can be a point-to-point motion, linear motion, spline motion etcetera. A combination of these points and motions augmented by triggering of I/O's to operate the attachments like the gripper, external sensors or the tools completes the application (Biggs, 2003).

Another popular way is offline-programming where a virtual work cell is programmed on a computing machine (Wittenberg, 1995). This program is then transferred to the real world after testing its operation in simulation. There is usually a touch-up required to adjust the program to the real world. The benefit of this type of programming is that it reduces the down-time for a work cell that is already used for production. Apart from the work done by researchers in the above two robot programming paradigms, there is also a considerable body of research that focusses on interaction techniques with robots (Haage, 2017). However, most research in these areas is still fundamentally based on the conventional method of storing points of interest and their associated motion profiles.

We would like to bring to notice as background those works that tried to use the constraint based programming approaches for robots. One such work is by Zanchettin (2016). who investigated on how to facilitate robot programming by combining constraint-based programming with vision. They made an experimental verification on a visually aided grasping task and proved the robustness of the application. Perzylo (2016) proposed a user-centric approach used to interpret under specified robot tasks, enabling communication on an abstract level. Such high level task description made the system amenable for users with little knowledge of robotics.

Bartels (2013) proposed robot motions that were defined as symbolic constraints using geometric features and the approach was validated by a pancake flipping robot. Makris (2014) combined speech and vision, resulting in user-friendly human interactions with a dual-arm robot for an assembly case in the automotive industry.

Programming by demonstration was combined with vision by Haage (2017), where the assembly of a cell phone was taught by demonstration and the vision system saved the movements of the operator and translated them into a robot program. Researchers have also come up with approaches in which skills are programmed offline and then fine-tuned online with vision (Matthaiakis, 2017) and haptics (Pedersen, 2016). Concerning constraint-based programming, various other frameworks exist (Sentis, 2006; Vanthienen, 2013; Mansard, 2009). The original idea goes back to Ambler (1975) who developed a method for inferring the position of objects by taking into account, the imposed, desired spatial relations between rigid bodies motivated by automatic assembly operations. Research on methods for motion planning in configuration space (Latombe, 1975) resulted in specifying the desired relative poses as an output of the application of various constraints (possibly conflicting), between object features. Samson (1991), introduced the task function approach that models a robot task as minimization of task functions. More recently, Mansard (2007) focused on solvers, that makes possible, fast reactive instantaneous control with inequality constraints. Their work (Mansard, 2012) spans, task prioritization, by using efficient solvers. In (De Schutter, 2007) a task specification approach based on constraints, denoted iTaSC was presented, which developed a systematic modeling procedure for geometrically complex robot tasks and closed the loop between modeling and control by supporting instantaneous equality constraints and instantaneous least-squares objective functions.

Framework Architecture

To realize the Human-Robot skill-based framework, a flexible architecture is developed, shown in Fig. 5. The centerpiece of this architecture is the middleware that allows a vendor-independent approach to robot programming. As mentioned before, the robots in industry can come from various OEM's. This is also the case with the various sensors that are required for execution of robot skills. Therefore this middleware is required that can launch components and serve as the communication protocol between the various software and hardware modules as shown in Fig. 3.

This middleware allows the addition of meta-layers that gives the ability to program the robot without concern to the underlying robot's manufacturer and programming environment. However, this still requires a driver to the robot. All those robots that have such an interface can reuse the application in a truly vendor

Figure 3. The role of the middleware and the component framework is to connect various software and hardware blocks

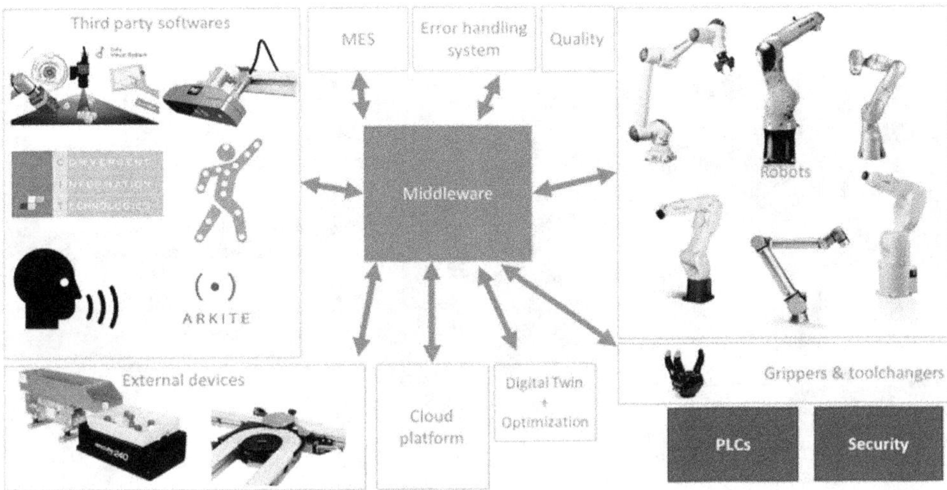

independent fashion. This particular aspect of the framework is explained in more detail in the description of process-executor.

The middleware that is realized in ROS connects to the implementation of various robot-skills. These skills can be invoked by the high-level human inputs. The input from human to invoke these skills can come from a Human-Machine Interface that allows a touch panel, speech recognition and teach by demonstration to program the application. The same inputs can be used to modify or reprogram. . The change of states in the FSM can be triggered by sensors, sub-task completion or human-input during the operation. An illustration is given in Fig. 4.

This description allows us to present the distributed components that are connected by the middleware to create this framework:

Figure 4. The role of FSM in the application development

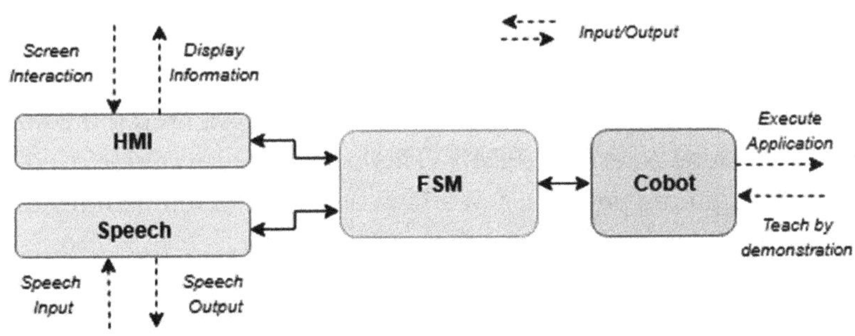

Figure 5. Distributed flexible assembly architecture, with the "process executor" being the central node, taking input from high-level sensing (speech, vision) and is responsible for executing the process by invoking appropriate instances of the different skills

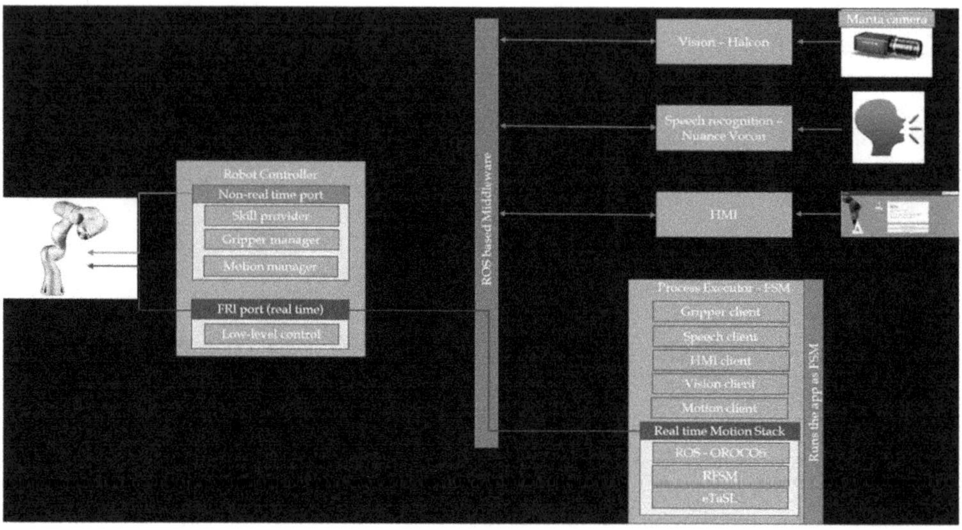

1. **HMI Component** for visualization of the task state and human-friendly programming.
2. **Vision Component** to identify and locate workpieces and tools.
3. **Speech Component** to recognize operator commands and give feedback.
4. **Teach by Demo** component to allow hand-guidance and record frames.
5. **Process Executor** to keep track of the current state: allowing to create new applications with various skills or execute a previously taught application.
6. **Skill Templates:** A low-level control system to compute real time motions with an option to invoke skills that are programmed with constraint-based programming approach.

A. Human-Machine Interface (HMI)

The first point of interaction for the human to start programming the application is the Human-Machine Interface (HMI). This is designed to guide even a novice through the programming procedure. It allows access to functionalities like:

1. Invoking teach by demonstration functionality.

2. Invoking programming by speech.
3. Selecting motion-profiles and parameterizing robot skills.
4. Store via-points for generating trajectories.
5. Storing trajectories with identifying names.
6. Combining stored trajectories into full-applications.
7. Building on top of previously stored applications.
8. Editing trajectories and thereby applications.
9. Viewing the status of the systems.

The HMI has access through the middleware to an SQL database where the data to accomplish the above tasks is stored. Next, it guides the user through the finite state machine which is presented in the next section. In other terms, it is a visual link with the process-executor.

When programming a new application for example, the HMI allows to visualize the different skills present in the application and the potential skills that can be added to the actual application. The order of the skills can also be modified through the HMI by drag and drop. The data is passed on to the process-executor as and when requires. Therefore the human, thanks to this user-friendly interface, stays continuously in loop and in control of the operation of the work-cell. Fig. 6 shows a snapshot of the HMI that is displayed to the operator.

The HMI is web-based which allows to use the power of all web-plugins such as bootstrap, Jquerry etc. This allows the HMI to run on a wide range of screen-sizes,

Figure 6. Human Machine Interface is the interaction tool that allows the human to intuitively program the work-cell, stay updated on the system status and edit/ create application data

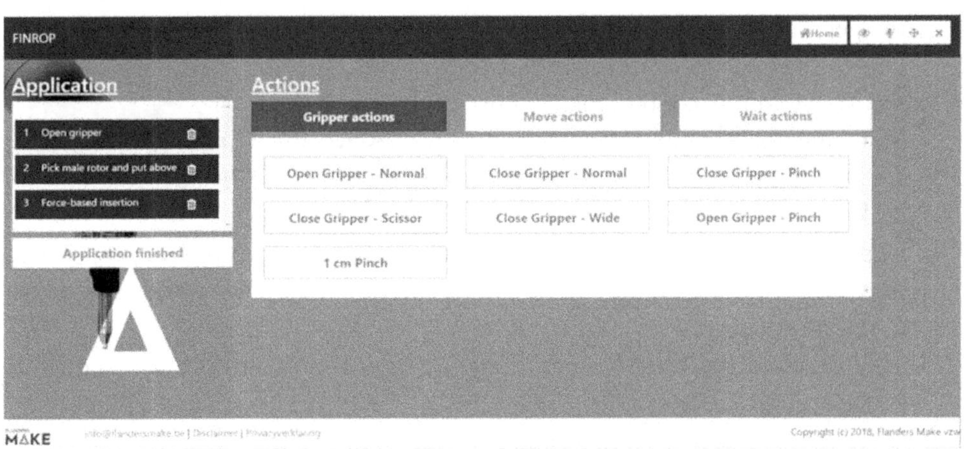

going from smartphones to tablets and computer screens. Moreover, implementing drag-and-drop features is also simplified.

B. Vision

To adapt to the demand of mass customization, machine vision offers valuable opportunities. The vision sub-system, as further described, supports the robot assembly by flexible feeding of the work cell and increased robustness. As such, we still use a kitting tray, where from the workpieces are picked, but the vision system allows it to be a low-precision, 3D-printed one, instead of a highly precise and expensive machined tray.

In order to meet the demands of a fast robot programming approach, the challenge is to identify and localize assembly parts with sufficient precision and within a reasonable development time, either by using shape-based model matching (trained via image templates) or CAD-based models of assembly parts. The use of CAD-based model(s) is beneficial for various reasons. On one hand, the CAD drawings can be compared with the real measurements to identify the position and orientation of each part in two- and/or three-dimensional perspective. Moreover, the system is able to create a check to verify if all parts are present, in order to start the assembly

Existing computer vision libraries, specifically HALCON™([Online]. Available: http://www.mvtec.com/products/halcon/) provide a fast and efficient development process with accurate results, ideal for highly customized products. To overcome undesired external light changes, reflections, and shadows in an industrial setting, an active diffuse light module is used. This creates homogeneous and stable lighting on the reflective metal assembly parts and increases contrast on curved edges and surfaces.

The toolchain used to extract object positions operates as follows:

1. Image(s) are acquired via a camera mounted above the assembly parts.
2. Due to the non-linear intrinsic response of the 2D camera, image(s) are first rectified to remove radial distortions.
3. Matching model(s), either trained via shape-based and/or CAD-based matching are used to locate multiple instances of all assembly part(s) under rotation and scaling.
4. Depth information is roughly defined within an initial calibration step where the height of the assembly table is identified and combined with the use of CAD information.
5. This removes the need for additional sensors like laser-distance finders or stereo configurations.
6. Once the assembly parts are detected, their individual poses are extracted.

Figure 7. Detecting compressor parts using machine vision

7. The vision ROS component broadcasts the pose of all recognized parts, relative to the camera position.

C. Speech-Recognition

To improve interaction with an operator and simplify the assembly application programming, a speech recognition and synthesis module is part of the framework. The main idea being that it allows hands-free interaction, which is an advantage for operators in a factory setting as their hands are typically busy e.g. doing hand-guiding, or while actually assembling products. As such, we foresee 2 different scenarios where speech is relevant;

- Programming the assembly process, namely selecting skills, instantiating their parameters and sequencing their order.
- Executing an assembly application, with commands like start, stop, go back etc.

Table 1. Programming by speech scenario, showing the teaching of a picking operation and a trajectory-following while grasping a rotor

Operator	Robot Dave
"Hey Dave"	"Hello"
"Create new application"	"Ok"
"Move to home position"	"Ok"
"Free motion"	"Ready to teach"
Hand guide to next waypoint	
"Waypoint"	"Ok"
. . . continue, until above tray:	
"End position"	"Trajectory stored"
"Pick small rotor"	"Pick action stored"
. . . teach move to fixture	
. . . add more skills, until:	
"Application ready"	"Application stored"

In Fig. 8, an example scenario is given of how the speech based robot programming is used in the presented framework. In addition, our focus being an industrially viable solution, requires that the speech module runs with:

- Low and predictable latency.
- Caters for industrial concerns about security and secrecy of information.
- Provides robustness against noise, and requires little training.

A survey to cater for these requirements ended up with relatively little candidates, as most solutions are cloud-based and a lot are still in development. We selected the Vocon Hybrid system ([Online]. Available: https://www.nuance.com/mobile/speech-recognition-solutions/vocon-hybrid.html), which offers an extensive programming SDK and runs offline. It is also multi-lingual, and its support for Dutch is naturally an advantage from the consortium's perspective.

D. Teach by Demonstration Functionality

The teach by demonstration functionality allows the human to grab the robot and exert force on its body to bring it to a certain pose/point in the workspace. The robot while executing this functionality stays in gravity compensation mode and reacts to the force exerted by offering no resistance to the human. Therefore it is a very intuitive way of bringing the robot to a point of interest and thereby storing this point for defining it as a way-point through which the robot end-effector should pass with a certain motion profile. This functionality is faster than the traditional method of generating motions via the teach pendant.

However, this functionality is available on limited robots. Modern collaborative robots often come with this functionality. Industrial non-collaborative robots with high pay-loads often lack this capability due to safety concerns. There have been works done that provide vision systems and attachments to robots that can provide them these capabilities. (Fryman, 2012).

In our case, the robot we used, a KUKA LBR iiwa, had this capability as an intrinsic feature. Therefore, full advantage was taken of this capability. The framework allows to invoke this ability and use it along with the speech recognition system. The result is a very user-friendly modality that humans can get acquainted to very quickly.

E. Process Executor

The process executor is first responsible for spawning the different ROS node involved in the middleware. Therefore, the process executor is be able to send the various commands (gripper, motions, vision, planner, ...) using their dedicated nodes. Secondly, this executor will set in place the Finite State Machine.

The Finite State Machine (FSM) is the brain of the application. It allows to keep track of the application current-state. This FSM is composed of various states linked together with transitions. The implemented FSM is shown in Figure 8. These transitions allow changes from state-to-state and are triggered by events coming either from the HMI or the speech module. Therefore, depending on the current state, a new one can be reached simply by saying an order out loud or by interacting with the HMI.

The role of the FSM is central to intuitive programming as it permits to create new applications in a simplified fashion. It allows access to previously taught applications whose relevant data is stored within the database. Creation of a new application requires the operator to communicate with the robot using speech. Saying "create new application" will start the programming process and thereby a name will be asked to be given to the application. Then the HMI will display the available skills (gripper, motion,..) that the operator can select in order to carry out actions of the new applications. Once the application is correctly appended with required skills, the operator can either execute the complete assembly, modify it by changing different steps in the process or delete it. A complete application can be saved in the database allowing to easily reuse or build upon it later.

When a complete application is called to be processed, the FSM enters in the 'Execute mode'. This is where it takes care of invoking the right skills at the correct

Figure 8. Example of the Finite State Machine (FSM). The process executor starts in the initial state of the FSM, i.e. "Waiting". To trigger the next state the application waits for a FSM event, e.g "Hi Dave"

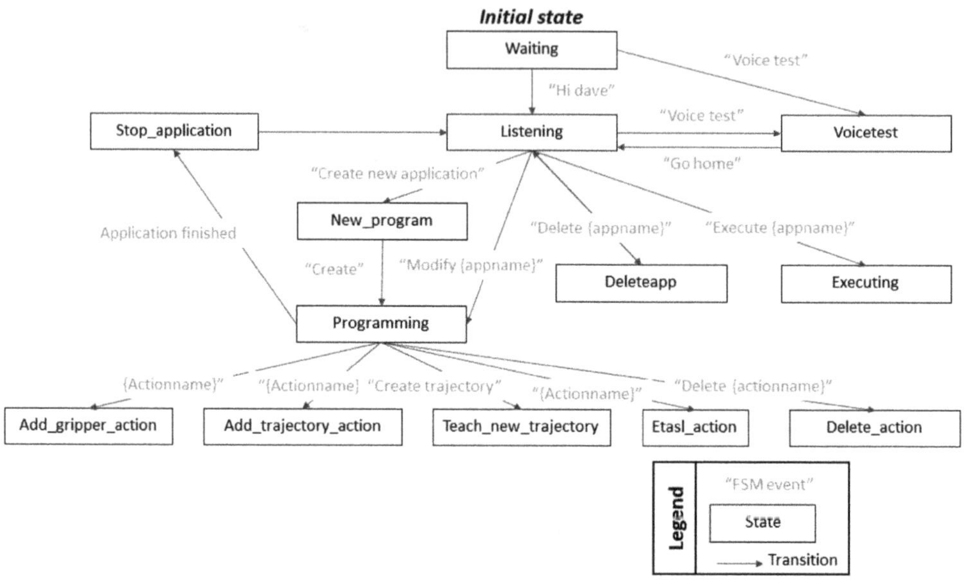

moment in time depending on the human/sensor/robot input it receives and the programmed process model. While performing the complete application, the operator is aware of the current progress of the application as well as the future steps involved in the application though the HMI. The FSM in 'Executing state' recuperates the skills to perform and sends its execution parameters to the robot in the workcell.

F. Skill Programming

In our assembly cell, we are using a Kuka LBR iiwa, using its Java-based Sunrise platform, fitted with a 3-finger Robotiq gripper. We built an extensible ROS driver, which serves a dual purpose. On one hand, it provides an interface for standard basic motion and grasping commands, (point-to- point move, open/close gripper, . . .). On the other hand, it allows to exploit specific features of the platform at hand by offering them as separate skills, which can then be invoked by the FSM. For example, it offers an interface to enable/disable the listening to motion commands over EtherCAT, provide gravity compensation, and access to the force recognizing features of the robot. It should be noted that the motions generated by eTaSL require a real-time interface with the robot, which in this setup is provided with EtherCAT (FRI port of Kuka LBR iiwa). It is necessary because eTaSL (or any other constrained based programming approach) optimizes the task at hand by taking the task state into the account at each instance and then provides the motion commands most optimal for task completion.

As previously described, we present two approaches of robot skill programming that address different tasks in varying situations. The first approach is via eTaSL meanwhile the second one is via Sunrise, the robot's own programming environment. The proposed system demonstrates the possibility of switching between them. In real industrial cases, this feature might be useful due to e.g. existing legacy of robot-centric infrastructure that is still required, thus preventing the use of vendor-independent framework for all tasks. In our case, apart from demonstrating the possibility of framework switch, the use of these two frameworks is also to exploit the best of both worlds. The Sunrise environment has access to the robot dynamical model and low level torque controller, hence making it better at performing fine manipulation tasks.

While eTaSL is more flexible in specifying and processing continuous motion such as spline trajectories. The eTaSL implementations have the further advantage of providing skills that are easy to adapt for further use. As the implementation remains the same, only the parameters need to be adjusted. Furthermore since the skills are executed in real-time and have the inbuilt capability of reacting to any unforeseen changes in the task due to the constant optimization, they are well-suited to take advantage of the sensor suite of modern robots. The eTaSL based robot skills integrate very well to the high-level interfaces like programming-by-demonstration, and speech etc. due to the inherent properties of constraint-based programming, where only the necessary constraints for task completion need to be specified and the rest of the details are left open for optimization in real time as per the state of the task. In the presented work eTaSL is used to create a skill that performs free space motion in Cartesian coordinates given the waypoints generated from kinesthetic teaching (also referred previously as Programming by Demonstration). The skill interpolates through the input points and generates the corresponding spline profile for each Cartesian axis.

The compressor assembly tasks were performed using insertion skill in addition to the free-space motion. The insertion skill was implemented both in the native Kuka Sunrise (Java) environment and eTaSL. For the Sunrise implementation Impedance control mode was used in the Cartesian space, enabling stable insertion for a tight tolerance assembly task. To improve robustness with respect to parts misalignment, a Lissajous-patterned jitter is superposed to the insertion force. The eTaSL implementation uses a real-time force based optimization to reach the point of interest which is the end of the cavity where the component needs to be inserted. For example the rotor that needs to be inserted in the housing.

Framework Validation

As stated earlier, to test the developed framework, a real industrial case of an "air compressor assembly" was chosen that is currently carried out by manual labour. The process is repetitive, dull and labour intensive.

A. Task Significance

The different components of an air compressor are shown in Fig. 2. They range from large, parametric objects, such as the housing and the rotors, to small o-rings and nuts. The assembly procedure, shown in Fig. 7, contains steps, like multiple rotor insertions, lid placement on top of rotor shafts etcetera that requires complex motions and robot sensitivity as shown in the attached video. These steps require a modern, 7-DOF robot equipped with force sensing. Even then, the programming of such complex steps takes a long time.

Scaling up or reusing the conventionally written program for other compressor models is arduous and requires a long reprogramming time. Therefore this application was considered perfect to showcase i) Quick programming of motions using intuitive inputs like speech, teach by demonstration and vision system, ii) reusable and scalable skills, iii) using middleware for an industrial application.

B. Task Description

The task consists of putting together the air compressor by following the steps shown in Fig. 9. The human operator starts the process by engaging the HMI to create a new application. The HMI allows the use of voice-recognition system to teach the robot the trajectory points from the kitting tray to the assembly fixture by demonstration. The entire assembly area depicting the objects of interests and the process of teaching is shown in Fig. 1. The significance of this teaching operation is to show that any operator without a knowledge of programming robots can intuitively program the robot motion and actions.

The voice-recognition system is engaged by a "Hey Dave" call. Once confirmation is received from the system, the operator orders the creation of a new application, whereby he proceeds to teach the trajectory via points from above the kitting tray to above the assembly fixture. The via points are stored by the voice commands as well. Once the trajectory teaching is done, the operator closes the teaching session by a voice command as well and receives the confirmation about the closure of session from the system._

The grasp points, for all the components are part of the workpiece metadata. Therefore in this application, the operator only teaches the trajectory from above the

Figure 9. Assembly sequence of the main parts. From left to right: unassembled - small rotor inserted - both rotors inserted - middle lid added - top lid added

kitting tray to the assembly fixture. The taught application is then made to run with a voice command, "run application". Subsequently, the robot proceeds to go above the kitting tray following the taught trajectory by invoking the spline skill. There it asks the vision system for the grasp point of the small rotor which is the first object to be inserted. In this application the sequence of workpieces to

be picked and inserted is hard coded. However, the framework allows to easily extend the application by specifying the object to be picked by voice command while teaching the trajectory.

The same is true for the skill to be performed at the end of the trajectory on the assembly fixture, which in this case is always an insertion operation, but can as easily be extended to order another skill, for example, "screwing". The grasp configuration of the gripper, for the object to be picked is part of the workpiece metadata. Therefore, when the robot has reached the grasp point, it invokes the gripping skill that grasps using the stored configuration for the workpiece. In case of the rotors, it is a "pinch grasp". The robot then proceeds back to the taught point above the tray and uses the spline to follow the trajectory to the assembly fixture, where it performs the rotor insertion in the respective housing cavity. The same process is repeated then for the big rotor, middle lid and the top lid until the compressor is assembled.

C. Results

The application showed very promising results as shown in the attached video. In the setup (with limited noise), speech recognition performed accurately for programming by voice and the spline skill programmed in a constrained-based way performed with robust results. The picking operations were reliable due to the feedback from the vision system. The insertion skill used was the one programmed in Kuka Sunrise (Java) to showcase the capability of the framework to deal with skills programmed using traditional methods. It performed the complicated operation of inserting the rotors with precision. The application will be further extended to perform the screwing operation to secure the inserted pieces. As the assembly should be flexible, the cobot will be using the gripper to take the nut runner. However, as currently available nut runners do not allow wireless remote starting, we are currently using one of the gripper's fingers to push the trigger.

Future Trends

We wish to further develop the framework to add more skills, such as screwing and gluing. We also plan on doing user-studies to quantify the performance of traditionally programmed robot skills versus constraint-based robot skills. Work is ongoing to make constraint-based programming more composable and require even less expert knowledge than currently necessary through an eTaSL skill programming template that will provide a graphical interface to create new skills. We tried to base ourselves on commercially available technologies whenever possible, to maximize the potential industry take up.

To create the presented unified framework, each of the components operate as a ROS node and sends appropriate custom messages. Through rosjava, Windows clients (for the speech, vision and robot) can be connected very easily to this middleware, albeit with limited functionalities (e.g. ROS actions are not supported) and there are issues when the ports are not properly closed, specifically on the robot platform. On the other hand, ROS still has various drawbacks for a practical application, such as the single-point-of-failure that is the roscore node and the fact that it has to run on Linux.

Therefore, work is also in progress on upgrading the current middleware from a ROS based version to a ROS2 based version. The initial results suggest that ROS2 is more robust and unlike ROS it does not have a single point of failure. It has a DDS core therefore it provides a more stable communications protocol. Moreover, the setting up of such a framework on a new system is a tedious task, as it requires a significant configuration effort. Using docker containers is a viable option to reduce this effort as they allow easy interconnection between various computing units be they Windows based or Linux based. The reusability is also easier when using docker as they save the user a ton of time that would otherwise be spent on configuration. Lastly it allows launching of all programs from a single PC.

Incorporating these upgrades will make the framework even more potent.

CONCLUSION

In this work, we took a step towards increasing the flexibility of assembly automation by making robot programming faster and intuitive. We developed a framework that uses eTaSL (a constraint-based programming language) to generate motions that supports complex online robot skills. The framework seamlessly integrates the conventional (Kuka sunrise) and non-conventional (eTaSL) motion stacks with a middleware that can be used to create robot applications via and HMI. This middleware further more connects the force-sensitive cobot in

a vendor-independent fashion, utilizes a versatile gripper, uses computer vision that can handle reflective pieces, provides speech recognition that frees up the hands of the operator and therefore provides a framework that makes programming robots quickly possible for anyone with rudimentary understanding of technology. The developed framework is validated with an actual industrial case of an air compressor assembly by Human-Robot Interaction.

REFERENCES

Aertbeliën, E., & De Schutter, J. (2014, September). eTaSL/eTC: A constraint-based task specification language and robot controller using expression graphs. In *2014 IEEE/RSJ International Conference on Intelligent Robots and Systems* (pp. 1540-1546). IEEE. 10.1109/IROS.2014.6942760

Ambler, A. P., & Popplestone, R. J. (1975). Inferring the positions of bodies from specified spatial relationships. *Artificial Intelligence*, *6*(2), 157–174. doi:10.1016/0004-3702(75)90007-7

Bartels, G., Kresse, I., & Beetz, M. (2013, October). Constraint-based movement representation grounded in geometric features. In *2013 13th IEEE-RAS International Conference on Humanoid Robots (Humanoids)* (pp. 547-554). IEEE. 10.1109/HUMANOIDS.2013.7030027

Biggs, G., & MacDonald, B. (2003, December). A survey of robot programming systems. In *Proceedings of the Australasian conference on robotics and automation* (pp. 1-3). Academic Press.

De Schutter, J., De Laet, T., Rutgeerts, J., Decré, W., Smits, R., Aertbeliën, E., ... Bruyninckx, H. (2007). Constraint-based task specification and estimation for sensor-based robot systems in the presence of geometric uncertainty. *The International Journal of Robotics Research*, *26*(5), 433–455. doi:10.1177/027836490707809107

Fryman, J., & Matthias, B. (2012, May). Safety of industrial robots: From conventional to collaborative applications. In *ROBOTIK 2012; 7th German Conference on Robotics* (pp. 1-5). VDE.

Haage, M., Piperagkas, G., Papadopoulos, C., Mariolis, I., Malec, J., Bekiroglu, Y., ... Tzovaras, D. (2017). Teaching assembly by demonstration using advanced human robot interaction and a knowledge integration framework. *Procedia Manufacturing*, *11*, 164–173. doi:10.1016/j.promfg.2017.07.221

Jäger, A., Moll, C., Som, O., Zanker, C., Kinkel, S., & Lichtner, R. (2015). *Analysis of the impact of robotic systems on employment in the European Union. Final report.* Luxembourg: Publications Office of the European Union.

Latombe, J. C. (1991). *Robot Motion Planning.* Kluwer. doi:10.1007/978-1-4615-4022-9

Makris, S., Tsarouchi, P., Surdilovic, D., & Krüger, J. (2014). Intuitive dual arm robot programming for assembly operations. *CIRP Annals*, *63*(1), 13–16. doi:10.1016/j.cirp.2014.03.017

Mansard, N. (2012, May). A dedicated solver for fast operational-space inverse dynamics. In *2012 IEEE International Conference on Robotics and Automation* (pp. 4943-4949). IEEE. 10.1109/ICRA.2012.6224851

Mansard, N., & Chaumette, F. (2007). Task sequencing for high-level sensor-based control. *IEEE Transactions on Robotics*, *23*(1), 60–72. doi:10.1109/TRO.2006.889487

Mansard, N., Khatib, O., & Kheddar, A. (2009). A unified approach to integrate unilateral constraints in the stack of tasks. *IEEE Transactions on Robotics*, *25*(3), 670–685. doi:10.1109/TRO.2009.2020345

Matthaiakis, S. A., Dimoulas, K., Athanasatos, A., Mparis, K., Dimitrakopoulos, G., Gkournelos, C., ... Angione, G. (2017). Flexible programming tool enabling synergy between human and robot. *Procedia Manufacturing*, *11*, 431–440. doi:10.1016/j.promfg.2017.07.131

Pan, Z., Polden, J., Larkin, N., Van Duin, S., & Norrish, J. (2010, June). Recent progress on programming methods for industrial robots. In *ISR 2010 (41st International Symposium on Robotics) and ROBOTIK 2010 (6th German Conference on Robotics)* (pp. 1-8). VDE.

Pedersen, M. R., Nalpantidis, L., Andersen, R. S., Schou, C., Bøgh, S., Krüger, V., & Madsen, O. (2016). Robot skills for manufacturing: From concept to industrial deployment. *Robotics and Computer-integrated Manufacturing*, *37*, 282–291. doi:10.1016/j.rcim.2015.04.002

Perzylo, A., Somani, N., Profanter, S., Kessler, I., Rickert, M., & Knoll, A. (2016, October). *Intuitive instruction of industrial robots: Semantic process descriptions for small lot production. In 2016 IEEE/RSJ international conference on intelligent robots and systems (iros)* (pp. 2293–2300). IEEE.

Samson, C., Espiau, B., & Borgne, M. L. (1991). *Robot control: the task function approach*. Oxford University Press, Inc.

Scioni, E., Decré, W., Aertbeliën, E., De Schutter, J., Tirmizi, A., & Witters, M. (2017). Fast and Intuitive Robot Programming: a Constraint-based Approach. ICRA.

Sentis, L., & Khatib, O. (2006, May). A whole-body control framework for humanoids operating in human environments. In *Proceedings 2006 IEEE International Conference on Robotics and Automation, 2006. ICRA 2006.* (pp. 2641-2648). IEEE. 10.1109/ROBOT.2006.1642100

Tirmizi, A., De Cat, B., Janssen, K., Pane, Y., Leconte, P., & Witters, M. (2019, May). User-Friendly Programming of Flexible Assembly Applications with Collaborative Robots. In *2019 20th International Conference on Research and Education in Mechatronics (REM)* (pp. 1-7). IEEE. 10.1109/REM.2019.8744135

Vanthienen, D., Klotzbu, M., De Schutter, J., De Laet, T., & Bruyninckx, H. (2013, November). Rapid application development of constrained-based task modelling and execution using domain specific languages. In *2013 IEEE/RSJ International Conference on Intelligent Robots and Systems*(pp. 1860-1866). IEEE. 10.1109/IROS.2013.6696602

Wittenberg, G. (1995). Developments in offline programming: An overview. *Industrial Robot: An International Journal*, *22*(3), 21–23. doi:10.1108/EUM0000000004186

Zanchettin, A. M., & Rocco, P. (2016, October). Robust constraint-based control of robot manipulators: An application to a visual aided grasping task. In *2016 IEEE/RSJ International Conference on Intelligent Robots and Systems (IROS)* (pp. 3634-3639). IEEE. 10.1109/IROS.2016.7759535

KEY TERMS AND DEFINITIONS

Collaborative Robots: A collaborative robot, also known as a cobot, is a robot that is capable of operating safely alongside human without any safety barrier separating the two. It has a complex sensor suite that not only allow safe operation but also makes it possible to assist humans in their tasks.

Docker: Docker is a tool that can package an application and its dependencies in a virtual container that can run on any Linux server. Docker containers are lightweight, a single server or virtual machine can run several containers simultaneously. It avoids the exercise to configure or reconfigure every time the application has to work on a new setup.

Human-Machine Interface: Also known as an HMI. An HMI is a software application that presents information to an operator or user about the state of a process, and to accept and implement the operators control instructions. Typically information is displayed in a graphic format (Graphical User Interface or GUI).

Manufacturing Execution Systems: A manufacturing execution system (MES) is an information system that connects, monitors and controls complex manufacturing systems and data flows on the factory floor. The main goal of an MES is to ensure effective execution of the manufacturing operations and improve production output.

Non-Sensitive Robot: A non-sensitive robot has no tactile sensing ability in its joint therefore it cannot determine if its body is in contact with anything external based on the interaction force.

Original Equipment Manufacturer: An original equipment manufacturer (OEM) is a company that produces parts and equipment that may be marketed by another manufacturer.

Product Lifecycle Management: In industry, product lifecycle management (PLM) is the process of managing the entire lifecycle of a product from inception, through engineering design and manufacture, to service and disposal of manufactured products.

Programmable Logic Controllers: A programmable logic controller otherwise known as a PLC is an industrial computer control system that monitors the state of the input devices and makes decisions based upon a custom program to control the state of the output devices.

Robot End-Effector: In robotics, an end effector is the device at the end of a robotic arm, designed to interact with the environment. The exact nature of this device depends on the application of the robot.

ROS: ROS is an open-source, meta-operating system for your robot. It provides the services you would expect from an operating system, including hardware abstraction, low-level device control, implementation of commonly-used functionality, message-passing between processes, and package management.

ROS2: ROS2 is a more robust incarnation of ROS. It avoids a single point of failure. It is based on a more stable communication protocol. Its architecture offers the ability to do real-time control of robotic devices.

Teach Pendent: Teach pendants are typically handheld devices and may be wired or wireless. A Teach Pendant is used to program various models of the robotic industrial machinery.

Chapter 6
Designing a Robot for Manufacturing Fiberglass Reinforced Plastic (FRP) Molded Grating

Marcos Vinícius Ramos Carnevale
Federal University of Rio de Janeiro, Brazil

Armando Carlos de Pina Filho
https://orcid.org/0000-0002-2622-1480
Federal University of Rio de Janeiro, Brazil

ABSTRACT

The use of robotics in the industrial environment has, in general, very similar goals. Because of productivity requirements, or due to reliability, industries have been constantly equipping their floor with robots. In that sense, the chapter observed—in a fiberglass company—the chance of using a robot to execute a boring and repetitive task. The task mentioned is, actually, the manufacturing of fiberglass reinforced plastic (FRP) molded grating. To confirm the possibility of using a robot to this job, a cost and time analysis was made about the whole molded gratings manufacturing process. Afterward, research about robotics was taken in parallel with the conception of the robot (named "roving-robot"). Calculations were made to the mechanical project of the robot. Applying computer-aided design (CAD), technical drawing and bill of materials were generated to permit the robot assembling. All of these project steps are presented in this chapter.

DOI: 10.4018/978-1-7998-1382-8.ch006

INTRODUCTION

This chapter presents a project developed after a real problem faced by a fiberglass manufacturer in Brazil. The firm had been undergoing lack of productivity during the fabrication of molded fiberglass gratings (FRP gratings, or fiberglass reinforced plastic gratings). From 2014 to 2015, the company could not reach the high market demand in the country. In that way, a solution had to be achieved considering the identified bottleneck: the number of molds; also, taking into account the very tight budget available. According to the supply chain analysts, a new mold would cost around U$43,000.00. Therefore, the use of robotics was suggested in order to reduce the task time of certain steps during the molded gratings production.

The main goal of the present chapter is to exhibit the mechanical conception of a robot capable of raising the FRP gratings productivity. Due to the shortage of money, it was used a low-cost and goal-focused mentality.

To familiarize the reader with specific subjects, the application of robotics in industrial environments will be briefly presented as background. In addition, the main focus of the chapter will be dedicated to summarize the main FRP products and their manufacturing methods.

After this, the reader will be capable of understanding the molded FRP gratings fabrication process. All the manufacturing steps will be described and analyzed in terms of time and cost. A specific model of grating will be selected to exemplify some calculations regarding the automation benefits. However, the estimations can be extended to any other similar product.

Subsequently, the reader will find some studies about the robotic branches. Between them, an introduction about this very recent science and the main classifications and components used to build a robot. Some specifications to the robotic solution proposed will be presented simultaneously to this review.

In the end of this chapter, the reader will find the conception of the robot shaped to solve the problem abovementioned. The knowledge of robotics already introduced helps the understanding of the choices made about mechanical elements. The robot concept will be fully presented by showing its computer aided design (CAD).

BACKGROUND

The Beginning of Robotics

The term "robot" has its origins at the Czech word "robota", whose literal translation is "slave" (Brum, 2016). Karel Capek, the playwright that wrote a science fiction narrative called "Rossum's Universal Robots", first used this word on 1921. Then, "robot" was spread out when the writer Isaac Asimov introduced the "Three Laws of Robotics" on 1942. Nevertheless, the ambition to make machines with human being movements can be observed since antiquity – either when the Egyptians built mechanical bodies in statue shapes, or when puppets could be driven by pulley systems in Ancient Greece.

The robot control by computers became possible only on 1948, after the invention of the transistor. Six years after, George Devol applied for the first patent of an industrial robot with digitally programmable operations. The Unimate had born, and was considered the first industrial robot commercialized, after being installed at General Motors on 1961. From then on, the use of robotics has grown really fast not only in industrial environments, but also in other sectors, like services, safety, health, entertainment and household tasks.

In that way, robotics has gradually become an interdisciplinary science. According to Mccomb (2011), it embraces engineering, electronics, psychology, sociology, physics, artificial intelligence, design, programming, mechanical architecture, among others.

Classification of Robots

In spite of being innumerable, some robot classifications are highly relevant to keep in mind while a project is being developed. Therefore, the majority of the robots can be sorted by autonomy, mobility and degrees of freedom (related with its cinematic structure).

The robot autonomy concerns its control system. It is very important to think about this attribute from the beginnings of the project – since, in the future, many choices will depend on this indication. In that sense, robots can be created to work:

- In a completely autonomous way.
- Controlled by a human being.
- Or partially controlled.

It is important to say that there is no better control system method; what exist are configurations better suitable to certain tasks (Mccomb, 2011).

The goal here is to develop a low-cost and agile robot capable of executing just one procedure – between two others – of FRP grids manufacturing. In that way, the gratings production would have alternated steps realized by the operators and the robot (as illustrated by Figure 1). Consequently, it is extremely important the interaction between the employees and the machine, so that each procedure can be made in the right time.

Figure 1. Division of tasks during the molded grids manufacturing
(produced by the authors)

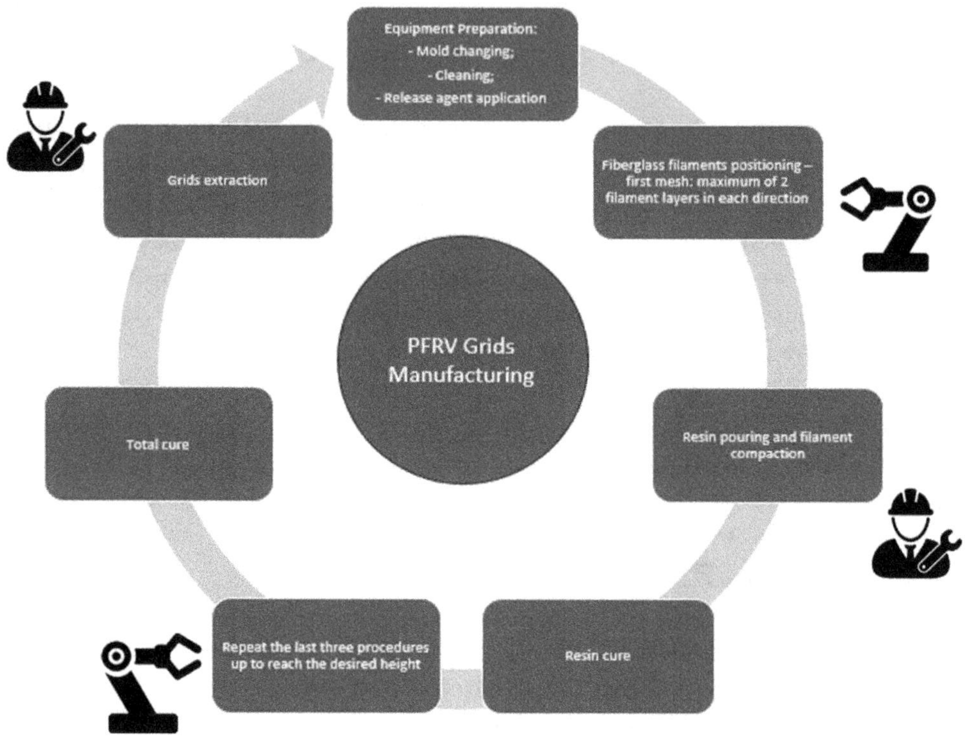

It follows, then, that the best solution to this project is a semi-automatic machine that can be partially controlled by the operator, because he will have to prepare the robot and finalize the manufacturing process. After the end of the braiding activity, the robot can indicate the employees that the further procedures can start. In addition, the braiding process has a complex setup that would be pretty difficult to automate: passing the fiberglass filaments through the aluminum pen and making the knot in one of the mold steel blocks.

As for mobility, it is possible to divide the robots between fixed and mobiles. In this project, it is easy to understand that the task to be automatized needs a high level of precision in a limited environment. The use of a mobile robot would not represent an advantage; so, it was decided to anchor the roving-robot to the molding table.

There are two main types of structures that can be used to design stationary robots: they can have a parallel or a serial construction. The parallel one is characterized by a closed chain mechanism, where there is at least one point connected to a fixed base by two or more independent kinematic chains (Malvezzi, 2008). Parallel robots have:

- Greater movement speed due to the best distribution of movable elements by each actuator.
- Greater rigidity, provided by the higher number of kinematic chains.
- Greater precision, since the actuators are not assembled in series – avoiding error propagations.

However, the closed chain brings to the mechanism some disadvantages, such as:

- Higher occupied space.
- Greater complexity to logically program the actuators.
- Great difficulty in calibration (Malvezzi, 2008).

In addition, it is noted that parallel robots end up having a higher cost when destined to perform simple tasks, if compared to the robots in series. This is due to the large number of actuators required for the movement of the also large number of mechanisms. For the project, the robot must perform the same movement the operators do, driving a guiding-pen between the blocks of the mold. This operation is quite simple because it is always rectilinear. Therefore, a serial mechanism is most appropriate for it.

In order to go ahead with the robot design, it is first necessary to analyze the demanded movements in a closer way. The fiberglass filaments are passed between the mold metal blocks with the aid of a pen (Figure 2). This movement is always straight; therefore, it is quite evident that all the required displacements are done in a Cartesian plane, in only two perpendicular directions defined here as X and Y. A third orthogonal direction, Z, may be necessary considering that the guiding-pen should be raised during the activity.

Presetting of the Roving-Robot Components

The first components studied for the roving-robot were the actuators. They can be classified according to the energy source they use for their operation. The main types

Figure 2. Aluminum guiding-pen
(produced by the authors)

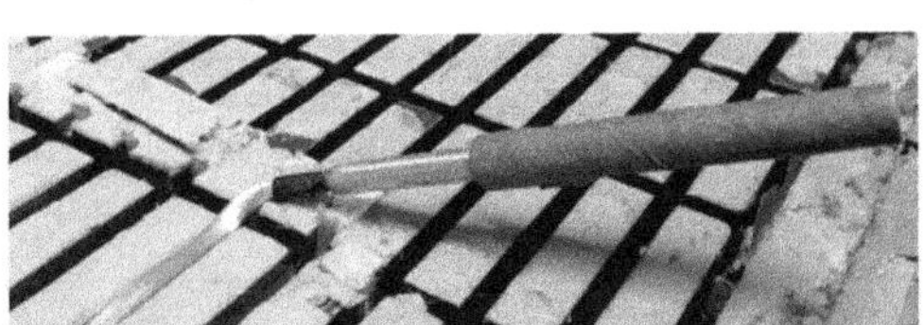

found in robotics are pneumatic, hydraulic and electric. Table 1 presents advantages and disadvantages of each actuator model.

The roving-robot must counts on actuators that allow:

- Accurate and reliable movements, so that the guiding-pen is properly positioned between the mold blocks.
- Fast movements, so that the used time of each table/mold assembly is reduced.
- Low cost.

It is concluded, therefore, that electric actuators are the most suitable for the project. This is because they are the only ones that provide the precision and reliability required. Added to this is the fact that no special devices are required, indispensable to other models.

More specifically, the motors that best meet the roving-robot project are the steppers. Servomotors could be used too, but this type of motor often requires a position encoder to keep track of the position of the motor shaft, especially if precise movements are required (Burris, 2019). Unlike AC and DC motors, stepper motors have as their main advantage their direct control through computers and microprocessors. The rotation of its axis is performed at discrete angular intervals (the “steps”) after receiving command pulses. Therefore, one can know which angular rotation was performed by the motor according to the number of pulses sent by the controller. This makes them ideal for control applications that use open mesh. Because of all this, the use of sensors will be first avoided in this project.

The operation of the robots is obtained by the control system that they use. Microcontroller options were sought in the market for the roving-robot project. The

Table 1. Advantages and disadvantages of actuators

Actuators	Pneumatic	Hydraulic	Electric
Energy	Compressed air (5-10 bar)	Mineral oil (50-100 bar)	Electric current
Options	Cylinders, flipper motors, piston motors	Cylinders, flipper motors, axial piston motors	Direct current, alternate current, stepper motor
Advantages	Light, cheap, quick movements	Strong, great power and speed, good weight-power ratio, stability with static loads	Precise, trustable, easy to control and to install, quiet, can be extremely quick
Disadvantages	Low accuracy, fragile, need special installation (filters, compressors, etc.), noisy	Low accuracy, fragile, need special installation (filters, compressors, etc.), very expensive	Can be expensive, depending on the model; limited power

(produced by the authors)

three main prototyping platforms found were Arduino, Intel Galileo and Raspberry. The first two boards are microcontrollers that always run the same software, while the latter is properly a computer that contains an operating system. As specified so far, only stepper motors will be used for the operation of the roving-robot. Considering that these motors require a simple hardware control, the Arduino was the most indicated board for this development; Arduino was also the most economical option found (Bruce, 2013).

The Arduino Uno model was selected to the current activity because it is accessible and has innumerable expansive modules (shields) that allow the easy control of stepper motors. In this sense, it will be used also the CNC Shield v3 – an Arduino Uno compatible board that facilitates the connection and control of three independent stepper motors (plus the clone command of one them).

Finally, to drive the motors, A4988 drivers will be adopted. These drivers have been specially developed to control small steps (micro stepping) of bipolar motors. They are the most widespread in the market for low cost applications. Still, they are quite affordable and can reach five resolutions, ranging from full-step to 1/16 of this (1/16-step).

Once concluded the presetting of some elements – such as structure, actuators and drivers – is necessary to define the problem to solve and the conditions to automatize, making possible the design of robot.

MAIN FOCUS OF THE CHAPTER

Contextualization: Factory Automation

According to ISO 10218 (International Organization for Standardization), an industrial robot can be defined as an "automatically controlled, reprogrammable, multipurpose manipulator, programmable in three or more axes, which can be either fixed in place or mobile for use in industrial automation applications". In spite of being a very broad and restrictive description, numerous authors consider as "robot" any machine that can execute tasks controlled by the human being. These tasks can be either remotely controlled or preprogrammed. With this last narrative, even not-autonomous mechanisms could be considered a robot (Bouteille, 2000).

Despite all the diversity of opinions, the purposes of applying robots in industrial environments converge in some few objectives. In that sense, the automation of a process seeks typically a cost reduction of products and services inside the factories. This reduction can be a consequence of many elements, like:

- Productivity increase.
- The possibility to reduce the number of employees.
- The optimization of raw material consumption.
- The opportunity of having a continuous production (uninterrupted).
- The savings from not training the operators.

In addition, factory automation can bring improvements on:

- Product quality.
- Reliability of methods and processes.
- Possibility to access harmful environments.
- The execution of repetitive, unpleasant and monotonous tasks.

Because of all this, it is easy to understand the accelerated growing of robotic applications in industrial environments. This is the central point of what is being called a new industrial revolution, characterized by modern and highly automatized factories (Heer, 2016). Also, this revolution has been allowing the return of the factories to countries considered developed, whose onerous labor costs had taken away the production from their cities. On the other hand, the robotics can help emerging countries in gaining productivity and diving, in a timely manner, into the new global tendencies of industry standards (Wentzel, 2016).

Composite Industry and the Fiberglass Reinforced Plastic (FRP) Material

Given all the advantages of task automation by robots, it was verified the opportunity to apply these concepts in one of the biggest FRP industries in Brazil. In order to preserve confidentiality, the company has asked to do not have its name cited in the present text.

The fiberglass reinforced plastic (FRP, also popularly known as just "fiberglass") is a composite material. A "composite" can be defined, in turn, as a multiphase substance – with significant proportionality of its formative phases – that compounds a blend with better characteristics than those of the individual components used alone (Kemerich et al., 2013). In the case of fiberglass, there are two key phases: the polymeric matrix and the reinforcement, where the matrix phase is a resin whose function is to bond the fiberglass fibers (reinforces). The mechanical resistance of a FRP component depends especially on the type, orientation, quantity and positioning of the fiberglass fibers.

The FRP has been showing itself as an excellent technologic alternative to aluminum, steel, concrete, PVC and wood. Its costs, moreover, are quite competitive when taken a long-term view. Among the biggest advantages of the FRP, there are: resistance to weather conditions and to chemicals; facility to assembly and to make adjustments; lightness; extended lifetime; low expansion coefficient and thermal conductivity; mechanical resistance.

Numerous manufacturing methods have been developed in order to reach massive production of composite materials. In the beginning, manual methods tended to predominate, as the lamination. In this process, layers of fiberglass films are usually positioned in a mold and impregnated by catalyzed resin. After the cure of the compound, the piece can be removed from the cast (Figure 3-I). Subsequently, the technology evolution brought methods that allowed the massive production of FRP parts. Among them, there are the pultrusion, the continuous lamination and the filament winding (Figure 3-II, III and IV, respectively).

All these methods are very efficient to get constant cross-section profiles, pipes and plates. However, they cannot be applied to obtain more specific products, as great-thickness plates, customized objects and floor gratings. In that sense, it was observed the opportunity to improve the production of molded FRP gratings by using

Figure 3. Main FRP fabrication processes
(produced/adapted by the authors)

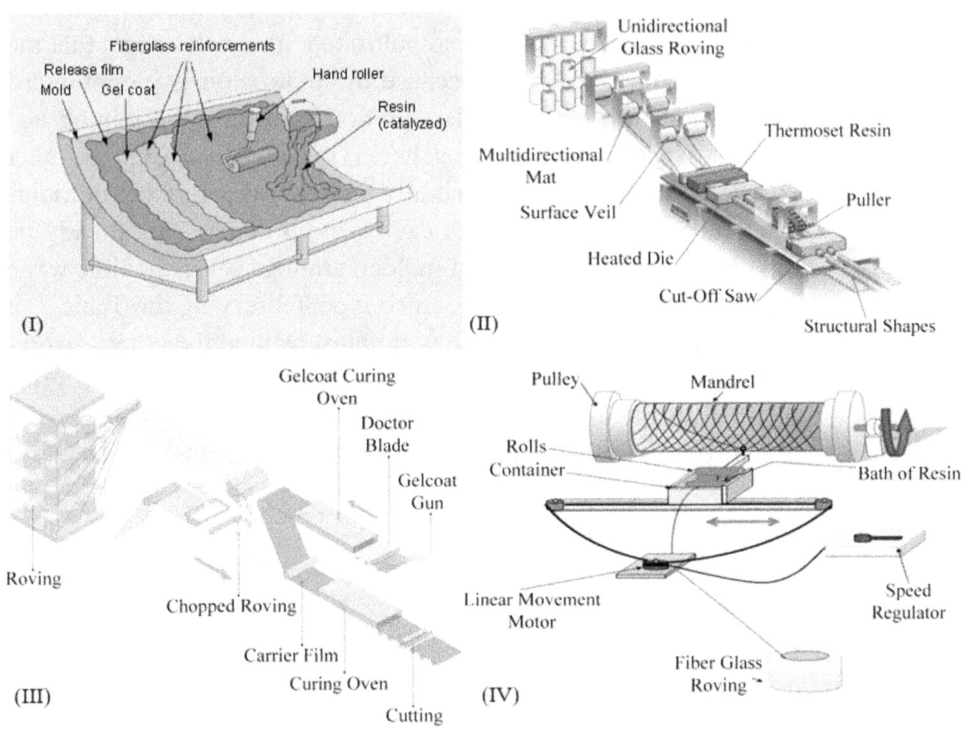

a robot. In spite of not being considered precisely a "lamination", the fiberglass gratings molding is a technique likewise manual, whose main stages are the fiberglass filament placement and the addition of catalyzed resin.

Molded Gratings

Floor gratings (also known as "floor grids") represent a quite important niche of the composite materials market. This is because these products gather all the advantages of fiberglass reinforced plastic against the steel – notably lightness, corrosion resistance and straightforward installation.

In almost all the countries, two types of FRP grids are available:

- Gratings mounted by pultruded profiles glued to each other.
- Molded/injected gratings.

Despite small differences, the manufacturing processes of opened-molding and injection result in very similar finished products (here treated as equivalents). Nowadays, the majority of the factories prefer the open molding method over the injection one, since the former allows a better distribution of the fiberglass filaments inside the cavities. In addition, the injection revealed itself a superfluous procedure, whose machinery requires higher spending with maintenance and skilled workforce.

Comparing both types of grid, molded and pultruded, it is well known that the latter can meet higher demands. This is because the pultrusion is a continuous fabrication process that achieves large scales; its limitations are the joining and gluing procedures (that is to say, the bottleneck here is the workforce). On the other hand, the molded grids manufacturing is generally limited by the number of molds available (because, once extracted from the cavity, the products can already be final finished). In that way, the production of molded gratings is less flexible when compared to the production of pultruded/mounted type. Observing the Table 2, it is easy to notice that it is not always that one type can substitute the other.

Then, one of the possibilities to meet higher demands of FRP molded grids production – without investing in costly molding equipment – could be the decrease of its manufacturing time, especially the reduction of the time when the cavities are being used.

Molded Gratings Manufacturing Process

In order to validate the assumption that the robotics could increase the molded grids production in the company, many analyzes were made regarding their manufacturing process steps.

Table 2. Comparison of the main models of FRP grids on the market

Characteristic	Molded Grid	Mounted Grid
Application	Indicated for traffic of people, vehicles and equipment with wheels/ castors.	Indicated for people traffic. Not recommended for vehicular traffic and wheel/ caster equipment.
Type of panels	Monolithic/mono-bloc panel, seamless and without glue.	Panel mounted by means of pultruded profiles with seams and glue.
Chemical Resistance (in addition to the manufacturing method, the type of resin is also a key factor in determining the resistance of the material against corrosion)	High chemical resistance due to the high percentage of resin in its composition (60 ~ 65%).	Moderate chemical resistance due to the lower percentage of resin (35 ~ 40%).
Mechanical Resistance	Good mechanical resistance, ideal for moderate spans.	Excellent mechanical resistance, recommended for larger spans.
Field Adjustment	Easy for field adjustment. The grids can be cut to meet the dimensions without any problem of finish or mechanical resistance loss.	Requires skilled labor for field adjustments. We recommend purchasing the trimmed panels from the factory.
Loading Bar	Support bar on all sides. That is, it can be supported in any direction.	It is necessary to identify and define the direction of the bearing bar. Can be supported only in one direction.
Weight	Higher weight when compared to mounted grids.	Low weight per square meter.
Main Markets	Chemical industries, paper and cellulose, water and sewage treatment plants.	Offshore platforms, confined rooms (that requires phenolic resin).

(produced by the authors)

This type of grating is produced inside molds (also known as cavities) that are assembled on a sort of table, as shown in the Figure 4. Each mold has steel blocks, whose cavities shape the grid pattern.

The company in question owns different types of tables and molds. The most common sizes of tables produce 2000x1000mm and 3660x1220mm gratings, while the variety of steel blocks dimensions allows the diversity of meshes. Therefore, each model of grid requires different amounts of raw material and workforce time. Aiming uniformity, this project was carried out by choosing one type of grid: the 38x38x38 with 2000x1000mm dimensions. The resultant panel – from now on “standard panel” – is the best-seller at the Brazilian market, and has 2000mm width, 1000mm length, 38mm height, and a 38mm square mesh.

Figure 4. FRP gratings table with mold
(produced by the authors)

The manufacturing of molded grids starts with a preparatory stage, when an operator changes the steel blocks (to obtain the right mesh) and cleans the table, if necessary. Then, he applies a release agent, so that the cured product can be easily removed from the cavity. The next step consists in placing the fiberglass filaments between the steel blocks, up to reach two layers of filament in each direction. This step is entirely manual, realized by an operator, and it is one of the most repetitive and boring tasks in the factory.

After this, the catalyzed resin is poured, and the operator uses a special tool to make the fiberglass compact inside the matrix, which is still liquid. If necessary, the process of filament positioning is repeated until obtained the desired height. Finally, after the complete cure of the resin, the grid is removed from the mold by extractor pins.

In that sense, ignoring the finishing steps (which have place outside of the molding equipment) and the extraction (that is very fast), it is possible to divide the gratings manufacturing in three main steps:

- Mold preparation.
- Fiberglass filaments positioning.
- Resin pouring and cure.

Consequently, a study was realized inside the company in order to understand the percentage of machine hour used by each of the steps previously mentioned. With that purpose, the time of each cycle was measured for all the grid models offered by the company. Table 3 illustrates the result obtained for two grid panels. It is highlighted that the standard panel (38x38x38 with dimensions 2000x1000mm) is present at the first column.

Specifically analyzing the filament positioning data, it is easy to notice that the greater the table dimensions, the greater the difficulty found by the operators to realize the task – consequently, the greater the time. This occurs because they have to change their position around the table many times during the work. In that sense, even for small tables manufacturing, the company decided to assign two operators to the task. Despite this fact, for all models of molded grids, the filament positioning is always the longer step. Yet, considering that the other procedures are less flexible to changes, the filament braiding has become the focus of the current text. Being a completely manual task, it showed itself as a good candidate to be improved by automation.

Fiberglass Filament Braiding

According to the fiberglass company, the total length of filament inside a standard panel reaches the 1,0km/m^2 ratio. This first explains why the braiding step represents a long time percentage during the grids manufacturing.

To start the filament positioning, the operator makes a knot in one of the steel blocks that are in the vertex of the table. Then, the procedure is realized with an "aluminum pen" that acts as a guide passing across the cavity shaped by the metallic blocks. Aluminum has been selected as the tool material because it has hardness lower than the rectified steel blocks one, preventing the mold damaging.

For the FRP gratings, the fiberglass filaments are the exclusive reinforcement component, having the function of ensuring the mechanical properties of the finished grid. Because of this, the way the filaments are positioned inside the resin defines

Table 3. Time to produce two different panels of FRP molded gratings

Manufacturing Step	38x38x38 Model, 2000x1000mm Panel	38x38x38 Model, 3660x1220mm Panel
Mold preparation	0,08h (7%)	0,10h (5%)
Filament braiding	0,73h (64%)	1,63h (79%)
Resin cure	0,33h (29%)	0,33h (16%)
Total time	1,15h (100%)	2,06h (100%)

(produced by the authors)

Table 4. Standard panel fabrication data

Data	38x38x38 Model, 3660x1220mm Panel
Filament total length (km)	1,9
Maximum number of filaments inside the guiding-pen	4
Number of filament layers	4
Manufacturing time (h)	1,15
Braiding process time (h)	0,73

(produced by the authors)

the quality of the product. The company's experience has shown that the mechanical resistance is greater for grids with filaments:

- Slightly stressed between the steel blocks.
- Braided also at the edges.
- With alternate layers in the directions parallel to the table edges. One layer consists of filaments braided at the X and Y directions. Each molded grating requires a different amount of filament to its best manufacturing.

For the standard panel here analyzed, up to four filaments can be placed inside the aluminum pen, in such a way that the operation is realized in a more efficient way, and without compromising the mechanical resistance of the grids. The Table 4 presents some important data regarding the standard panel manufacturing.

Hence, an important calculation to be made is the average speed that the operators can braid the filaments into the cavities (V_{braid}):

$$V_{braid} = \frac{Total\ Filament\ Length}{Number\ of\ filaments\ by\ pen \times Time} \quad (1)$$

This expression results in about 180mm/s (or 10,85m/min) average speed for the standard panel.

Conditions to Automatize

The first boundary condition to automatize the filament braiding is represented by the average speed of the fiberglass positioning inside the mold. To meet the target, the robot must be capable of performing a speed greater than the one measured for two operators working in each mold (0,19m/s, according to the practical observation). In addition, it is already known that the company does not have a high budget for this production line. Then, the costs of components and the installation cannot exceed the price of a new mold machine. The robot must also execute its task at the tables and molds already presents at the factory. In short, the final solution must be simple, cost-effective and perform considerable speeds.

Thinking about the physical obstacles, it is important to notice that the grids manufacturing consists in three main procedures; and just one of them is being automatized. Thus, the robot cannot hamper the work done by the operators during the two other steps. Another condition is to consider that the working environment is vulnerable to catalyzed resin spill.

Having all the project requirements and limitations, it is possible now to start studying the proper solutions to design the robot components.

SOLUTIONS AND RECOMMENDATIONS

Initial Approach

The roving-robot is a project that must be adapted to the tables and the molds already available at the factory. In that way, the first step done was the design of these elements. In many cases when the mechanical design of complex components is needed, a cloud of points can be first generated by 3D scanners, and then corrected using a CAD software. However, considering that the table geometry is quite simple, the modeling required just few measurements with a caliper and a metric tape. While taking the measures, it was realized that the supplier has probably used high dimensional and parallelism tolerances. Considering this, it was defined that the robot has to be field assembled.

Having the mold design, it was possible to draft the roving-robot, with a structure that can be divided in 3 main parts:

- Y axis: structure that allows the movement along the direction parallel to the bigger side of the table (2000mm).
- X axis: structure that allows the movement along the direction parallel to the smaller side of the table (1000mm).
- Z axis: structure with movement orthogonal to both axis previously defined (that allows the lifting of the aluminum pen).

Figure 5 illustrates the sketch made before the robot design. This draft helped to guide the selection of the mechanical elements and the electric actuators most suitable to the robot. In that way, it is easy to notice that the proposal was to use

Figure 5. Project draft
(produced by the authors)

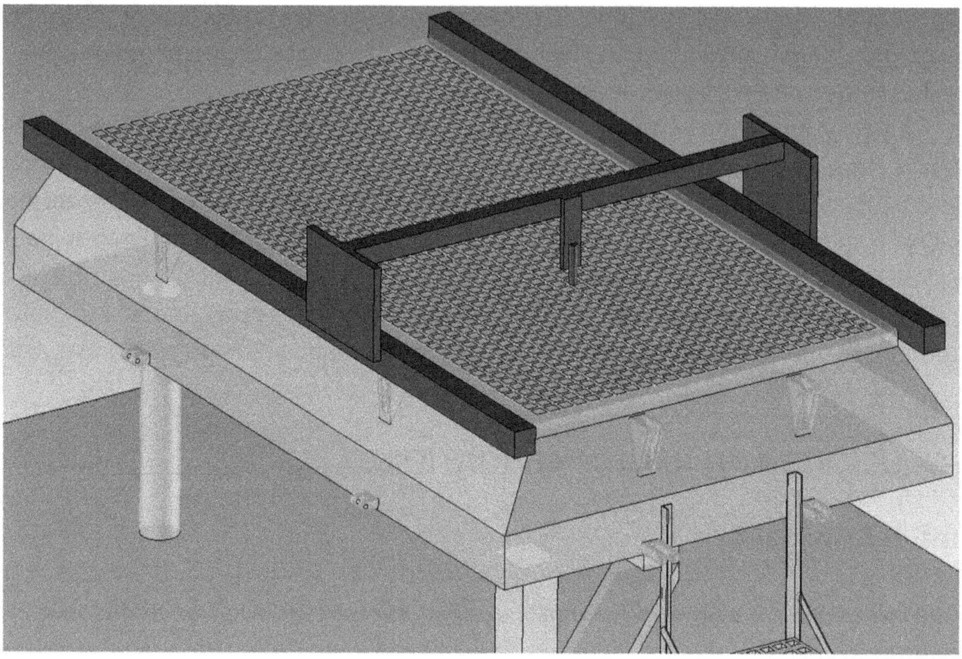

three-associated axis to compose a Cartesian system that coordinates the movements. It is expected that four stepper motors are needed to drive the robot: two for the Y, one for the X and one for the Z axis. To start the calculations, the total mass of the system was estimated in 5kg. This is the mass loading the Y motors. For the X motor, in turn, the prediction was a maximum load of 2kg.

Displacement Speed

In order to go ahead with the project, it was essential to set a speed target to the roving-robot. That is because high velocities can be achieved, but increasing the costs of the components needed to buy. So, instead of early limiting the budget, it was agreed with the fiberglass company the target of reducing by half the time required to position the filaments into the mold cavities. Consequently, the time-target had become 22 minutes (considering that the operators are able to realize the task in 43.8 minutes).

The robot speed cannot be calculated considering an ordinary average velocity. This assertion is true because stepper motors, even more than other actuators, need well-defined profiles for acceleration and deceleration. In case this does not occur, there will be always the risk of step losses (slipping) during speed variations – due to the system inertia (Quinones, 2017). The majority of stepper motor controllers use two variants to drive the actuator: the SPS (step-per-second) and the SPS variation in time. These parameters define angular velocity and angular acceleration, respectively.

Knowing all this, a speed curve can be divided into three main parts: acceleration, cruise speed and deceleration. The total number of steps required for the movements can be allocated in these three time-parts. The total number of steps, in turn, can be calculated using the controller micro stepping, the stepper motor resolution and the physical dimensions of the mechanisms. On the other hand, the cruise speed and the steps division depend on both, the controller and the system inertia.

In the case of the roving-robot, three are the main movements in X and Y (Figure 6):

- 27x 2000mm courses on the Y direction, per filament layer.
- 53x 1000mm courses on the X direction, per filament layer.
- 78x 38mm course tunings to change the mold aisle, in both directions, per filament layer.

Therefore, the time division for each movement type was made – in order to meet the 22-minute target agreed with the company. For this, the time distribution was realized according to the method explored by QUINONES (2017), allocating 20% of each rout for acceleration, 60% for cruise speed (constant), and the other 20% for deceleration. It is also reasonable to consider the Y speed 50% higher than the X one, because the first axis will count on two-stepper-motor torques (needed for the system acceleration).

For the 38mm-tunings, an interval of 0,5s for each tuning is suitable with the motor specifications. The total number of tunings for the standard grid panel corresponds to 312. Table 5 shows the number of movements in each direction, as well as their distances.

So, the remaining time to the X and Y movements is:

$$t_{xy} = 1314 - 0,5 \times 312 = 1158s \tag{2}$$

Figure 7 illustrates the speed profile in both directions.

In that way, using the area below the curves, it is possible to calculate the cruise speed to meet the target defined by the company (by solving the three-equation system below):

$$\Delta S_x = 1000mm = V_{cz} \times 0,8t_x \tag{3}$$

Figure 6. Displacements need to position the filaments
(produced by the authors)

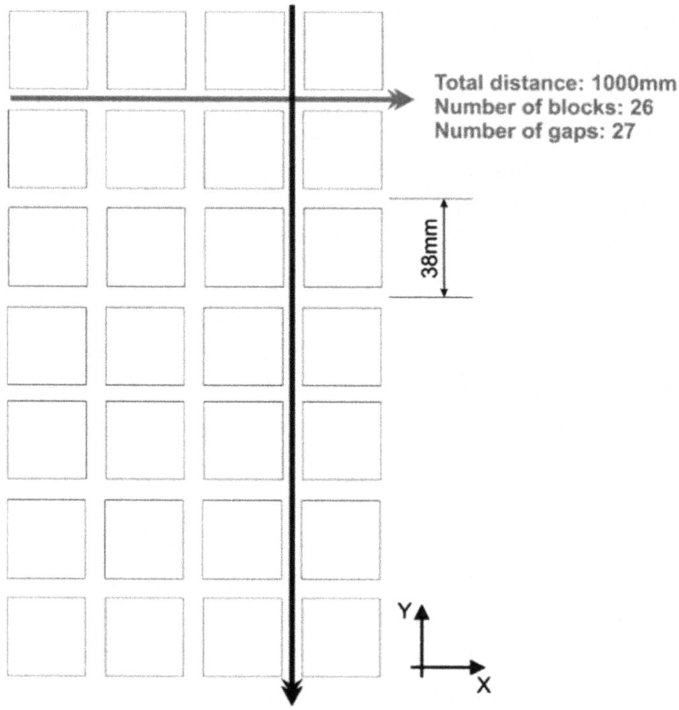

$$\Delta S_y = 2000mm = 1,5V_{cz} \times 0,8t_y \quad (4)$$

Table 5. Number of movements required

Movement	Displacement	Number of Movements by Layer	Total Number of Movements for the Standard Panel (Four Layers)	Total Displacement for the Standard Panel
X-Axis	2000	27	108	216
Y-Axis	1000	53	212	212
Adjusts	38	78	312	11,856

(produced by the authors)

Figure 7. Speed profiles for X and Y-axis
(produced by the authors)

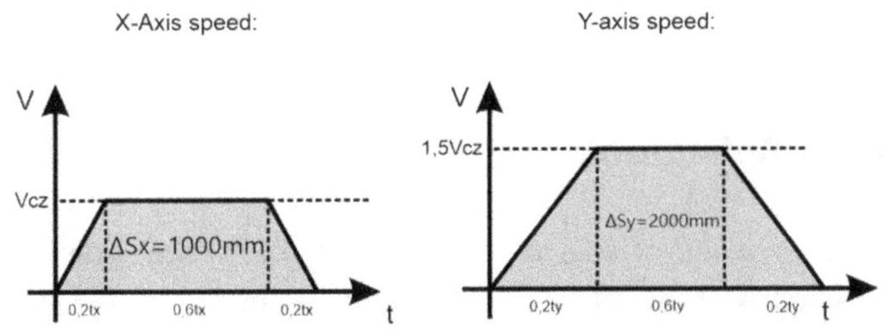

$$t_{xy} = 19,4s = 108 \times t_y + 212 \times t_x \quad (5)$$

Replacing t_x and t_y into the equation (5):

$$V_{cz} \cong 384mm / s \quad (6)$$

$$1,5V_{cz} \cong 576mm / s \quad (7)$$

At this point, it is already known the speed that each stepper has to perform. Time and accelerations can be obtained by using simple kinematic equations:

$$t_X = \frac{\Delta S_x}{V_{cz} \times 0,8} = \frac{1000mm}{382mm / s \times 0,8} \cong 3,25s \quad (8)$$

$$t_y = \frac{\Delta S_y}{1,5V_{cz} \times 0,8} = \frac{2000mm}{573mm / s \times 0,8} \cong 4,34s \quad (9)$$

$$a_X = \frac{V_{cz}}{0,2 \times t_x} = \frac{382mm / s}{0,2 \times 3,27s} \cong 590mm / s^2 \quad (10)$$

$$a_y = \frac{1,5V_{cz}}{0,2 \times t_y} = \frac{573mm / s}{0,2 \times 4,36s} \cong 664mm / s^2 \quad (11)$$

Stepper Motor Selection

The stepper motors are classified according to their size by the American regulation NEMA (Associated Manufacturer of Electrical Supplies). A NEMA 17 stepper motor, for instance, is characterized by flange size 1.7 inches. Nowadays, the most used steppers are models NEMA 17, 23 and 34. Many motor constructors respect this regulation, each of which with its own specific product.

Because of the A4988 driver choice (selected to not encumber the project), the stepper motor selection is restricted by the maximum electric current supported by these controllers – around 2.0A. That said, the actuators NEMA 23 from the Neoyama supplier, model AK23/15F6FN1.8, were appointed for this project. They request 2.1A when connected in the phase series bipolar, having a torque of 15kgf.cm. The small-exceeded current of 0.1A will not be a problem if the steppers work in fair conditions (far from their operational limit). Now, it is necessary to verify if the motors can meet the linear motion system calculated before. The selected drivers will need to be changed if the project is undersized, in order to select motors that are more powerful. In contrast, if they are oversized, the downgrade to a NEMA 17 can be considered.

Linear Motion on X and Y

Looking again at the draft made for the roving-robot, the reader can notice that one of the most important decisions to the project is the type of mechanism that will allow its linear motion. In other words, how to reach a translational motion from the rotational movement of the motor shafts? This is important mainly for the X and Y directions, where the displacements are bigger and can have a greater impact on the filament-positioning time. According to Bruzzone (2004), the main mechanisms to perform this function are:

- Screw shaft and nut.
- Pulley and belt.
- Gear and rack.
- Gear and chain.
- Friction wheel.

Considering the accuracy needed, it was ruled out the possibility of using the gear/chain mechanism and the friction wheels. The former, besides requiring higher costs with maintenance and lubrication, allows clearance. In addition, their transmission produces noises, cannot absorb vibrations, and is not indicated for environments with emission of particulate matters (Bruzzone, 2004) – existing at the factory because of the FRP profiles cutting. The latter, on the other hand, cannot achieve high precision and accuracy levels when used in an open loop system. That is because its inability of correcting (or forecasting) any eventual slippage without using sensors. For the same reason, only toothed belts (timing belts) will be taken into account for selection, being dismissed the flat and the V-belts.

It is in that sense that, looking at similar mechanical projects, there is a normal usage of the remaining mechanical elements: screw shaft and nut; timing belts and pulley; or gear and rack. Table 6 illustrates a guideline for helping the initial approach.

Initially, all the three remaining mechanical solutions can achieve good speeds and accuracy. However, it is important to highlight that the table presents estimated maximum and minimum values for these two variables. Of course, the real number will substantially depend on how much the company is willing to invest on them. Therefore, the further analyses will compare the cost-benefit of each solution. First, these elements will be sized and have their costs estimated. Then, the one with lowest cost and easiness to buy, which can meet the project goals (speed and accuracy), will be selected to the roving-robot.

Linear Movements by Screw Shafts and Nuts

A threaded bar can generate linear movements from its rotation. According to Gordo & Ferreira (2012), these components can have different kinds of threads, as square, ACME and metric. The square one is normally used on heavy-duty equipment. So, for the robot drive, it is possible to use either ACME or metric screw shafts.

Table 6. General characteristics of linear transmission methods

	Screw Shaft and Nut	Gear and Rack	Timing Belts and Pulley
Route	Limited	Unlimited	Limited
Maximum speed (m/s)	2	5	10
Power	High	High	Low
Precision (mm)	0,01/0,001	0,1	0,1/0,01
Cost	High	Medium	Low

(Righettini, 2010)

To specify the best threaded bar that can be employed in this project, it is important to understand its terminology:

- The pitch diameter is the average between the major and minor diameter.
- The pitch is the distance between 2 crests (p).
- The lead is the axial advance during one complete turn (d_p).

In addition, the threads can have the number of start greater than one. This allows a greater axial advance for each completed turn. If the thread number of start is n_e , the lead can be calculated by the equation:

$$d_p = n_e p \tag{12}$$

Using a direct coupling between the stepper motor and the bar (without speed reducer), the number of turns that the motor has to perform (N) to achieve the linear displacement ΔS is:

$$N = \frac{\Delta S}{d_p} = \frac{\Delta S}{n_e p} \tag{13}$$

The motor rotation frequency to cover the desired path in the Δt time interval can be obtained by the equation:

$$f = \frac{N}{\Delta t} = \frac{\Delta S}{n_e p \times \Delta t} \tag{14}$$

Then, the number of revolutions per minute (RPM) needed are:

$$RPM = 60 \times f = 60 \times \frac{\Delta S}{n_e p \times \Delta t} \tag{15}$$

It is possible to observe that revolution per minute number is inversely proportional to the thread pitch and the number of start of the screw shaft.

It was found that the price of recirculating ball screws were four times greater than the metric ones. In that way, the initial calculations were made using metric threads, despite of their low efficiency. Before defining the bar diameter, it was necessary to calculate the pitch to obtain the needed speed.

The largest pitches available on retail stores were 20mm, with a single start number. Therefore, the motor RPM to achieve the 573mm/s cruise speed would be:

$$RPM = 60 \times \frac{576}{1 \times 20} \cong 1729 \tag{16}$$

For a particular spinning, the number of pulses per second (PPS) can be obtained from the expression:

$$PPS = \frac{RPM \times PPR}{60} \tag{17}$$

where PPR is the number of pitches per revolution – calculated from the product of the pitches needed by the motor to complete one revolution and the number of micro steps:

$$PPR = P_{motor} \times n_{micro} \tag{18}$$

Considering that micro stepping will not be used ($n_{micro} = 1$, because the screw shaft system would already provide a good resolution) and selecting a motor NEMA 23:

$$P_{motor} = 200 \tag{19}$$

$$PPS = \frac{1719 \times 200}{60} \cong 5764 \tag{20}$$

However, it was realized that this number of pulses per second is too high. Making an extrapolation at the torque-PPS graphic, the torque tends to be zero.

At this point, the alternatives are:

- Select more powerful stepper motors and drivers that can assist them.
- Order special threaded shafts that can provide longer axis advances (option that would increase the costs up to 400%).
- Try to use other linear positioning mechanisms.

The last alternative was chosen after consulting some stepper motor suppliers. That is because either the use of special screw shafts or the selection of more powerful motors would increase a lot the robot cost.

Linear Movements by Timing Belts

Belts are flexible transmission elements used in numerous applications, as linear displacement of objects from the torque of the motors. With the exception of the toothed model (timing belts), the other types of belts perform a friction-conditioned motion transmission. For this reason only toothed belts are being considered for the roving-robot.

According to Gates (2017), there are two main configurations of timing belts and pulleys for linear positioning mechanisms. The first one uses two pulleys: a driving pulley, coupled to the motor; and a driven one, whose function is only to allow the belt stretching, so that the movement can be realized. Figure 8-I illustrates not only this mechanism, but also the magnitude of the tensile forces in each part of the belt. Note that the mechanism needs a belt with the ends attached, composing a closed geometry.

The second configuration for linear positioners is represented by Figure 8-II. In this case, the belt has an open geometry, being tensioned between two side supports (stationary). In addition, the mechanism uses only one toothed pulley (coupled to the motor) and two bearings, with the role of increasing the contact angle between the transmission elements. Unlike the first arrangement, this alternative keeps the belt static while the pulley moves the positioner. In addition, the latter case makes it easier to pretension the belt, avoiding the need of tensioning pulleys.

Considering the simplicity of the second pulley-belt arrangement, this option was selected for the project – hoping to obtain a low-cost assembly and fabrication mechanism.

When a loaded system is driven by belts, there is always a difference in tensions on the entering and leaving sides of the belt. The entering side has a greater tension, while the leaving one becomes loose. The Figure 8-II illustrates the tensions T_1 (tight) e T_2 (slack). The difference between these two forces is called "effective tension", represented by T_e:

$$T_e = T_1 - T_2 \tag{21}$$

The first step in the selection of timing pulleys and belts is the choice of the most suitable tooth profile. There are two main tooth profiles for synchronizing belts: trapezoidal and curvilinear. The latter have numerous advantages over the former,

Figure 8. Linear positioning configurations with belts and pulleys
(Gates, 2017)

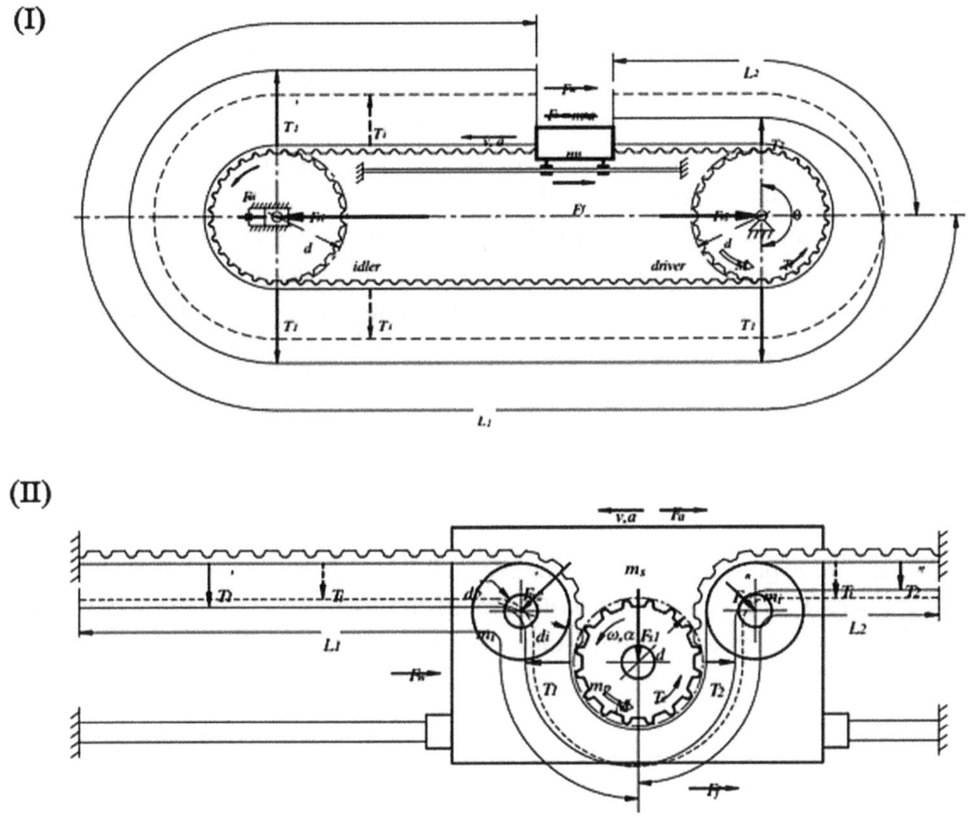

according to a technical article published by Stock Drive Products (2017). Thus, curvilinear teeth:

- Have a greater depth: avoiding that the movement "skips" a tooth, and reducing the probability of relative position loss.
- Have a lighter construction.
- Have a greater contact area between belt and pulley.
- Allow a more uniform stress distribution.
- Are cheaper (since a thinner belt can withstand higher loads).

Because of the superiority of belts with curvilinear profile teeth, many of them are also known as HTD belts (High Torque Drives). However, there is also a variation of the HTD type, produced by Gates, which was found to be even more efficient and with superior accuracy in relation to the normal HTD products. These belts, named GT2, were designed specifically for linear motion transmission (Stock Drive Products, 2017). Their rounded teeth have been modified to form profiles that guarantee a smooth and accurate accommodation in the pulley grooves, avoiding slacks in the assembly. This model is also the best one for mechanisms that need to alternate the movement direction (Reprap, 2017), being widely used in printers, plotters and laboratory equipment. This is precisely the case of the roving-robot; in this way, GT2 belts will be employed for its construction.

The minimum diameter of the toothed pulley must now be validated, so that the 573mm/s speed can be performed by the robot (calculations that were also realized for the threaded bars). Considering r_p the pulley pitch radius and v the robot linear speed, the number of revolutions per minute (RPM) needed is:

$$RPM = \frac{60 \times v}{2\pi r_p} \tag{22}$$

The available pulleys for purchasing that could match the robot design (restricted by the assembly space) have a primitive radius between 10 and 15mm. Therefore, considering the worst case – 10mm, since the amount of RPM is inversely proportional to the radius – the number of revolutions per minute, according to equation (E22) would be:

$$RPM = \frac{60 \times 576mm / s}{2\pi \times 10mm} \cong 550 \tag{23}$$

In order to calculate the number of pulses per second it is necessary to know the number of micro steps that the system will use. This number can be determined by the required tolerance and the resolution of the mechanism being modeled. Thus, the resolution R_{es} of the robot depends on the angle of the motor pitch ($\alpha = 1,8°$, according to its datasheet), the pitch radius of the pulley and the amount of micro steps used:

$$R_{es} = \frac{2\pi r_p \times \alpha}{360° \times n_{micro}} \tag{24}$$

Neglecting other errors, the number of required micro steps can be obtained by equating the resolution of the mechanism with the gap between the aluminum pen and the mold blocks (1.16mm, after made the measurements).

$$n_{micro} = \frac{2\pi \times 10mm \times 1,8^{\circ}}{360^{\circ} \times 1,16mm} = 0,27 \tag{25}$$

The number of micro steps resulted smaller than one; this means that even using $n_{micro} = 1$ the system would be capable of reaching a resolution that meets the positioning gap of the pen (remembering that all the other positioning errors have been ignored). For $n_{micro} = 1$ the equation (E24) would then become:

$$R_{es} = \frac{2\pi \times 10 \times 1,8^{\circ}}{360^{\circ}} = 0,31mm \tag{26}$$

Therefore, not using micro steps, the number of steps per second is given again by equations (17) and (18):

$$PPS = \frac{550 \times 200}{60} \cong 1834 \tag{27}$$

From stepper motor datasheet, it can be realized that this value of pulses per second is within the stepper operative range. Now it is important to verify if the available torque meets the accelerations and the frictional forces of the design. Before, however, it was chosen to the project an aluminum toothed pulley with a 12.74mm pitch radius, 40 teeth and 2mm pitch – low cost component that can be easily found in the market and fits in the physical spaces available for the robot designing.

The torque required to operate the mechanism must be calculated for both the Y-axis and the X-axis. This is because, although the Y-direction exhibits greater acceleration and greater displacement mass, it will count on two stepper motors for the movement execution.

Starting the verification for the Y-axis, the torque of the chosen linear positioner system is given by:

$$M = T_e \times r_p \times FS \tag{28}$$

where FS is the securing factor and T_e (the effective tension) is a summation of the terms below:

- Force required to accelerate the system (F_a).
- Friction force between guides and bearings (F_f).
- Forces external to the system (F_w).
- Inertial force for bearings and pulley acceleration (F_{ai}).

The only significant external forces to the system would be caused by possible shocks between the guiding aluminum pen and the transverse filaments previously positioned in the mold. Since these shocks are difficult to measure, an extra safety factor will be used for the engine torque, while the term F_w will be ignored. In addition, friction forces between the guides and the linear bearings will be neglected at this point. As the mass of the bearings and the pulley (whose diameter will not be bigger than 30mm) are relatively small, they can also be discarded for the pre-calculation. Thus, the belt effective tension is approximately equal to the force required for the acceleration of the masses:

$$T_e = F_a \tag{29}$$

The accelerating force of the system, in turn, is extracted by Newton's second law. The mass to be moved in Y was estimated in 5kg; considering two motors:

$$F_a = \frac{m \times a}{2} \tag{30}$$

$$F_{ay} = \frac{m_y \times a_y}{2} = \frac{5kg \times 0{,}664m / s^2}{2} = 1{,}66N \tag{31}$$

Expressed in the same unit used by the manufacturer of the selected stepper motor, where g is the gravity acceleration:

$$F_{ay} = \frac{1{,}64N}{g} = \frac{1{,}64N}{9{,}8m / s^2} = 0{,}17kgf \tag{32}$$

$$T_{ey} = F_{ay} = 0{,}17kgf \tag{33}$$

The securing factor FS will be a multiplication of the securing factor recommended by the motor supplier ($FS_{stepper} = 2$) and a factor due to the shock:

$$FS = FS_{stepper} \times FS_{choque} \tag{34}$$

For numerous mechanical calculations, it is recommended to use shock factors between 2.0 and 2.3. Since the magnitude of the contact between the pen and the transverse filaments is not known, the more conservative value was used. Therefore:

$$FS = FS_{stepper} \times FS_{choque} = 2 \times 2,3 = 4,6 \tag{35}$$

Then, the required torque for the Y-axis will be:

$$M_y = T_{ey} \times r_p \times FS = 0,17kgf \times 1,274cm \times 4,6 = 0,99kgf.cm \tag{36}$$

Repeating the same calculations for the X-axis, but considering only one motor and its corresponding acceleration:

$$T_{ex} = F_{ay} = 0,17kgf \tag{37}$$

$$M_x = 1,06kgf.cm \tag{38}$$

It is noted that this value is well below the motor torque curve for the necessary number of PPS calculated. Thus, it is possible to increase the resolution of the system using a larger number of micro steps, or even increase the speed of the motors. These analyzes, however, will be more useful after the robot assembly. While the robot works, some unpredictable variables (such as the shock between the guiding-pen and the positioned filaments) will be understood more clearly.

Like any belt-driven mechanism, the pre-tension to which it is subjected (T_i) must be determined. This initial tension is necessary to prevent the loose side of the belt from causing positioning losses. The required pre-tension depends on the configuration of the robot, the load conditions and the pre-tensioning method. Practical experience has shown that the ideal pre-tensioning for linear positioners is between 100% and 120% of the effective tension value T_e, whereas slack side the tension T_2 has a magnitude between 10% and 30% of T_e (Gates, 2017).

The best way to validate the pitch number – previously chosen by the market facility – is to use the belt technical information. Doing this, it was realized that the 2mm pitch is well adjusted to the project.

Lastly, also the belt width can be defined according to the supplier instructions. They often inform the maximum torque values to which the belts can be subjected – according to the pulley and the maximum RPM of the system. After some analyzes, the 9mm width GT2 belts were selected; they are one of the most widespread in the market, widely used in printers, plotters and CNC machine. The total cost of a set with pulleys, bearings and timing belts was estimated in around U$23.00.

Linear Movements by Gear and Rack

Given the feasibility of using belts at a low cost, it became interesting – first of all – a price analysis of the gears and racks set. In fact, the calculations of torque and rotation speed for gears are very similar to those made with toothed pulleys.

After a price quotation, it was clearly perceived that the gear and rack mechanism would have a much higher cost when compared to the pulley and timing belt one: around U$152.00. It was considered not only unitary price of the elements, but also the freight expenses.

In this way, the project had continuity with the use of timing belts and pulleys.

Base Designing

Knowing which mechanical elements would be present in the project, an "outside-in" modeling had been started. That is, the first analyses were about the robot's positioning and support manner in relation to the table and the mold.

For the Y-axis movement, linear aluminum rails and open pillow blocks were selected to guide the motion. These rails will be fixed on stainless steel 2.1/2'' L profiles. The latter, in turn, will be welded to the table, as illustrated by Figure 9.

In addition, flanges were required for the belt tensioning. In order to pre-tension the belts, fastening systems with two plates, bolts and nuts were also used. The tightening of the bolts approximates the plates (where the belts are attached) from the bulkhead, as shown in Figure 10.

Material of Machined Parts

According to previous calculations, it was concluded that the mass of the moving elements is the variable with the greatest impact on the torque required by the motors. The larger the mass of the system moved, the lower the acceleration and, consequently, the lower the speed reached. In this way, the structure of the robot

Figure 9. Linear guides assembly for movement on the Y-axis
(produced by the authors)

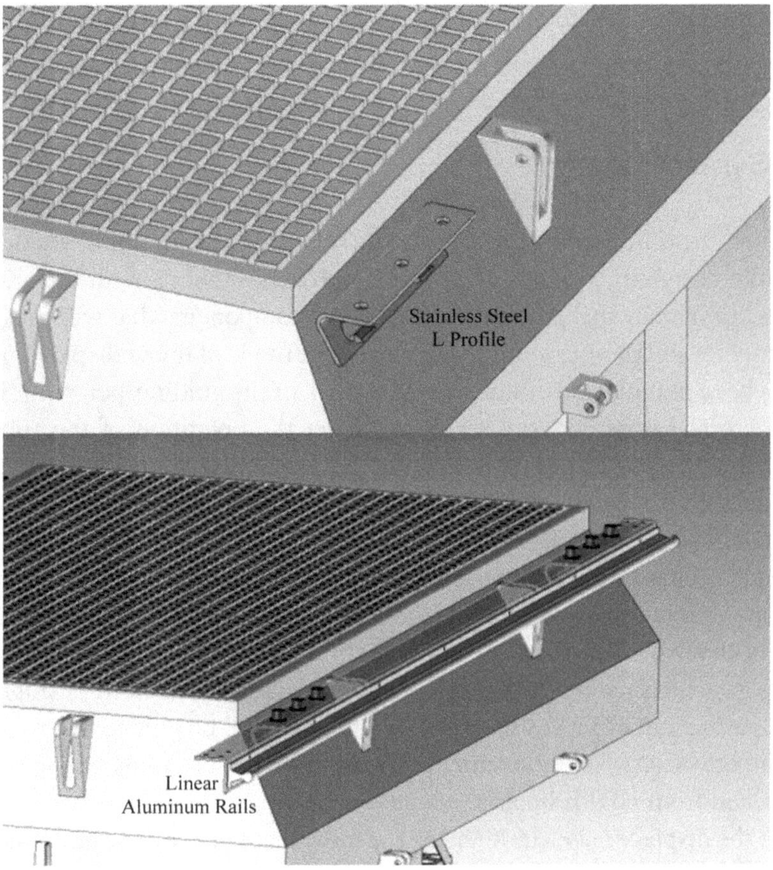

Figure 10. Belt tensioning system
(produced by the authors)

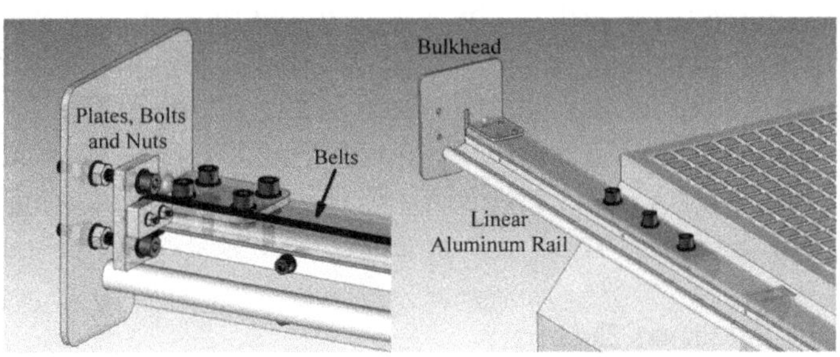

must be constructed, preferably, from low specific mass materials. As an alternative to aluminum, whose density reaches 2.70g/cm^3, the application of high density polyethylene (HDPE), with only 0.95g/cm^3, was seen as a good solution. HDPE is a rigid and easily machinable material. It has adequate mechanical resistance to the low efforts required by the project.

Z-Axis System Design

Given the physical limitations of the robot in relation to the mold, it was decided to continue the designing by an "inside-out" approach – that is, with the creation of the Z-axis. Two main analyzes were done on the components that will move in this axis: the method of fixing the guiding-pen; and the mode of linear displacement in Z.

As has been previously commented, the shock of the guiding-pen with filaments in the transversal direction can seriously affect the operation of the robot. This problem can be avoided, naturally, with the implementation of more complex control and sensor systems. However, in order to avoid additional costs, a purely mechanical solution was sought. The idea is represented by Figure 11, where it can be noticed a mechanism of springs and joints allowing a movement similar to that of a gyroscope (but in only two axes). If there is any impact on the tip of the guiding-pen, the mechanism allows it a slight inclination. In this way, the tip of the pen has an elevation, which makes it possible to jump small obstacles. It is important to note that, getting out of its vertical position, the pen will be guided by cross-shaped slots in the direction of the movement. The entire structure must be manufactured and machined from HDPE boards.

Unlike the displacements in X and Y, the movement in Z will be quite short (with a maximum range of 40mm) and parallel to the gravity acceleration. The mass of the moving elements is also much lower. In addition, there is a physical limit to be considered: the entire structure of the robot cannot exceed the borders of the mold, since the factory counts on several tables arranged next to each other.

In this way, a movement realized by a trapezoidal threaded bar was considered pertinent to the movements in Z. The length needed is around 100mm, and there are not so many restrictions regarding its pitch and its diameter. This is because the forces that oppose the motion are really small. For this same reason, the stepper motor used on the Z-axis will be a NEMA 17. The chosen threaded bar has 2mm of pitch and 8mm of diameter. The Z movement system is shown in Figure 12.

X and Y-Axis System Design

The displacement system in the X direction was previously defined by using timing belts and toothed pulleys. To support the structure of the X-axis and to guarantee the

Figure 11. Guiding-pen assembly
(produced by the authors)

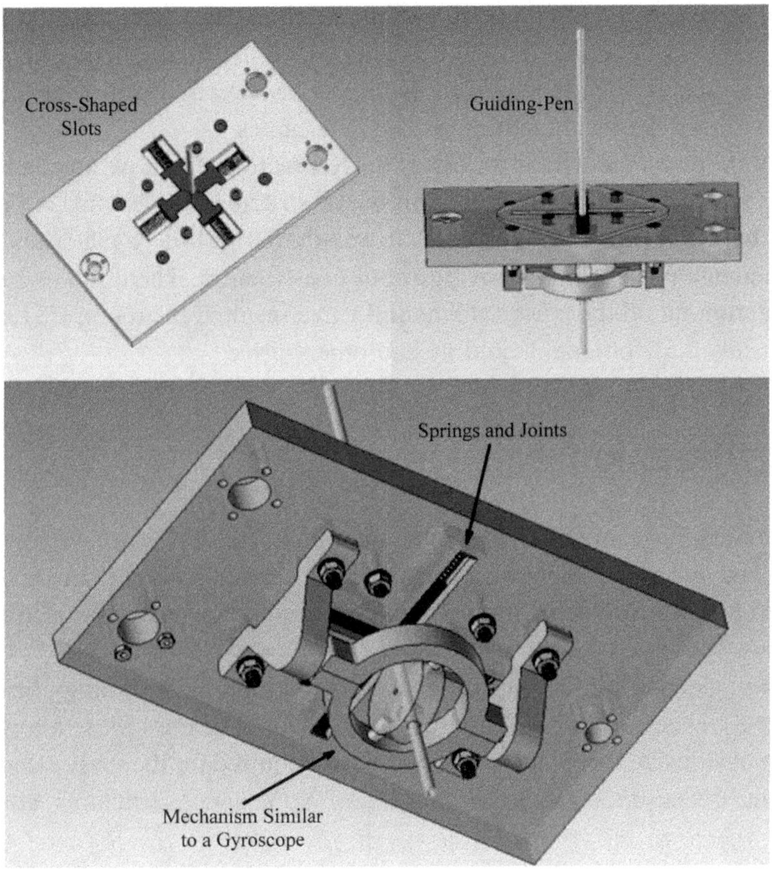

Figure 12. Z displacements: guides and linear bearings employed
(produced by the authors)

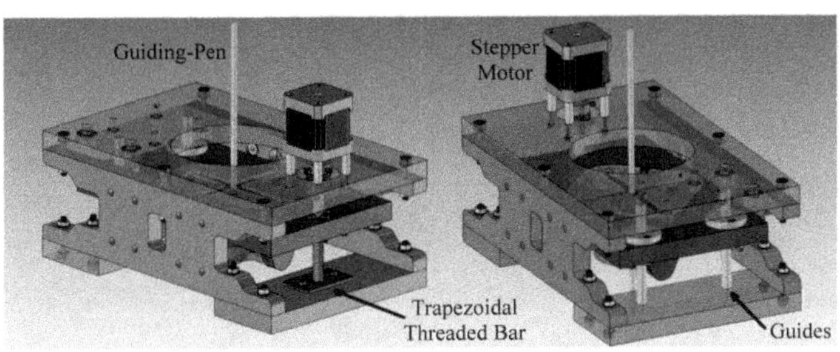

rectilinear motion, linear guides of 16mm diameter were chosen. In this case, the selected pillow blocks will be closed. Figure 13 shows not only these linear bearings, but also the NEMA 23 motor coupled to the toothed pulley and the belt. It is also possible to observe the pairs of ball bearings which the function is to increase the angle of contact between the belt and the pulley, as already mentioned. The entire structure will also be manufactured with HDPE plates.

Finally, the linear guides and the belt of the X direction will move together with the HDPE plates that will be driven in Y. This whole structure is presented in Figure 14.

With the aid of the CAD software, it was drafted a bill of materials with all the components present in the roving-robot construction. Then, it was realized a price quotation for all the parts, which total value resulted in around U$1,000.00 – approximately 2.5% of a new molding equipment price.

FUTURE RESEARCH DIRECTIONS

As commented throughout the text, some minor adjustments would be more useful to be realized during the robot's assembly – to be carried out in the field. However, in order to avoid errors, some future studies to complement this project might be interesting.

In this sense, one can develop an exclusive work on the metrology involved in the roving-robot construction. With this study, it would be possible to know exactly which dimensional and geometric tolerances are required for the fabrication of each component. Because the robot is composed of X, Y, and Z structures, attached to

Figure 13. Drive system in the X direction: pillow blocks, linear guides, timing belt and pulley
(produced by the authors)

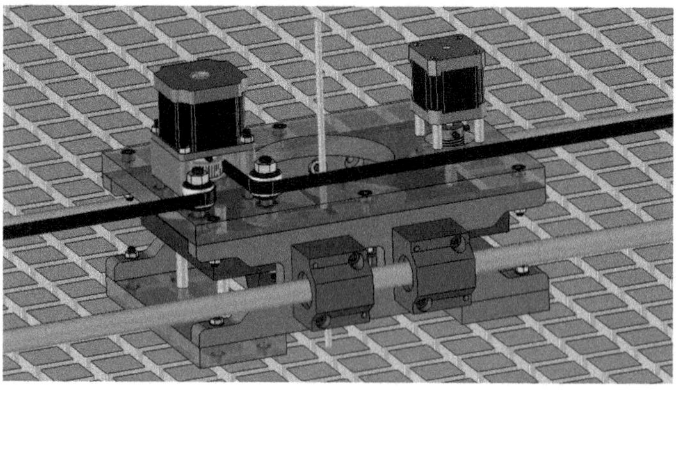

Figure 14. Roving-robot final assembly
(produced by the authors)

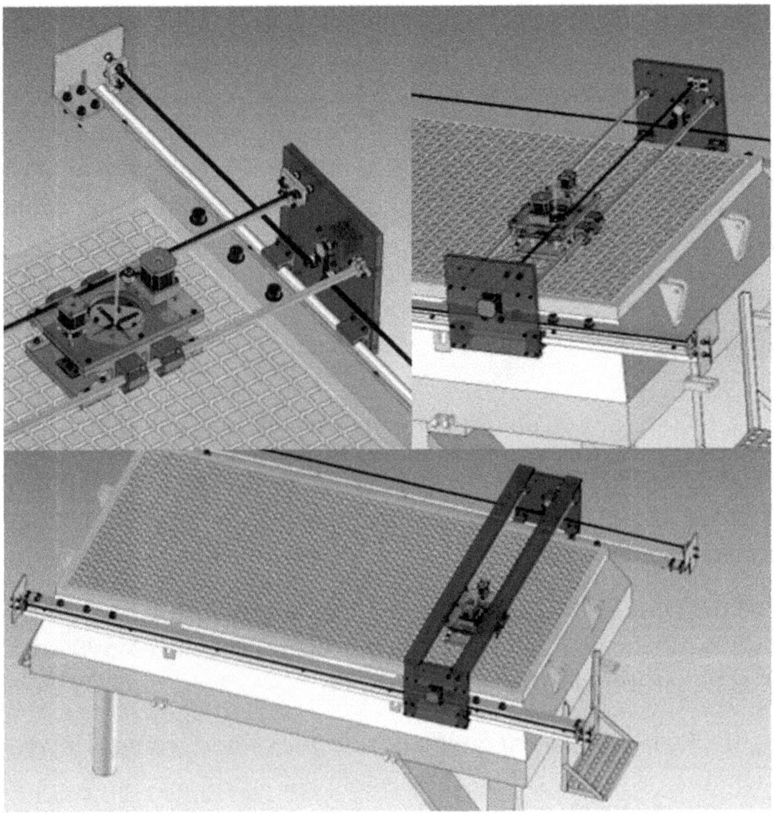

each other, each dimensional error would propagate during the motions – added also to possible errors coming from the stepper motors.

Another suggestion is to deepen the study of solutions to avoid shocks between the guiding-pen and the filaments already positioned in the mold. The current project took a completely mechanical approach to the subject; however, it is possible to carry out analyzes in the scope of control and automation, with the use of sensors.

CONCLUSION

This chapter presented the design and the specifications of a robot capable of positioning fiberglass filaments inside FRP gratings molds. This operation is currently carried out by operators, being an extremely monotonous and repetitive task, besides representing an inadequate bottleneck in the factory. The time spent on positioning the filaments may be reduced by half with the automation, increasing the productivity of this industrial line. In addition, robots can work for hours on end, and will save time spent on employees training. A division of tasks between robot and operators is required; however, the number of employees may also be reduced.

This text aimed to navigate along the reasoning line used to project a robot. It is hoped that the reader could understand the step-by-step presented, having now a better understanding of robots mechanical designing. The same reasoning used here can be applied and adapted to a variety of other mechanical projects – contributing to a faster and most effective approach.

REFERENCES

Bouteille, D. (2000). *Les automatismes programmables*. Toulouse: Cepadues.

Bruce, J. (2013, August 21). *Home Automation with Raspberry Pi and Arduino*. Retrieved from http://www.makeuseof.com/tag/how-to-build-home-automation-system-raspberry-pi-and-arduino

Brum, M. (2016, December 22). *Introduction to Educational Robotics*. Retrieved from http://www.educacional.com.br/upload/dados/materialapoio/124590001/8214768/Rob%C3%B3tica%20Educativa.pdf

Bruzzone, L. (2004). *Meccanica applicata alle macchine*. Genova: Università degli Studi di Genova.

Burris, M. (2019, May 17). *Choosing Between Stepper Motors or Servo Motors*. Retrieved from https://www.lifewire.com/stepper-motor-vs-servo-motors-selecting-a-motor-818841

Gates. (2017, January 3). *Timing belt theory*. Retrieved from http://www.gatesmectrol.com/mectrol/downloads/download_common.cfm?file=belt_theory06sm.pdf&folder=brochure

Gordo, N., & Ferreira, J. (2012). *Machine Elements*. São Paulo: Telecurso 2000 Profissionalizante.

Heer, C. (2016, August 18). *Survey: 1.3 million industrial robots to enter service by 2018*. Retrieved from http://www.ifr.org/news/ifr-press-release/survey-13-million-industrial-robots-to-enter-service-by-2018-799/

Kemerich, P., Piovesan, M., Bertoletti, L., Altmeyer, L., & HohmVorpagel, T. (2013, January-April). *Glass Fiber: Characterization, Disposal and Environmental Impact Generated*. Retrieved from https://periodicos.ufsm.br/reget/article/viewFile/7590/pdf

Malvezzi, F. (2008, February 27). *Evaluation of kinematic performance of a tridimensional parallel mechanism.* Retrieved from http://www.teses.usp.br/teses/disponiveis/3/3151/tde-08012008-085930/pt-br.php

Mccomb, G. (2011). *Robot builder's bonanza.* TAB.

Quinones, J. (2017, January 20). *Applying acceleration and deceleration profiles to bipolar stepper motors.* Retrieved from http://www.ti.com/lit/an/slyt482/slyt482.pdf

Reprap. (2017, January 27). *Choosing Belts and Pulleys.* Retrieved from http://reprap.org/wiki/Choosing_Belts_and_Pulleys

Righettini, P. (2010). *Progettazione funzionale di sistemi meccatronci: Introduzione al modulo.* Bergamo: Università degli Studi di Bergamo.

Stock Drive Products. (2017, January 27). In *Handbook of timing belts and pulleys.* Retrieved from: http://www.sdp-si.com/d265/html/d265t003.html

Wentzel, M. (2016, January 22). *Fourth Industrial Revolution: how Brazil can prepare for the Economy of the Future.* Retrieved from http://www.bbc.com/portuguese/noticias/2016/01/160122_quarta_revolucao_industrial_mw_ab

ADDITIONAL READING

Budynas, R. G., & Nisbett, J. K. (2015). *Shigley's Mechanical Engineering Design.* McGraw-Hill.

Ceccarelli, M. (2012). *Service Robots and Robotics: Design and Application.* IGI Global. doi:10.4018/978-1-4666-0291-5

Craig, J. J. (2017). *Introduction to Robotics: Mechanics and Control.* Pearson.

Habib, M. K. (2015). *Handbook of Research on Advancements in Robotics and Mechatronics.* IGI Global. doi:10.4018/978-1-4666-7387-8

Nehmzow, U. (2003). *Mobile Robotics: A Practical Introduction.* Springer-Verlag. doi:10.1007/978-1-4471-0025-6

Riascos, L. A. M. (2010). *Fundamentos de robótica: manipuladores e robôs móveis.* Pleiade.

Rivin, E. I. (1988). *Mechanical Design of Robots.* McGraw-Hill.

Siegwart, R., & Nourbakhsh, R. (2004). *Introduction to autonomous mobile robots.* MIT Press.

KEY TERMS AND DEFINITIONS

Automation: The use of computerized and/or mechanical systems in order to optimize productive processes.

Cartesian Robot: Robot whose three main control axes are linear and form a right angle between them; these kind of robots move in a straight line instead of rotating.

Fiberglass Reinforced Plastic (FRP): Composite material formed of a polymer matrix (resin) reinforced with fiberglass fibers.

FRP Molded Gratings: Gratings made of fiberglass reinforced plastic with the use of a molding equipment. These grids have bi-directional strength and a high corrosion resistance, being a great alternative to metallic products.

Roving-Robot: Name of the robot projected to automate part of the FRP molded gratings production.

Stepper Motor: Brushless DC electric actuator, capable of converting digital pulses into the mechanical rotation of its own shaft; widely used in projects that demand high precision and accuracy.

Timing Belt: Toothed belt capable of transmitting movements in a synchronized way when connected to toothed pulleys.

Chapter 7
Modeling, Model Reduction, and Control of a Hands-Free Two-Wheeled Self-Balancing Scooter

Qiong Li
Purdue University, USA

Wangling Yu
Purdue University Northwest, USA

H. Henry Zhang
Purdue University, USA

ABSTRACT

Designing a two-wheeled self-balancing scooter involves in the synergistic approach of multidisciplinary engineering fields with mutual relationships of power transmission, mass transmission, and information transmission. The scooter consists of several subsystems and forms a large-scale system. The mathematical models are in the complex algebraic and differential equations in the form of high dimension. The complexity of its controller renders difficulties in its realization due to the limit of iteration period of real time control. Routh model reduction technique is employed to convert the original high-dimensional mathematical model into a simplified lower dimensional form. The modeling is derived using a unified variational method for both mechanical and electrical subsystems of the scooter, and for the electronic components equivalent circuit method is adopted. Simulations of the system response are based on the reduced model and its control design. A prototype is developed and realized with Matlab-Labview simulation and control environment.

DOI: 10.4018/978-1-7998-1382-8.ch007

INTRODUCTION

In general, the mechanical engineering design focuses on the analysis of motions, forces or torques to figure out its variables such as displacements, velocities or forces. The mechanical products were developed with simple feedforward and feedback control systems by using analog/digital devices to manipulate designed variables to achieve certain control functions and performances. With the introduction of digital computers into their control systems, the industrial products became multidisciplinary, and their comprehensive functionalities were executed with mechanical and electrical systems and achieved much higher precision of their motions. Starting in early 1980s, more and more industrial products have been migrating to mechatronic products. Nowadays, mechatronic products possess smart mechanisms driven by local actuators, which are integrated with sensors, and operated by embedded microprocessors coded with artificial intelligence, operational laws, as well as arithmetic and logic computations. The mechatronic systems have functionalities via the task-driven reconfigurable and flexible precision mechanisms. (Zhang, Nawrocki and Li, 2018)

As for the formulation of mechatronic systems, generally speaking, a machine set is mainly composed of two categories: power-generating machine and power-consuming machine. The power electronics delivers the manipulated input of energy flow into the power-generating machine, such as a DC motor, while the drivetrain builds the connection between the power-generating machine and power-consuming machine, obtaining the desired energy output measured by corresponding sensors. Basically, three types of energy flows are handled: electrical energy, mechanical energy and hydraulic energy. If this system is integrated with information processing subsystem bridging the input and output, automatically handling data based on the manipulated variables and measured variables in a feedback or feedforward way, a mechatronic system is constructed.

The terminology mechatronics was coined by Tetsuro Mori in 1969, by combining the mecha- from mechanical and –tronics from electronics to describe the electronic control systems for mechanical factory equipment built by Yaskawa Electric Corporation. Different literatures gave several definitions for this new terminology, and all of them pointed out that mechatronics is of multidisciplinary engineering. To our best understanding, it is more proper to interpret mecha- as mechanism instead of mechanical for the term mechatronics. The first generation of mechatronic systems, such as anti-lock braking systems, cameras, printers and disk storage, were developed by integrating the mechanical system, sensors, actuators and microcontrollers. The vital element of such a system is integration, consisting of both horizontal integration from various disciplines and vertical integration from design to manufacturing. Thus, a simultaneous engineering is recommended for mechatronic system design, and an overall integrated multidisciplinary modeling has to take place, in order to create sophisticated control functions.

As a multidisciplinary product, the mechatronic system normally consists of a number of subsystems of different engineering fields with controllers coded in the embedded microprocessor. Modeling of mechatronic system is an essential step in the product development, assuring their dynamic responses meet the design requirements and specifications. In this chapter, the modeling method relies on a unified framework for the multidisciplinary system developed in recent years. The network-based modeling approach was employed in mechanical engineering due to its analogy to electrical circuitry, (Zhang, Nawrocki and Li, 2018) while the variational modeling method for mechanical engineering, such as Lagrangian and Hamiltonian function, was extrapolated to the cases of electrical network due to their metaphorically mutual analogy. The network representation of subsystems formulates the basics of modeling the mechatronic systems.

The mathematical model of the entire mechatronic system results in a large scale modeling of complex form and high dimension, whose real-time control design is challenging and full of complexity. (Isermann, 2003; Giurgiutiu and Lyshevski, 2009) Furthermore, as a simultaneous engineering design, system optimization is required to find out the best way to realize all the design specification from various disciplines. For the optimal control of mechatronic system, a reduced model that inherits the dominant features and performance of the original high dimension model is desirable, and can be derived using model order reduction techniques. (Kim, Kim, and Kwak, 2005; Nasrallah, Michalska and Angeles, 2007; Hu and Tsai, 2008; Li and Zhang, 2019)

MATHEMATICAL MODELING

In this section, the large scale mathematical model of the hands-free two-wheeled self-balancing scooter as a real-time mechatronic system is established, based on a unified framework of variational method for both mechanical and electrical subsystems. As a naturally unstable system, with the rider standing on it, the two-wheeled self-balancing scooter without handle for moving forwards/backwards, and turning left/right control is treated as an inverted pendulum on a moving platform. The important electric and electronic components out of the mechatronic device, such as sensors, electric motors, as well as other essential components, such as tires, are modeled based on their algebraic and dynamic equations.

Variational Method for Multidisciplinary System

The systematic modeling methods, including network method, which is rooted in electrical circuit studies, variational method, which stems from Hamilton and Lagrange equations of motion, and bond graph methods, which graphically illustrate the system energy interactions, are based on energy analysis. Each of them has its own uniqueness. Network method establishes systems by connecting the power flow of energy sources. Based on the inter-connective constraints, such as flow continuity constraint and energy compatibility constraint, it builds up different constraint matrices to complete the state space modeling of systems. Bond graph method is very similar to network method. However, it presents the interactive energy relationships in energy bonds to couple the energy ports of basic components. Variational method has advantages over the aforementioned methods in the case in which the complicated interconnectivity of system constraints is dealt, and in the solution procedure of variational method there is no need to explicitly formulate the tedious flow continuity constraint and energy compatibility constraint. It is a suitable method to handle multidisciplinary energy couplings for the physical system with lumped parameters. Among the three methods, variational method is the only one directly describing systems with algebraic forms through the use of the techniques from Hamilton and Lagrange to present the internal connections and constraints, so as to form the dynamical modeling. (Li, 2015)

For a physical system, its energy flow at inlet / outlet ports is created by a couple of variables, i.e., effort variable and flow variables, whose product is the power. The system components can be categorized as energy sources, energy stores, energy dissipaters, energy converters and energy couplers. Energy sources include flow sources and effort sources, energy stores perform flow stores and effort stores, energy dissipaters are dampers and friction components, and through energy converters and energy couplers no power is stored or dissipated. The multidisciplinary system modeling of the two-wheeled self-balancing scooter can be developed based on the unified framework from Lagrange's' equations and variational approach. This method is the most applicable approach for modeling the dynamics of the mechatronic system.

Modeling A Two-wheeled Self-balancing Scooter

Per the above stated energy based variational method, the mechanical subsystem can be modeled as follows:

Tractive force F_t from the wheel is the scooter's driving force. (Vora et al., 2014) Given the wheel radius r_w, the motor torque T_{tm}, it is described as a linear algebraic equation of the tractive force F_t and the wheel torque T_t provided by the electric motor.

$$F_t = T_t / r_w = \frac{T_{tm} \cdot \zeta \cdot \eta_t}{r_w} \tag{1}$$

where ζ is the gear ratio of the electric motor, and η_t is the mechanical efficiency of the geartrain connected to the motor shaft, and in most cases, the mechanical efficiency η_t is between 90% - 95$\%$.

The elastomeric tire of the scooter wheel has rolling resistance during motions. The rolling resistance torque $T_r = F_P \cdot l_{oo'}$, where F_P is the force from the integral of the distributed normal pressure, and the offset distance $T_r = F_P \cdot l_{oo'}$ is determined by the hysteresis of the rubber tire. The rolling resistance force F is

$$F = \frac{T_r}{r'} = F_P \cdot \frac{l_{oo'}}{r'} = F_P \cdot C_0 \tag{2}$$

where r' is the effective radius of the wheel when the scooter is loaded with the rider, and C_0 is the defined rolling resistance coefficient,

$$C_0 = F = \frac{T_r}{r'} = F_P \cdot \frac{l_{oo'}}{r'} = F_P \cdot C_0 .$$

Given the road grade angle β, the gravity dragging component force F_{PT} is

$$F_{PT} = F_P \sin \beta \tag{3}$$

Derive the Lagrangian of the equivalent inverted pendulum as follows: (Li, 2015; Yu and Chen, 2018)

$$\begin{aligned} L = {} & \frac{1}{2} m_b \left[\dot{x}^2 + (r_b \dot{\theta})^2 + 2 r_b \dot{x} \dot{\theta} \cos \theta \right] + \frac{1}{2} I_b \dot{\theta}^2 + m_w \dot{x}^2 \\ & + I_w \frac{\dot{x}^2}{r_w^{\,2}} - 2 m_w g r_w \cos \beta - m_b g \left(r_b \cos \beta \cos \theta + r_w \cos \beta \right) \end{aligned} \tag{4}$$

The dynamics of the mechanical scooter system is

$$\left(I_b + 2m_b r_b^2\right)\ddot{\theta} + 2m_b r_b \cos\theta\, \ddot{x} - 2m_b r_b \left(\sin\theta\right)\dot{x}\dot{\theta} - m_b g r_b \sin\theta \cos\beta = 0 \tag{5}$$

$$m_b \ddot{x} + \left(m_b r_b \cos\theta\right)\ddot{\theta} - \left(m_b r_b \sin\theta\right)\dot{\theta}^2 + 2m_w \ddot{x} + 2I_w \frac{\ddot{x}}{r^2} = F_T \tag{6}$$

Equations (5) and (6) can be linearized and presented in state space form.

As a part of the electrical subsystem, motor drive for the motor speed or torque control via manipulating the duty cycle of the PWM signal is simplified as the H-bridge circuit of four n-channel enhancement type MOSFETs. Furthermore, since only two MOSFETs in the H-bridge circuit are turned on simultaneously, and one of the two is associated to the ground, the motor drive modeling is equivalently reduced to the modeling of n-channel MOSFET. The current of the motor drive is an algebraic equation (Li, 2015):

$$i_D = k_n' \frac{W}{L}\left[\left(V_{GS} - V_t\right)V_{DS} - \frac{1}{2}V_{DS}^2\right]$$

if $V_{GS} < V_t$, OFF;

if $V_{GS} \geq V_t$, ON. (7)

where, the process transconductance parameter $k_n' = \mu_n C_{ox}$ is the product of electron mobility μ_n and oxide capacitance C_{ox}; MOSFET is featured with channel width W, channel length L, and aspect ratio of W to L. MOSFET is cut off if the gate-to-source voltage V_{GS} is less than the threshold voltage V_t; elseif the gate-to-source voltage V_{GS} is larger than the threshold voltage V_t, MOSFET is turned on and work in the triode region.

The dynamic model of electric motor has two equations: the electrical portion and the mechanical portion. The former is Kirchhoff's voltage law applied to the armature circuit: I_a is the current flowing through the armature circuit, U_a is the average PWM voltage supply to the armature, k_a is the back electromotive force constant, R_a is the resistance in the armature (see Equation (8)); while the latter is Newton's second law applied to the mechanical rotor/stator mechanisms of the electric motor: J is the moment of inertia of the rotor, ω_r is the angular speed of the rotor, and T_L is the load torque applied to the motor shaft (see Equation (9)).

$$\frac{dI_a}{dt} = -\frac{R_a}{L_a} I_a - \frac{k_a}{L_a} \omega_r + \frac{1}{L_a} U_a \tag{8}$$

$$\frac{d\omega_r}{dt} = \frac{k_a}{J} I_a - \frac{B_m}{J} \omega_r + \frac{1}{J} T_L \tag{9}$$

Sensors used in the mechatronic scooter can also be modeled. There is a sensor fusion of accelerometer and gyro (see Figure 1). As an inertial measurement unit (IMU), the accelerometer's model is a damped mass-spring system: its natural frequency $\omega_n = \sqrt{\frac{k}{M}}$ is determined by *k* spring of stiffness *k* and proof-mass *M*, and *b* is damper coefficient, and the damping ratio is $\zeta = \frac{b}{2\sqrt{kM}}$.

$$M\frac{d^2x}{dt^2} + b\frac{dx}{dt} + kx = -M\frac{d^2y}{dt^2} \tag{10}$$

where x is the relative displacement of the proof-mass, and y is the displacement of the accelerometer casing mounted onto the scooter platform.

The gyroscope is the device for detecting the rotation around the central axes of the platform between two wheels. Per the principles of Coriolis force, the microgyroscope senses the angular rate via its vibrating element in the two degree of freedom spring-mass-damping system with the air damping coefficients c_x and c_y, the spring rates k_x and k_y. The motion dynamics of the gyroscope in the driving direction x and sensing direction y is presented in Equations (11) and (12), given the external force $F(t) = F_0 \sin \omega_D t$.

$$m_p \frac{d^2x}{dt^2} + c_x \frac{dx}{dt} + k_x x = F_0 \sin \omega_D t \tag{11}$$

$$m_p \frac{d^2y}{dt^2} + c_y \frac{dy}{dt} + k_y y = 2m_p \frac{dx}{dt} \Omega \tag{12}$$

Figure 1. Sensor fusion of accelerometer and gyroscope
(Li, 2015)

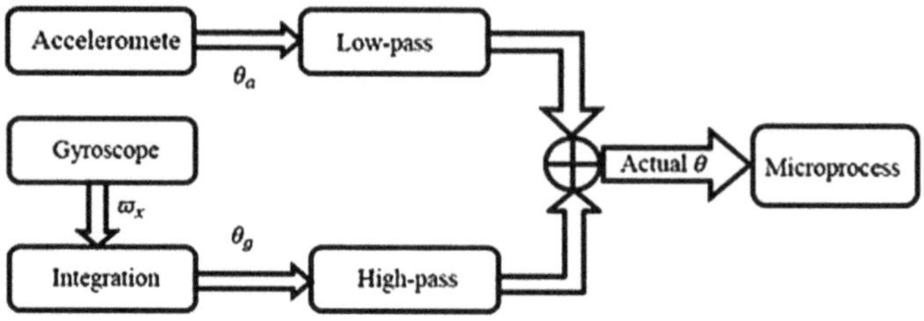

Other components, such as electromagnetic relay and optical encoder, also have been mathematically modeled. Utilizing the equivalent circuit methods and the algebraic relationship of the conversion between digital electric signal and analog electric signal, microprocessor and its DAC and ADC can be mathematically modeled.

With the above preparation, the holistic configuration of the feedback control of the two-wheeled self-balancing scooter is illustrated in Figure 2.

The algebraic equations define the relationships of the state variables, and the dynamic equations of the mechatronic subsystems describes the characteristics of the entire multidisciplinary system. The standard form of state space function is presented as follows:

$$\dot{x}\left(\mathrm{t}\right)=Ax\left(t\right)+Bu\left(t\right) \tag{13}$$

Figure 2. Architecture and control of the two-wheeled self-balancing scooter
(Li, 2015)

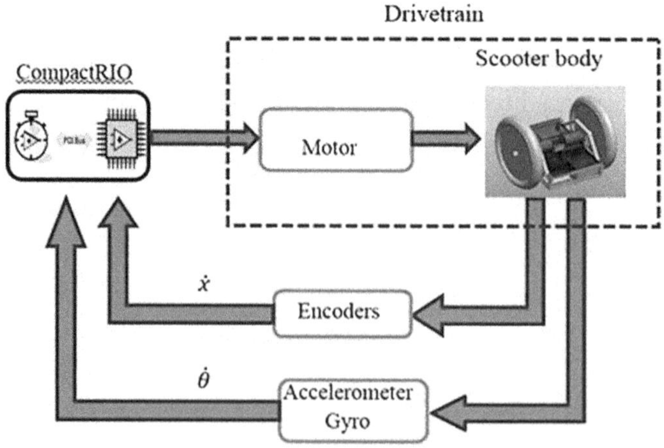

$$y(t) = Cx(t) + Du(t) \tag{14}$$

where

$$A = \begin{bmatrix} 0 & 0 & 0 & \dfrac{\dot{\theta}_0^2 - \dfrac{2m_b r_b \dot{x}_0 \dot{\theta}_0 + m_b g r_b \cos\beta}{I_b + 2m_b {r_b}^2}}{\dfrac{2m_w + 2I_w \frac{1}{r^2} + m_b}{m_b r_b} - \dfrac{2m_b r_b}{I_b + 2m_b {r_b}^2}} & 0 & 0 \\ 1 & 0 & 0 & 0 & 0 & 0 \\ 0 & 0 & 0 & \dfrac{\dot{x}_0 \dot{\theta}_0 + \frac{g}{2}\cos\beta - \dfrac{m_b r_b \dot{\theta}_0^2}{2m_w + 2I_w \frac{1}{r^2} + m_b}}{\dfrac{I_b + 2m_b {r_b}^2}{2m_b r_b} - \dfrac{m_b r_b}{2m_w + 2I_w \frac{1}{r^2} + m_b}} & 0 & 0 \\ 0 & 0 & 1 & 0 & 0 & 0 \\ 0 & 0 & 0 & 0 & -\dfrac{R_a}{L_a} & -\dfrac{k_a}{L_a} \\ 0 & 0 & 0 & 0 & \dfrac{k_a}{J} & -\dfrac{B_m}{J} \end{bmatrix}$$

$$B = \begin{bmatrix} 0 & 2\dfrac{\zeta \cdot \eta_t}{r_w} \dfrac{-\dfrac{1}{m_b r_b}}{\dfrac{2m_w + 2I_w \frac{1}{r^2} + m_b}{m_b r_b} - \dfrac{2m_b r_b}{I_b + 2m_b {r_b}^2}} & (2m_w + m_b) \dfrac{-\dfrac{1}{m_b r_b}}{\dfrac{2m_w + 2I_w \frac{1}{r^2} + m_b}{m_b r_b} - \dfrac{2m_b r_b}{I_b + 2m_b {r_b}^2}} \\ 0 & 0 & 0 \\ 0 & 2\dfrac{\zeta \cdot \eta_t}{r_w} \dfrac{-\dfrac{1}{2m_w + 2I_w \frac{1}{r^2} + m_b}}{\dfrac{I_b + 2m_b {r_b}^2}{2m_b r_b} - \dfrac{m_b r_b}{2m_w + 2I_w \frac{1}{r^2} + m_b}} & (2m_w + m_b) \dfrac{-\dfrac{1}{2m_w + 2I_w \frac{1}{r^2} + m_b}}{\dfrac{I_b + 2m_b {r_b}^2}{2m_b r_b} - \dfrac{m_b r_b}{2m_w + 2I_w \frac{1}{r^2} + m_b}} \\ 0 & 0 & 0 \\ \dfrac{1}{L_a} & 0 & 0 \\ 0 & \dfrac{1}{J} & 0 \end{bmatrix}$$

$$C = \begin{bmatrix} 1 & 0 & 0 & 0 & 0 & 0 \end{bmatrix}$$

$D = 0$

The two-wheeled self-balancing scooter is treated as an illustrative example of large scale system, even though its dimension is not high enough to be of large scale system. But it is typical since it contains both mechanical system and electrical system with multiple inputs of angular speed, tilt angle and wheel speed. The mathematical modeling and model reduction techniques can be demonstrative in the entire procedure for system modeling, simulation and control design.

MODEL REDUCTION

The large scale system is high in dimension, and its computational efforts for system modeling and analysis are time consuming, even the realization of high dimensional control design is impractical, especially in real-time control. The characteristics of the large scale system lays in its hierarchical structures and decentralized control. The system characteristic matrix can be decoupled into a number of subsystems, and most of them are MIMO systems.

Any modeling task always needs to make a compromise between simplicity and accuracy. For large scale system, their complicated mathematical model commands heavy computational efforts, making it is hard to work with their original form. Simplifying the complex model without losing its main characteristics and performance is a practical demand in the engineering practice of mechatronics.

Quite a few model reduction methods are developed: for example, (1) Davison technique (Davison, 1966), (2) Chidambara technique (Janardhanan, 2013), and (3) Marshall technique (Janardhanan, 2013) belong to one category, and they are essentially based on analyzing the dominant characteristics of the original system, and keeping the desirable pole-zero on the complex plane to formulate simpler model and produce proper responses. Other approaches of model reduction and their criteria are different, for example, (1) Aggregation methods (Linhart and Skogestad, 2012), (2) Frequency domain based methods, including moment matching (Cunedioglu, Mugan, and Akcay, 2006), (3) Pade approximation methods (Paraskevopoulos, 1980; Yadav and Prajapati, 2012), (4) Routh approximation techniques and continued fraction method (Wang, Li and Wang, 2011), and (5) norm based methods (Kumar and Nagar, 2014). Their model reduction is achieved by balanced truncation, by impulse/step error minimization, or using Wilson's technique, and pole placement techniques. In this chapter, since the state space model is at hand, Routh approximation technique is employed to illustrate the model order reduction procedure for original full order model of the two-wheeled self-balancing scooter. The procedure is that applying the Routh approximation technique to obtain the reduced order model, which needs first to develop a Routh-like table based on the original full order model, then construct the desired reduction by matching the coefficients of its Routh table of the reduced order model with that of the Routh table of the original system. One of the advantages Routh approximation techniques has is that the system stability is inherited.

For the matrices in the state space presentation of Equations (13) and (14), the transfer function can be derived. Equation (15) is the transfer function for continuous time system, while Equation (16) is for discrete time system.

$$H\left(s\right) = C\left(sI - A\right)^{-1} B + D \tag{15}$$

$$H\left(z\right) = C\left(zI - A\right)^{-1} B + D \tag{16}$$

There are two concepts associating with the system, which are time moments m_i and Markov parameters h_i. Equation (17) shows the expression for time moments m_i. And the Markov parameters h_i for continuous time system, can be derived from Equation (18) and equation (19).

$$m_i = \int_0^\infty t^{i+1} h\left(t\right) dt \tag{17}$$

where $h\left(t\right) = Ce^{At}B$ is the impulse response of the system.

$$h_i = CA^{i-1}B \tag{18}$$

$$H\left(s\right) = D + \sum_{i=1}^{\infty} h_i s^{-i} \tag{19}$$

with constraints

$$h\left(0\right) = \left[Ce^{At}B\right]_{t=0} = CB\,,\ \left[\frac{d^i h\left(t\right)}{dt^i}\right]_{t=0} = \left[\frac{d^i}{dt^i}\left[CAe^{At}B\right]\right]_{t=0} = CA^iB\,.$$

The numerical simulation is performed using the scooter parameters summarized in Table 1.

Given

$$m_f = 0.0792 \times 0.75 \times 24 \times 20 = 28.5120 lb ,$$

$$I_H = I_G + m_H \cdot (Z_h / 2)^2 = \frac{1}{3} m_H (Z_h / 2)^2 + m_H \cdot (Z_h / 2)^2 ,$$

Table 2 lists some calculated values of the terms in the mathematical model of the two-wheeled self-balancing scooter, that reflect the influences of rider's body, scooter frame and two electric motors with geartrains.

The corresponding Routh table (see Table 3) for the reduced model of two wheeled self-balancing scooter is summarized as follows.

Table 1. Scooter parameters

Average height of human body Z_h	*1.75 m*=5.7414698 *ft*
Height of aluminum frame Z_f	*0.75in=0.0625 ft*
Width of aluminum frame X_f	*20 in=1.6666667ft*
Length of aluminum frame Y_f	*24 in=2 ft*
Radius of wheel r_w	*8 in=0.6666667 ft*
Density of aluminum D_f	*2.7 g/cm³ =0.0972lb/in³*
Weight of wheels m_w	10 lb
Weight of motor m_m	15 lb
Average weight of human body m_H	150 *lb*
the moment of inertia for the wheels I_w	$I_w = \frac{1}{2} m_w r^2 = 0.5 \times 10 \times 0.6666667 \times 0.6666667 = 2.2222 lb \cdot ft^2$
Distance between center of mass m_b and center of wheel r_b	$Z_h/2$

Table 2. Some terms

$2m_w + 2I_w\frac{1}{r^2} + m_b = M$	$2 \times 10 + 2 \times \frac{320}{8^2} + 208.5120 = 238.5120$
$I_b + 2m_b{r_b}^2 = M_b$	$1661.7 + 2 \times 208.5120 \times \left(\frac{5.7414698}{2}\right)^2 = 5098.4$
$m_b r_b = T_b$	$208.5120 \times 5.7414698 = 1197.2$
$2m_w + m_b = W$	$2 \times 10 + 208.5120 = 228.5120$
$\frac{\zeta \cdot \eta_t}{r_w} = R$	$20 \times \frac{0.95}{0.6666667} = 28.5$

SIMULATION AND EXPERIMENTS

Based on the simplification of the scooter mechanical structure, in which the scooter is treated as an inverted pendulum structure system in mathematical modeling. In such modeling method, the tires of the scooter are considered as rigid objects without deformation, and all the dynamics in electrical subsystem is a very short transient in comparison to the mechanical system. Various control methods have been proposed, such as fuzzy control, Lyapunov control and linear state feedback control. Generally speaking, these conventional methods focus on modeling of the main mechanical parts and corresponding control design so as to compensate the imperfection of the modeling process.

Numerical simulation of the reduced model and its simplified control design is employed to estimate and evaluate the system responses and performances. Once the satisfactory results are achieved, the hands-free two-wheeled self-balancing scooter is prototyped, and the test runs of the final design of two-wheeled self-balancing scooter are conducted to verify the accuracy of the modeling.

Table 3. Routh table

	$a_0^0 = 1.6628e+18$ $a_0^1 = 9.2541e+18$	$a_2^0 = 0$ $a_2^1 = 0$
$\alpha_1 = \frac{a_0^0}{a_0^1} = 0.1797$ $\alpha_2 = a_0^1/a_0^2$	$a_0^2 = a_2^0 - \alpha_1 a_2^1 = 0$ $a_0^3 = a_2^1 - \alpha_2 a_2^2$	$a_2^2 = a_4^0 - \alpha_1 a_4^1$ $a_2^3 = a_4^1 - \alpha_2 a_4^2$

The ideal design is mainly on logic level. However, designing a complicated mechatronic system such as the hands-free two wheeled self-balancing scooter, requires a lot of experience in manufacturing, power supply and split, DAQ, and other aspects of energy transmission, mass transmission and information transmission. During the prototyping procedure, interfacings between sensors, drives, actuators, and microcontroller as well as rechargeable battery sets are carefully considered for better efficiency and accuracy. PWM frequency and electrical motor drive also carefully selected for the electric motors.

As shown in Figure 3(a), the internal arrangement of the NI cRio controller, motor drives, geartrain integrated electric motors, accelerometer and gyroscope integration, optical encoder housed inside of electric motor, and their electric power supply in the hands-free two-wheeled self-balancing scooter. The wheels are connected to the motor through geartrain on the central line of the scooter platform. Two types of batteries are used, rechargeable battery in the middle of the drivers supplies the power in order to drive the system, but sensors are powered by 5 volt battery. Figure 3(b) shows the test run of the hands-free two wheeled self-balancing scooter.

The programming of the mechatronic system for the two-wheeled self-balancing scooter is developed in Labview software. Figure 4 to Figure 5 shows the FPGA program panel and the scooter host program panel, which integrate the information transmission between the sensors and its FPGA control modules. Figure 6 demonstrates the Labview program for optical encoder that measures the motor angular position and speed.

Figure 3. Two-wheeled self-balancing scooter

Figure 4. Scooter FPGA program panel in Labview
(Li, 2015)

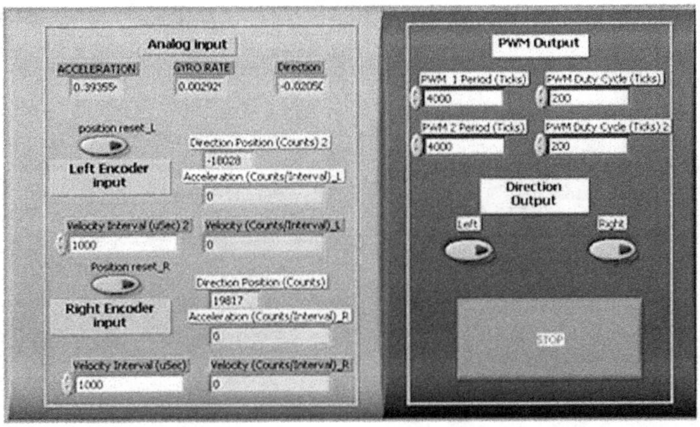

The Labview control program is uploaded to the cRio microprocessor for the offline control of the hands-free two-wheeled self-balancing scooter, and the rider tests run on the prototype. (see Figure 3(b)) The experimental results demonstrate its decision making ability on the direction control and moving control from the rider's Tai Chi changing poses. It is also noticed that the response time of the physical protype and its control realization is a little longer than the predicted by the numerical simulation.

CONCLUSION

This work uses the hands-free two-wheeled self-balancing scooter to demonstrate the design procedure of new century industrial product: from large scale modeling to model reduction, control design, and simulation as well as its prototyping. This is a design practice of analytical mechatronics, and the prototyping of a smart inverted pendulum type mechatronic device involves in multidisciplinary aspects of sensing, actuating, mechanism developing, microcontroller programming, control design, and their interfacings.

As another goal of this research, it can be concluded that, from the numerical simulations of Matlab-Labview software, and test runs of the real-time hands-free two-wheeled self-balancing scooter, the controller design based on reduced order model is feasible and sufficient. As one of the developed model reduction techniques, Routh model reduction technique is an effective way for mechatronic system model reduction.

Figure 5. Scooter host program panel
(Li, 2015)

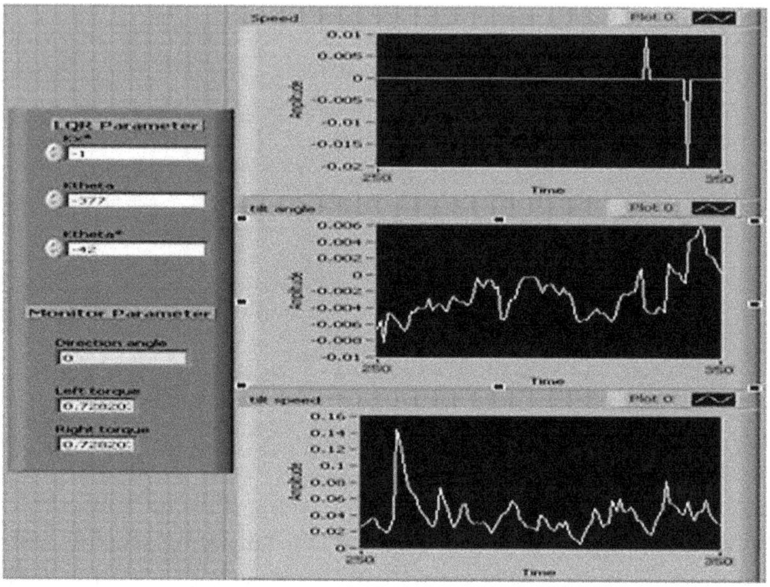

REFERENCES

Cunedioglu, Y., Mugan, A., & Akcay, H. (2006). Frequency domain analysis of model order reduction techniques. *Finite Elements in Analysis and Design*, *42*(5), 367–403. doi:10.1016/j.finel.2005.08.005

Davison, E. J. (1966). A method for simplifying linear dynamic systems. *IEEE Transactions on Automatic Control*, *11*(1), 93–101. doi:10.1109/TAC.1966.1098264

Giurgiutiu, V., & Lyshevski, S. E. (2009). *Micromechatronics: Modeling, Analysis, and Design with Matlab*. CRC Press.

Hu, J., & Tsai, M. (2008). Design of robust stabilization and fault diagnosis for an auto-balancing two-wheeled cart. *Advanced Robotics*, *22*(2-3), 319–338. doi:10.1163/156855308X292600

Isermann, R. (2003). *Mechatronic systems fundamentals*. Springer.

Janardhanan, S. (2013). *Model Order Reduction and Controller Design Techniques*. Academic Press.

Kim, Y., Kim, S. H., & Kwak, Y. (2005). Dynamic analysis of a nonholonomic two-wheeled inverted pendulum robot. *Journal of Intelligent & Robotic Systems, 44*(1), 25–46. doi:10.100710846-005-9022-4

Kumar, D., & Nagar, S. K. (2014). Model reduction by extended minimal degree optimal Hankel norm approximation. *Applied Mathematical Modelling, 38*(11-12), 2922–2933. doi:10.1016/j.apm.2013.11.012

Li, Q. (2015). *Large Scale Modeling, Model Reduction and Control Design for A Real-time Mechatronic System*. Purdue University.

Li, Q., & Zhang, H. H. (2019, May). Model Reduction of A Real Time Multidisciplinary Mechatronic System. In *20th International Conference on Research and Education in Mechatronics (REM 2019)*. University of Applied Sciences Upper Austria. 10.1109/REM.2019.8744102

Linhart, A., & Skogestad, S. (2012). An Aggregation Model Reduction Method for One-Dimensional Distributed Systems. *AIChE Journal. American Institute of Chemical Engineers, 58*(5), 1524–1537. doi:10.1002/aic.12688

Nasrallah, D., Michalska, H., & Angeles, J. (2007). Controllability and posture control of a wheeled pendulum moving on an inclined plane. *IEEE Transactions on Robotics, 23*(3), 564–577. doi:10.1109/TRO.2007.898953

Paraskevopoulos, P. N. (1980). Padé-type order reduction of two-dimensional systems. *IEEE Transactions on Circuits and Systems, CAS-27*(5), 413–416. doi:10.1109/TCS.1980.1084833

Vora, A., Wu, H., Wang, C., Qian, Y., Shaver, G., Motevalli, V., . . . Zhang, H. (2014). Development of a SIL, HIL and Vehicle Test-bench for Model-based Design and Validation of Hybrid Powertrain Control Strategies (SAE 2014-01-1906). *SAE Annual Meeting*.

Wang, Z., Li, L., & Wang, X. (2011). Modification Algorithm on Routh-Pade Model Reduction of Interval Systems. In ICIC 2011, LNCS 6838 (pp. 701–704). Springer. doi:10.1007/978-3-642-24728-6_94

Yadav, P. K., & Prajapati, N. L. (2012). One Dimensional Routh Pade Approximantsfor Two Dimensional Systems using Genetic Algorithm. *International Journal of Advanced and Innovative Research*, 576-580.

Yu, W., & Chen, H. (2018). Analog Controller Design for Mechatronic Rotary Inverted Pendulum (Part 1). *2018 ASEE Mid-Atlantic Section Spring Conference*.

Zhang, H. H., Nawrocki, R., & Li, Q. (2018, July). On Basics and Applications of Multidisciplinary Engineering and Technology Education. *3rd International Contemporary Educational Research Congress.*

Chapter 8
An Efficient Learning of Neural Networks to Acquire Inverse Kinematics Model

Fusaomi Nagata
Sanyo-Onoda City University, Japan

Maki K. Habib
https://orcid.org/0000-0001-8088-8043
The American University in Cairo, Egypt

Keigo Watanabe
Okayama University, Japan

ABSTRACT

In this chapter, effective learning approach of inverse kinematics using neural networks with efficient weights update ability has been presented for a serial link structure and industrial robot. Generally, in making neural networks learn a relation among multi inputs and outputs, a desired training data set prepared in advance is used. The training data set consists of multiple pairs of input and output vectors. The input layer receives each input vector for forward computation, and it is compared with the yielded vector from the output layer. The time required for the learning process of the neural networks depends on the number of total weights in the neural networks and that of the input-output pairs in the training data set.

DOI: 10.4018/978-1-7998-1382-8.ch008

INTRODUCTION

Since the planning and the control of a serial link structure and of industrial robots are directly affected by the method used to solve its inverse kinematic problem, it is important and an essential requirement for such structures to have the capability to develop a highly accurate and efficient solutions to the inverse kinematic problem. The inverse kinematics problem is defined as how to map a task coordinate system from Cartesian space to joint space given desired task space information for trajectory tracking (Xu et al, 2017; Carlos et al (2018); Zhou et al, 2018).

The authors have the experience of designing and implementing a neural network based technique with the industrial robot PUMA560. The trained neural network could acquire the desired relation between the contact force with an object and the velocity (Nagata & Watanabe, 2002). This chapter introduces the design of a neural network based approach with efficient learning algorithm to solve the inverse kinematics problem of a serial link structure.

For example, when designing a leg module with a serial link structure as shown in Fig. 1 for a multi-legged robot used in mechatronics education, first of all, its inverse kinematics problem must be solved. Up to now, biomimetic artificial neural network approaches have been devoted to robotic inverse kinematics problems. The neural network technique mimicking the human grain is known as one of the important results in biomedical engineering. In this first section, several related researches that challenged to solve inverse kinematics problems of manipulators or mechatronics systems are surveyed. Mao and Hsia proposed a neural network approach to solve the inverse kinematics of redundant robot manipulators in an environment with obstacles (Mao & Hsia, 1997), in which only the knowledge of the robot forward kinematics functions was required and the neural network was trained in the inverse modeling manner. Tejomurtula and Kak (1999) proposed a solution of inverse kinematics concerning a simple two link manipulator. It was reported that the proposed neural network method yielded multiple and precise solutions suitable for real-time applications. Xia and Wang proposed a new recurrent neural network called the dual network (Xia & Wang, 2001). It was shown that the dual network was capable of asymptotic tracking for the motion control of kinematically redundant manipulators.

Koker et al. (2004) proposed a neural network based inverse kinematics solution for a three joints manipulator, in which the designed neural networks gave reliable three angles according to a given position in Cartesian coordinates. Then, neural networks and genetic algorithms were fused together to solve the inverse kinematics problem of a Stanford robotic manipulator with six-DOFs while minimize the error at the end effector (Koker, 2013). Hoang et al. proposed a differential kinematics algorithm to generate omnidirectional walking trajectory of a leg based on back-

Figure 1. Leg module with a four-DOFs serial link structure for mechatronics education

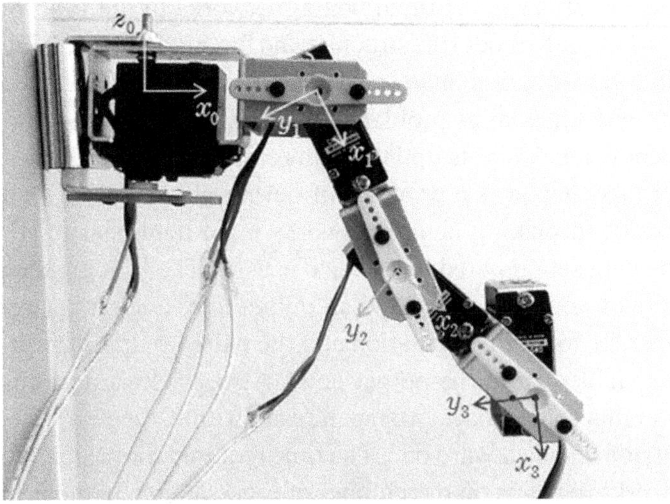

stepping control using Lyapunov stability. Simulation results for walking motion of one leg of the 6LR were shown to prove the effectiveness and applicability of the proposed controller (Hoang et al, 2014). Aggarwal et al. presented a non-conventional technique for solving the inverse kinematics problem using artificial neural networks. A feed forward multi-layer neural network was applied to this research. The inverse kinematic solution for a PUMA 560 robot was developed by training the neural network with the end-effector's Cartesian co-ordinates and its corresponding joint angles.

As for other types of approaches, Duka designed a feed-forward neural network to solve the inverse kinematics problem of a three-link planar manipulator (Duka, 2014), and then presented adaptive neuro-fuzzy inference system (ANFIS) that could produce the solution to the inverse kinematics problem for a planar robot, in which the ANFIS learned the mapping between the end-effector's position/orientation and three joint angles with acceptable accuracy (Duka, 2015). Maeda et al. designed a position control scheme for actual robot systems using high dimensional neural networks, in which complex-valued neural network and quaternion neural network learned the inverse kinematics of the robot systems. Two-dimensional SCARA robot and three-dimensional robot were well controlled using the inverse kinematics (Maeda et al, 2014). Furthermore, recently, Almusawi et al. presented a novel inverse kinematics solution for a DENSO robotic arm based on artificial neural network architecture, in which, it is described that the novelty of the proposed neural network is the inclusion of the feedback of joint angles as well as the desired position and orientation to the input layer of the neural network (Almusawi et al., 2016).

An expected the effect of artificial neural networks in the field of biomedical engineering is to be able to model the structure and functionalities of biological neural networks. As introduced earlier, many researches using neural networks tried to cope with robotic inverse kinematics problem, however, efforts on the improvement of learning efficiency, i.e., weights updating times, have been hardly seen. The main contribution of this chapter is a proposal of developing effective weight updating method. Generally, in making neural networks learn nonlinear relations suitably, desired training data set prepared in advance is used. The training data set consists of multiple pairs of an input vector and an output one. Each input vector is given to the input layer for forward calculation and the paired output vector is compared with the vector yielded from the output layer. Also, backward calculation means updating the weights using a back propagation algorithm. One cycle consists of one forward calculation and backward one. The time required for the learning process of the neural networks depends on the number of total weights in the neural networks and the one of the input-output pairs in the training data set. This chapter describes neural networks with efficient weights update ability in order to effectively learn the inverse kinematics of a leg kit with multi-DOFs (Nagata et al, 2013; Nagata et al, 2015). In the proposed learning process, after certain number of iteration, input-output pairs having the worse errors are extracted from the original training data set and form a new temporary set. Note that an iteration of learning uses all pairs in the training data set. Then, from the following iteration, the temporary set is applied instead of the original set. In this case, only pairs with worse errors are used for updating weights until the mean value of errors reduces to a desired level. Once the learning is conducted using the temporary training data set, the original set is applied again instead of the temporary set. It is expected by alternately giving the two kinds of training data sets the convergence time can be efficiently reduced. The effectiveness is proved through simulation experiments using a kinematic model of a leg with four-DOFs. Additionally, when the tip of the leg with a serial link structure is controlled kinematically in Cartesian space, an inverse Jacobian matrix is needed to generate velocities in joint space. However, the calculation of the inverse Jacobian is not easy. Hence, neural networks based inverse kinematics is used to implement the function of Jacobian to overcome the complexity. Then, the proposed learning method is applied to the inverse kinematics problem of an industrial robot with six-DOFs (Nagata & Watanabe, 2016). The effectiveness is further proved through simulation experiments using the kinematic model of an industrial robot.

LEG WITH FOUR-DOFS AND ITS FORWARD KINEMATICS

Generally, multi-legged robot has multiple legs with a serial link structure. Figure 2 shows an example of a leg with four-DOFs, which is used for graduation study of undergraduate students (Nagata, 2013). Five coordinate systems

$$\Sigma_k - x_k y_k z_k (0 \leq k \leq 4)$$

are assigned at each joint. The position of the arm tip in base frame $\Sigma_0 - x_0 y_0 z_0$ is defined with $\boldsymbol{x} = [x\ y\ z]^T$. Table 1 tabulates the Denavit-Hartenberg (DH) notation extracted from the leg with four joints. a is the link length which is the distance between two adjacent z-axes measured along x-axis, α is the link twist angle between two adjacent z-axes measured around x-axis, d is the link offset which is the distance between two adjacent x-axes measured along z-axis and θ is the joint angle between two adjacent x-axes measured around z-axis. Actually, Fig. 2 shows the initial pose of the leg with the angles of

$$\boldsymbol{\theta} = [\theta_1\ \theta_2\ \theta_3\ \theta_4]^T = [0\,0\,0\,0]^T.$$

The homogeneous transform ${}^{k-1}\boldsymbol{T}_k$ using the four parameters is written by

$$ {}^{k-1}\boldsymbol{T}_k = Rot(z,\theta)Trans(0,0,d)Trans(a,0,0)Rot(x,\alpha) = \begin{pmatrix} C_\theta & -S_\theta C_\alpha & S_\theta S_\alpha & aC_\theta \\ S_\theta & C_\theta C_\alpha & -C_\theta S_\alpha & aS_\theta \\ 0 & S_\alpha & C_\alpha & d \\ 0 & 0 & 0 & 1 \end{pmatrix} \tag{1}$$

where S_θ and C_θ denote $\sin\theta$ and $\cos\theta$, respectively. ${}^0\boldsymbol{T}_4$ is obtained by

$$ {}^0\boldsymbol{T}_4 = {}^0\boldsymbol{T}_1 {}^1\boldsymbol{T}_2 {}^2\boldsymbol{T}_3 {}^3\boldsymbol{T}_4 = \begin{pmatrix} R_{11} & R_{12} & -S_{\theta 1} & x \\ R_{21} & R_{22} & C_{\theta 1} & y \\ R_{31} & R_{32} & 0 & z \\ 0 & 0 & 0 & 1 \end{pmatrix} \tag{2}$$

where

$$R_{11} = C_{\theta 4}(C_{\theta 1}C_{\theta 2}C_{\theta 3} - C_{\theta 1}S_{\theta 2}S_{\theta 3}) + S_{\theta 4}(C_{\theta 1}S_{\theta 2}C_{\theta 3} - C_{\theta 1}S_{\theta 3}S_{\theta 4}) \tag{3}$$

$$R_{12} = S_{\theta 4}(C_{\theta 1}S_{\theta 2}S_{\theta 3} - C_{\theta 1}C_{\theta 2}C_{\theta 3}) - C_{\theta 4}(C_{\theta 1}S_{\theta 2}C_{\theta 3} + C_{\theta 1}C_{\theta 2}S_{\theta 3}) \tag{4}$$

$$R_{21} = C_{\theta 4}(S_{\theta 1}C_{\theta 2}C_{\theta 3} - S_{\theta 1}S_{\theta 2}S_{\theta 3}) - S_{\theta 4}(S_{\theta 1}S_{\theta 2}C_{\theta 3} + S_{\theta 1}C_{\theta 2}S_{\theta 3}) \tag{5}$$

$$R_{22} = S_{\theta 4}(S_{\theta 1}S_{\theta 2}S_{\theta 3} - S_{\theta 1}C_{\theta 2}C_{\theta 3}) - C_{\theta 4}(S_{\theta 1}S_{\theta 2}C_{\theta 3} + S_{\theta 1}C_{\theta 2}S_{\theta 3}) \tag{6}$$

$$R_{31} = S_{\theta 4}(S_{\theta 2}S_{\theta 3} - C_{\theta 2}C_{\theta 3}) - C_{\theta 4}(S_{\theta 2}C_{\theta 3} + C_{\theta 2}S_{\theta 3}) \tag{7}$$

$$R_{32} = S_{\theta 4}(S_{\theta 2}C_{\theta 3} + C_{\theta 2}S_{\theta 3}) + C_{\theta 4}(S_{\theta 2}S_{\theta 3} - C_{\theta 2}C_{\theta 3}) \tag{8}$$

$$\begin{aligned} x = {} & l_1 C_{\theta 1} + l_2 C_{\theta 1}C_{\theta 2} + l_3 C_{\theta 1}(C_{\theta 2}C_{\theta 3} - S_{\theta 2}S_{\theta 3}) \\ & + l_4 C_{\theta 1}(C_{\theta 2}C_{\theta 3}C_{\theta 4} - S_{\theta 2}S_{\theta 3}C_{\theta 4} - S_{\theta 2}C_{\theta 3}S_{\theta 4} - C_{\theta 2}S_{\theta 3}S_{\theta 4}) \end{aligned} \tag{9}$$

$$\begin{aligned} y = {} & l_1 S_{\theta 1} + l_2 S_{\theta 1}C_{\theta 2} + l_3 S_{\theta 1}(C_{\theta 2}C_{\theta 3} - S_{\theta 2}S_{\theta 3}) \\ & + l_4 S_{\theta 1}(C_{\theta 2}C_{\theta 3}C_{\theta 4} - S_{\theta 2}S_{\theta 3}C_{\theta 4} - S_{\theta 2}C_{\theta 3}S_{\theta 4} - C_{\theta 2}S_{\theta 3}S_{\theta 4}) \end{aligned} \tag{10}$$

$$z = -l_2 S_{\theta 2} - l_3(S_{\theta 2}C_{\theta 3} + C_{\theta 2}S_{\theta 3}) - l_4 C_{\theta 4}(S_{\theta 2}C_{\theta 3} + C_{\theta 2}S_{\theta 3}) + l_4 S_{\theta 4}(S_{\theta 2}S_{\theta 3} - C_{\theta 2}C_{\theta 3}) \tag{11}$$

Figure 2. Kinematics model of the leg with four-DOFs

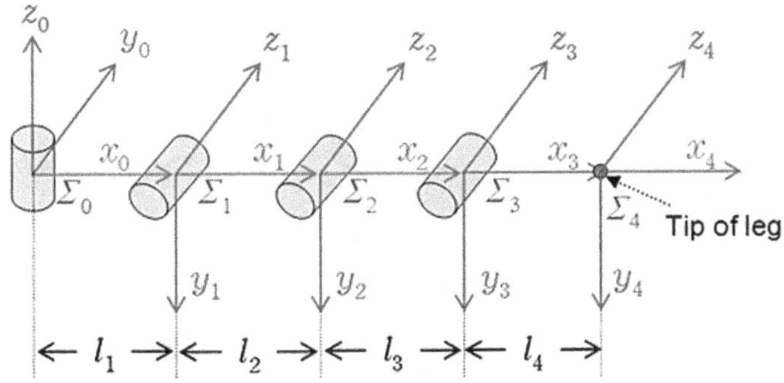

DESIGN OF NEURAL NETWORK-BASED INVERSE KINEMATICS

An original mobile robot for mechatronics education can be designed with the leg module shown in Fig. 1. In this chapter, Eq. (2) for $\boldsymbol{\theta} \in \Re^4 \rightarrow \boldsymbol{x} \in \Re^3$ is called the forward kinematics $\boldsymbol{x} = fkine(\boldsymbol{\theta})$ and can be analytically calculated. On the contrary $\boldsymbol{x} \in \Re^3 \rightarrow \boldsymbol{\theta} \in \Re^4$ is to be the inverse kinematics $\boldsymbol{\theta} = ikine(\boldsymbol{x})$. It is not easy but complex to obtain the analytical solutions of the inverse kinematics. In this section, the mapping of inverse kinematics is tried to be acquired in the neural networks shown in Fig. 3. The number of the hidden layers and that of neurons are not important for researching the weights update method proposed in the next section. This means that they are out of evaluation, so that two hidden layers and 30 neurons were tentatively set. After here, the neural networks are called the NN.

Table 1. Denavit-Hartenberg notation designed for the leg with four joints

k	a	α	d	θ
1	l_1	$-\frac{\pi}{2}$	0	θ_1
2	l_2	0	0	θ_2
3	l_3	0	0	θ_3
4	l_4	0	0	θ_4

Figure 3. Neural networks with three inputs of $\boldsymbol{x} \in \Re^3$ and four outputs $\boldsymbol{\theta}_{nn} \in \Re^4$

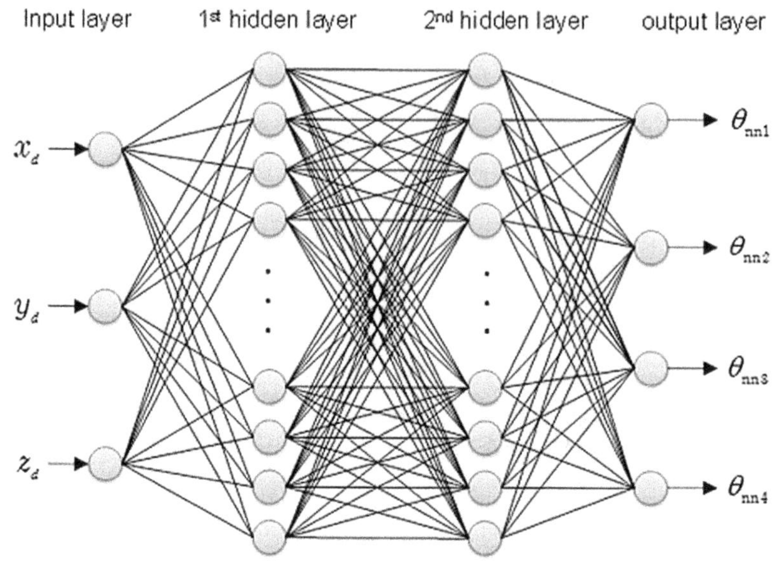

Neural Networks Leaned With Regularly Prepared Training Data Set

Figure 4 shows training data set

$$\boldsymbol{x}_j \in \Re^3 (1 \le j \le 100)$$

composed of 100 samples representing the relations $\boldsymbol{x}_j = fkine(\boldsymbol{\theta}_j)$ between $\boldsymbol{\theta}_j \in \Re^4$ in joint space and $\boldsymbol{x}_j \in \Re^3$ in Cartesian space, which are randomly generated by using the forward kinematics. The implicit relation of inverse kinematics that Fig. 4 has, is tried to be learned with the NN illustrated in Fig. 3. The NN has four layers where the input layer has three units of $\boldsymbol{x}_j$ and the output layer has four units of $\boldsymbol{\theta}_j$. The first and second hidden layers have thirty units, respectively.

After passing enough learning process using the training data set, the performance of the trained NN was evaluated using the test data set along the rectangle path in Fig. 4. Figure 5 shows the result, in which markers '+' are the test data set

$$\boldsymbol{x}_{dj} \in \Re^3 (1 \le j \le 120)$$

Figure 4. Training data set $\boldsymbol{x}_j \in \Re^3$ representing the relation $\boldsymbol{x}_j = fkine(\boldsymbol{\theta}_j)$ between $\boldsymbol{\theta}_j \in \Re^4$ in joint space and $\boldsymbol{x}_j \in \Re^3$ in Cartesian space, which are prepared randomly

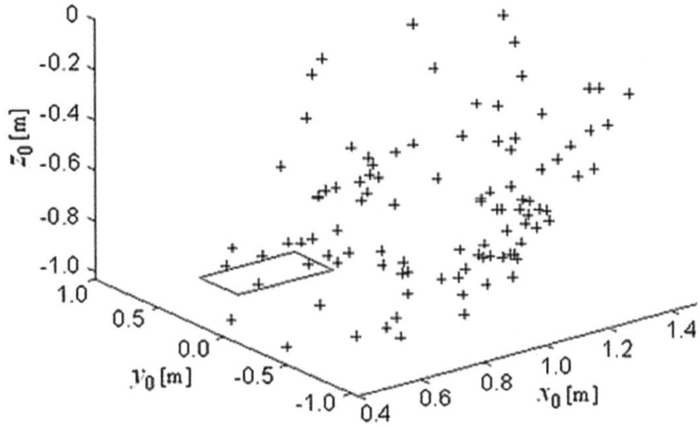

for the input layer of the NN. Also 'x' are the output of forward kinematics $fkine(\boldsymbol{\theta}_{nnj})$. Note that $\boldsymbol{\theta}_{nnj}$ is the output from the NN in response to $\boldsymbol{x}_{dj}$. It is recognized from the results that the NN learned with the training data set shown in Fig. 4 performs a passable generalization for inverse kinematics but some small errors are observed.

Neural Networks Learned With Randomly Prepared Training Data Set

In this subsection, the performance of the NN learned with regularly prepared training data set (60 pairs of $\boldsymbol{x}_j$ and $\boldsymbol{\theta}_j$) is evaluated. Figure 6 shows the training data set representing a path consisting of $\boldsymbol{x}_j (1 \leq j \leq 60)$, i.e., the sampled number for training data set is 60. The NN shown in Fig. 3 was trained with the set until the error satisfactorily converged. Figure 7 shows the performance result of the trained NN. Marker '+' in the upper part of the figure are test data set $\boldsymbol{x}_{dj} (1 \leq j \leq 300)$ for the NN. Also, 'x' in the lower part of the figure are the output of forward kinematics $fkine(\boldsymbol{\theta}_{nnj})$. Note that $\boldsymbol{\theta}_{nnj}$ is the output from the forward calculation of the NN when $\boldsymbol{x}_{dj} (1 \leq j \leq 300)$ is given to the input layer. It is recognized from the result that the NN learned with the training data set shown in Fig. 6 performs a desirable generalization for inverse kinematics.

Figure 5. Markers '+' are the test data set $\boldsymbol{x}_{dj}$ for learning of the NN. Also 'x' are the output of forward kinematics $fkine(\boldsymbol{\theta}_{nnj})$. Note that $\boldsymbol{\theta}_{nnj}$ is the output from the forward calculation of the NN when $\boldsymbol{x}_{dj}$ is given to the input layer

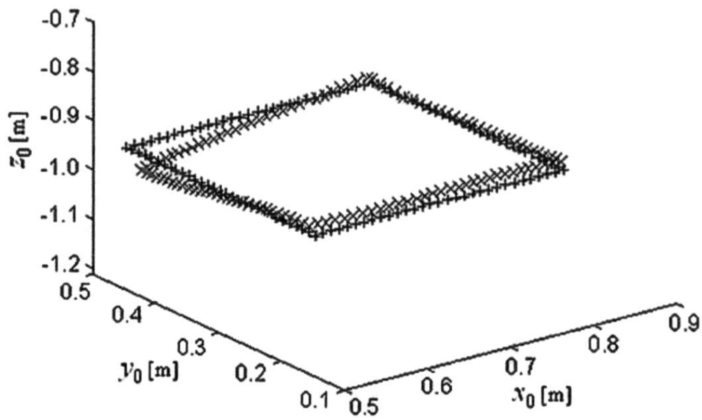

EFFICIENT LEARNING ABILITY

In the Case of Conventional Learning Method

When the back propagation algorithm is used for adjusting weights in neural networks, it is serious problem that much time is required for satisfactory convergence. Here, an efficient training method is introduced by using the case shown in Fig. 6. First of all, error E_i for criterion at the i-th learning procedure was defined by

Figure 6. Training data set representing the relation between $\boldsymbol{\theta}_j \in \Re^4$ and $\boldsymbol{x}_j \in \Re^3$, which are prepared regularly

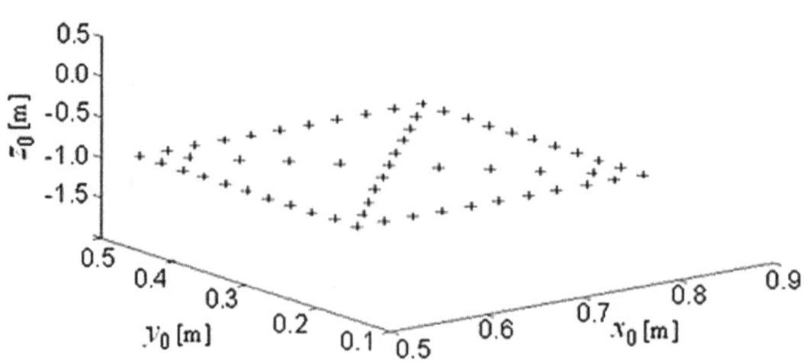

Figure 7. Marker '+' in the upper part of the figure are the test data set $\boldsymbol{x}_{dj}$ *for NN. Also 'x' in the lower part of the figure are the output* $\boldsymbol{x}_j$ *of forward kinematics* $fkine(\boldsymbol{\theta}_{nnj})$*. Note that* $\boldsymbol{\theta}_{nnj}$ *is the output from the forward calculation of the NN when* $\boldsymbol{x}_{dj}$ *is given to the input layer*

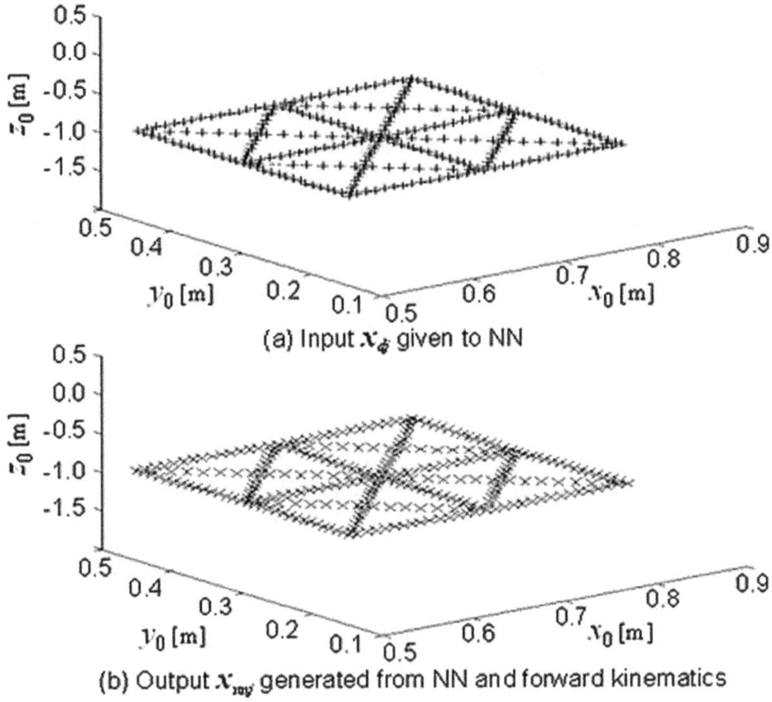

$$E_i = \bar{e} = \frac{1}{60}\sum_{j=1}^{60} e_j \tag{12}$$

where e_j is the error in case that the *j*-th sample, i.e., a pair of $\boldsymbol{x}_j$ and $\boldsymbol{\theta}_j$ in training data set is given to input and output layers, which is obtained by

$$e_j = \sqrt{\sum_{k=1}^{4}(\theta_{jk} - \theta_{\mathrm{nn}jk})^2} \tag{13}$$

where $\boldsymbol{\theta}_j \in \Re^4$ and $\boldsymbol{\theta}_{nnj} \in \Re^4$ are the training vector for the output layer and the actual output from the NN, respectively. k $(1 \le k \le 4)$ is the number of each joint.

j $(1 \leq j \leq 60)$ is the sampled number in the training data set. The learning, i.e., the tuning of weights, was continued until the following condition is satisfied.

$$E_i < E_d \tag{14}$$

where E_d is the maximum allowable error to be set in advance. In the conventional learning method, all samples in the training data set, i.e., sixty pairs, were sequentially given for updating weights. Figure 8 shows the learning result, in which sixty times of weights updating by using a back propagation algorithm were conducted in one learning procedure on the horizontal axis. In other words, one learning procedure means one iteration using sixty pairs. Also note that one weights updating procedure consists of a pair of a forward propagation to calculate the error and a back propagation to update all weights based on the error. It was confirmed that totally 22,810 × 60 = 1,368,600 times of weights updating were conducted and the total calculation time was 8,351 seconds. In this case, the desired maximum allowable error E_d was set to 0.02. The specification of CPU was Intel(R) Core(TM) i3 560 3.33 GHz.

In the Case of Proposed Efficient Learning Method

On the other hand, in our proposed method, a new training data set consisting of pairs of input $\boldsymbol{x}_l$ and output $\boldsymbol{\theta}_l$, that have not been well trained, is extracted from the original training data set. The condition of the extraction is given by

Figure 8. An example of learning result using a conventional method, in which the desired maximum allowable error E_d is set to 0.02

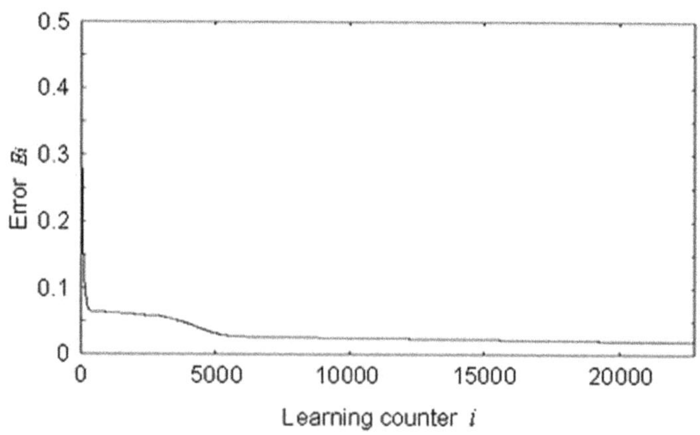

$$e_l > \bar{e} + e_{\mathrm{sd}} \tag{15}$$

where

$$e_l = \sqrt{\sum_{k=1}^{4} (\theta_{lk} - \theta_{\mathrm{nn}lk})^2} \tag{16}$$

$$e_{\mathrm{sd}} = \sqrt{\frac{1}{60} \sum_{j=1}^{60} (\bar{e} - e_j)^2} \tag{17}$$

where e_{sd} is the standard deviation of e_j $(1 \le j \le 60)$. The pairs of $\boldsymbol{x}_l$ and $\boldsymbol{\theta}_l$ satisfying Eq. (15) are extracted and form a new training data set with less number of pairs than that of the original training data set, i.e., 60. When this new training data set is applied instead of the original set, the error $\tilde{E}_i$ at the i-th learning procedure, i.e., iteration, is calculated by

$$\tilde{E}_i = \frac{1}{m} \sum_{l=1}^{m} e_l \tag{18}$$

where m is the number of pairs extracted from the original set based on Eq. (15). In the proposed method, E_i and $\tilde{E}_i$ are alternately used for the evaluation of convergence at the i-th learning procedure. Note that the extracted training data set does not increase the mean errors E_i and $\tilde{E}_i$. E_i and $\tilde{E}_i$ are criteria used to know the progress of training. Figure 9 illustrates the concept of the proposed learning process, in which the original training data set and the extracted one are alternately applied for 200 times of iterative learning, respectively. The vertical axis illustrates the number of the pairs in the training data set. The learning process using the extracted training data set works as mini-batch gradient descent used in deep learning techniques. In the mini-batch gradient descent and during each iteration samples randomly extracted from original training data set are used to update the weights. The number of samples is designated as mini-batch size. On the other hand, in the proposed learning process, m samples with larger errors are extracted and forms a temporary subset for weights updating.

Then, the proposed method was applied to the same problem explained in the previous section to evaluate its effectiveness. Figure 10 shows the learning result

Figure 9. Image of the proposed learning process, in which original training data set and extracted one are alternately applied to the NN for 200 times of iterative learning, respectively

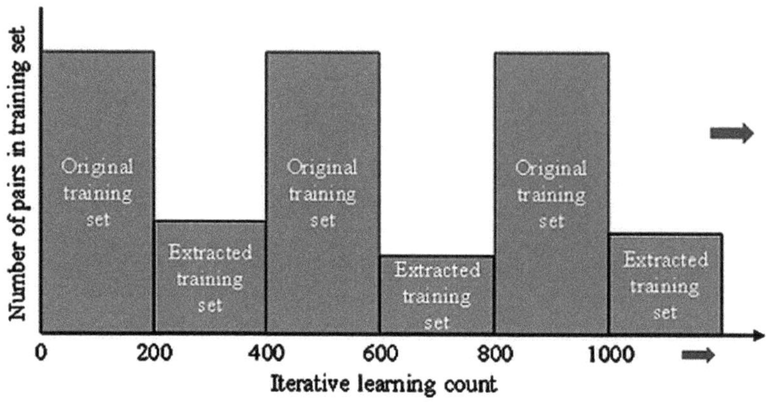

until E_i reached to 0.02, in which sequential 100 times of E_i and 1,000 times of $\tilde{E}_i$ are used alternately and repeatedly. It is observed that although the iterative learning counter is almost the same as 22,000 times shown in Fig. 8, total times of weights updating were reduced to 650,220 and the calculation time largely decreased to 4,096 seconds.

KINEMATIC CONTROL OF THE LEG MODULE

Jacobian of the Leg Module

When the tip of a leg module is controlled kinematically in Cartesian space, the Jacobian matrix is required. By differentiating Eqs. (9), (10) and (11) with respect to time, the following relation is obtained.

$$\begin{pmatrix} \dot{x} \\ \dot{y} \\ \dot{z} \end{pmatrix} = \begin{pmatrix} J_{11} & J_{12} & J_{13} & J_{14} \\ J_{21} & J_{22} & J_{23} & J_{24} \\ 0 & J_{32} & J_{33} & J_{34} \end{pmatrix} \begin{pmatrix} \dot{\theta}_1 \\ \dot{\theta}_2 \\ \dot{\theta}_3 \\ \dot{\theta}_4 \end{pmatrix} \quad (19)$$

where

Figure 10. Learning result using the proposed method, in which original training data set and extracted one are alternately applied to the NN for 100 and 1000 times of iterative learning, respectively

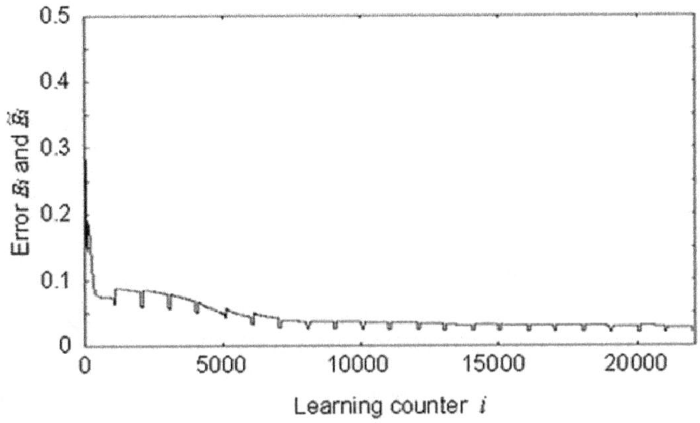

$$\begin{aligned} J_{11} &= -l_1 S_{\theta1} - l_2 S_{\theta1} S_{\theta2} - l_3 S_{\theta1}(C_{\theta2}C_{\theta3} - S_{\theta2}S_{\theta3}) \\ &\quad - l_4(S_{\theta1}C_{\theta2}C_{\theta3}C_{\theta4} - S_{\theta1}S_{\theta2}S_{\theta3}C_{\theta4} - S_{\theta1}S_{\theta2}C_{\theta3}S_{\theta4} - S_{\theta1}C_{\theta2}S_{\theta3}S_{\theta4}) \end{aligned} \tag{20}$$

$$\begin{aligned} J_{12} &= -l_2 C_{\theta1} S_{\theta2} - l_3 C_{\theta1}(S_{\theta2}C_{\theta3} + C_{\theta2}S_{\theta3}) \\ &\quad - l_4(C_{\theta1}S_{\theta2}C_{\theta3}C_{\theta4} + C_{\theta1}C_{\theta2}S_{\theta3}C_{\theta4} + C_{\theta1}C_{\theta2}C_{\theta3}S_{\theta4} - C_{\theta1}S_{\theta2}S_{\theta3}S_{\theta4}) \end{aligned} \tag{21}$$

$$\begin{aligned} J_{13} &= -l_3 C_{\theta1}(S_{\theta2}C_{\theta3} + C_{\theta2}S_{\theta3}) \\ &\quad - l_4(C_{\theta1}S_{\theta2}C_{\theta3}C_{\theta4} + C_{\theta1}C_{\theta2}S_{\theta3}C_{\theta4} + C_{\theta1}C_{\theta2}C_{\theta3}S_{\theta4} - C_{\theta1}S_{\theta2}S_{\theta3}S_{\theta4}) \end{aligned} \tag{22}$$

$$J_{14} = -l_4(C_{\theta1}S_{\theta2}C_{\theta3}C_{\theta4} + C_{\theta1}C_{\theta2}S_{\theta3}C_{\theta4} + C_{\theta1}C_{\theta2}C_{\theta3}S_{\theta4} - C_{\theta1}S_{\theta2}S_{\theta3}S_{\theta4}) \tag{23}$$

$$\begin{aligned} J_{21} &= l_1 C_{\theta1} + l_2 C_{\theta1} C_{\theta2} + l_3 C_{\theta1}(C_{\theta2}C_{\theta3} - S_{\theta2}S_{\theta3}) \\ &\quad + l_4(C_{\theta1}C_{\theta2}C_{\theta3}C_{\theta4} - C_{\theta1}S_{\theta2}S_{\theta3}C_{\theta4} - C_{\theta1}S_{\theta2}C_{\theta3}S_{\theta4} - C_{\theta1}C_{\theta2}S_{\theta3}S_{\theta4}) \end{aligned} \tag{24}$$

$$\begin{aligned} J_{22} &= -l_2 S_{\theta 1} S_{\theta 2} - l_3 S_{\theta 1}(S_{\theta 2} C_{\theta 3} + C_{\theta 2} S_{\theta 3}) \\ &\quad - l_4 (S_{\theta 1} S_{\theta 2} C_{\theta 3} C_{\theta 4} + S_{\theta 1} C_{\theta 2} S_{\theta 3} C_{\theta 4} + S_{\theta 1} C_{\theta 2} C_{\theta 3} S_{\theta 4} - S_{\theta 1} S_{\theta 2} S_{\theta 3} S_{\theta 4}) \end{aligned} \tag{25}$$

$$\begin{aligned} J_{23} &= -l_3 S_{\theta 1}(S_{\theta 2} C_{\theta 3} + C_{\theta 2} S_{\theta 3}) \\ &\quad - l_4 (S_{\theta 1} S_{\theta 2} C_{\theta 3} C_{\theta 4} + S_{\theta 1} C_{\theta 2} S_{\theta 3} C_{\theta 4} + S_{\theta 1} C_{\theta 2} C_{\theta 3} S_{\theta 4} - S_{\theta 1} S_{\theta 2} S_{\theta 3} S_{\theta 4}) \end{aligned} \tag{26}$$

$$J_{24} = -l_4 (S_{\theta 1} S_{\theta 2} C_{\theta 3} C_{\theta 4} + S_{\theta 1} C_{\theta 2} S_{\theta 3} C_{\theta 4} + S_{\theta 1} C_{\theta 2} C_{\theta 3} S_{\theta 4} - S_{\theta 1} S_{\theta 2} S_{\theta 3} S_{\theta 4}) \tag{27}$$

$$J_{32} = -l_2 C_{\theta 2} - l_3 (C_{\theta 2} C_{\theta 3} - S_{\theta 2} S_{\theta 3}) + l_4 (S_{\theta 2} C_{\theta 3} S_{\theta 4} + C_{\theta 2} S_{\theta 3} S_{\theta 4} - C_{\theta 2} C_{\theta 3} C_{\theta 4} + S_{\theta 2} S_{\theta 3} C_{\theta 4}) \tag{28}$$

$$J_{33} = -l_3 (C_{\theta 2} C_{\theta 3} - S_{\theta 2} S_{\theta 3}) + l_4 (S_{\theta 2} C_{\theta 3} S_{\theta 4} + C_{\theta 2} S_{\theta 3} S_{\theta 4} - C_{\theta 2} C_{\theta 3} C_{\theta 4} + S_{\theta 2} S_{\theta 3} C_{\theta 4}) \tag{29}$$

$$J_{34} = -l_4 (C_{\theta 2} C_{\theta 3} C_{\theta 4} - S_{\theta 2} S_{\theta 3} C_{\theta 4} - S_{\theta 2} C_{\theta 3} S_{\theta 4} - C_{\theta 2} S_{\theta 3} S_{\theta 4}) \tag{30}$$

Here, a constraint condition is considered. The orientation condition of the leg tip is fixed to the ground as given by

$$\theta_2 + \theta_3 + \theta_4 = c (c = \text{const.}) \tag{31}$$

Hence,

$$\dot{\theta}_2 + \dot{\theta}_3 + \dot{\theta}_4 = 0 \tag{32}$$

By reflecting the content of Eq. (32) into Eq. (19),

$$\begin{pmatrix} \dot{x} \\ \dot{y} \\ \dot{z} \\ 0 \end{pmatrix} = \begin{pmatrix} J_{11} & J_{12} & J_{13} & J_{14} \\ J_{21} & J_{22} & J_{23} & J_{24} \\ 0 & J_{32} & J_{33} & J_{34} \\ 0 & 1 & 1 & 1 \end{pmatrix} \begin{pmatrix} \dot{\theta}_1 \\ \dot{\theta}_2 \\ \dot{\theta}_3 \\ \dot{\theta}_4 \end{pmatrix} \tag{33}$$

is obtained, in which Jacobian $\boldsymbol{J}(\boldsymbol{\theta})$ is defined as

$$\boldsymbol{J}(\boldsymbol{\theta}) = \begin{pmatrix} J_{11} & J_{12} & J_{13} & J_{14} \\ J_{21} & J_{22} & J_{23} & J_{24} \\ 0 & J_{32} & J_{33} & J_{34} \\ 0 & 1 & 1 & 1 \end{pmatrix} \tag{34}$$

In order to conduct the kinematic control, the desired relative position Δx should be described in Cartesian space and be transformed into relative angle $\Delta\boldsymbol{\theta}$ in the joint space using the inverse Jacobian $\boldsymbol{J}^{-1}(\boldsymbol{\theta})$. However, since the calculation of $\mathbf{J}^{-1}(\boldsymbol{\theta})$ is considerably complicated, an easy method to generate $\Delta\boldsymbol{\theta}$ with the NN shown in Fig. 3 is introduced.

Kinematic Control Without Inverse Jacobian

It is assumed that $\boldsymbol{\theta}(k)$ and $\boldsymbol{x}(k)$ are the position vectors in joint and Cartesian spaces, respectively, so that the relation is represented with the forward kinematics function as

$$\boldsymbol{x}(k) = fkine\{\boldsymbol{\theta}(k)\} \tag{35}$$

where k denotes the discrete time. The desired movement $\Delta\boldsymbol{x}_d(k)$ in Cartesian space is written by

$$\Delta\boldsymbol{x}_d(k) = \boldsymbol{x}_d(k) - \boldsymbol{x}_d(k-1) \tag{36}$$

To conduct the above motion, the following movement in joint space can be generated using the trained NN.

$$\Delta\boldsymbol{\theta}_d(k) = ikine_{\_nn}\{\boldsymbol{x}_d(k)\} - ikine_{\_nn}\{\boldsymbol{x}_d(k-1)\} \tag{37}$$

where the function

$$\boldsymbol{\theta}_{nn}(k) = ikine_{\_nn}\{\boldsymbol{x}_d(k)\}$$

means the mapping of inverse kinematics through the NN shown in Fig. 3. Through the proposed procedure, the desired movement $\Delta\boldsymbol{\theta}_d(k)$ for kinematic control can be generated without the need to use $\boldsymbol{J}^{-1}(\boldsymbol{\theta})$.

INDUSTRIAL ROBOT MOTOMAN WITH SIX-DOFS

Figure 11 shows an industrial robot Motoman with six-DOFs. Six coordinate systems $\Sigma_k - x_k y_k z_k$ $(0 \le k \le 5)$ are assigned at each joint. The position of the arm tip viewed in base frame $\Sigma_0 - x_0 y_0 z_0$ is defined with $\boldsymbol{x} = [x\ y\ z]^T$. Table 2 tabulates the Denavit-Hartenberg notation (DH) extracted from the robot (Gan 2013). a is the link length which is the distance between two adjacent z-axes measured along x-axis, α is the link twist angle between two adjacent z-axes measured around x-axis, d is the link offset which is the distance between two adjacent x-axes measured along z-axis and θ is the joint angle between two adjacent x-axes measured around z-axis.

$$^{k-1}\boldsymbol{T}_k = Rot(z,\theta)Trans(0,0,d)Trans(a,0,0)Rot(x,\alpha) = \begin{bmatrix} C_\theta & -S_\theta C_\alpha & S_\theta S_\alpha & aC_\theta \\ S_\theta & C_\theta C_\alpha & -C_\theta S_\alpha & aS_\theta \\ 0 & S_\alpha & C_\alpha & d \\ 0 & 0 & 0 & 1 \end{bmatrix} \tag{38}$$

where, e.g., the homogeneous transform Rot(z, θ) conducts the rotation of θ [rad] to right screw direction around z-axis; Trans(0, 0, d) translates to z-direction with d [mm]; S_θ and C_θ denote sinθ and cosθ, respectively. Hence, ${}^0T_6$ is obtained by

$$^{0}\boldsymbol{T}_6 = {}^{0}\boldsymbol{T}_1\,{}^{1}\boldsymbol{T}_2\,{}^{2}\boldsymbol{T}_3\,{}^{3}\boldsymbol{T}_4\,{}^{4}\boldsymbol{T}_5\,{}^{5}\boldsymbol{T}_6 = \begin{bmatrix} R_{11} & R_{12} & R_{13} & x \\ R_{21} & R_{22} & R_{23} & y \\ R_{31} & R_{32} & R_{33} & z \\ 0 & 0 & 0 & 1 \end{bmatrix} \tag{39}$$

3×3 rotational matrix $\boldsymbol{R}$ in ${}^0\boldsymbol{T}_6$ has to be transformed into roll, pitch and yaw angles in kinematic simulation (Corke 1996). As is known, the inverse kinematics

Figure 11. Six-DOFs Industrial robot Yaskawa Motoman applied to mold polishing process

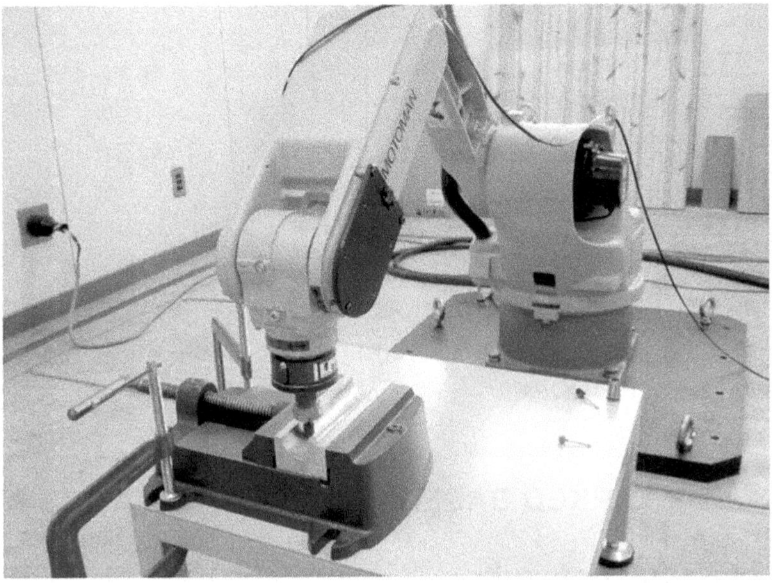

Figure 12. Neural network with five inputs of $\mathbf{x}_d$ and six outputs of $\boldsymbol{\theta}_{nn}$. In simulations, the number of the hidden layers and that of neurons are tentatively set to 2 and 20, respectively

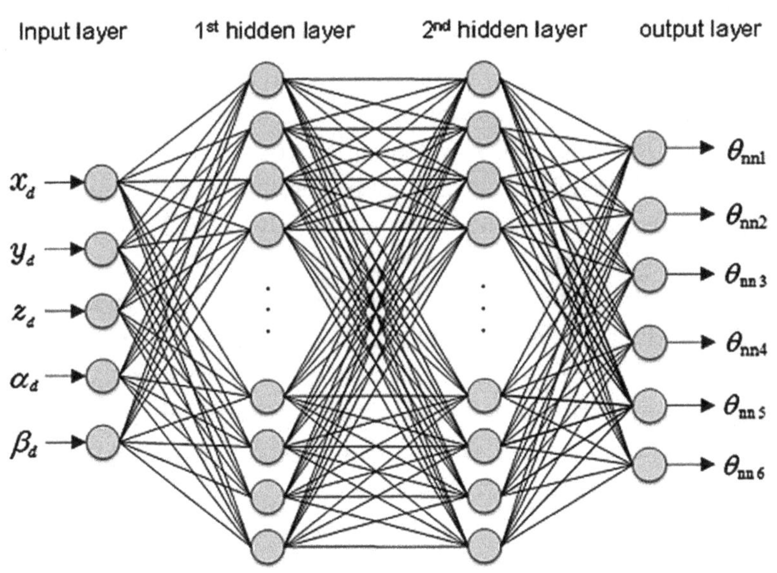

Figure 13. Image of desired points along a spiral path prepared for training data set

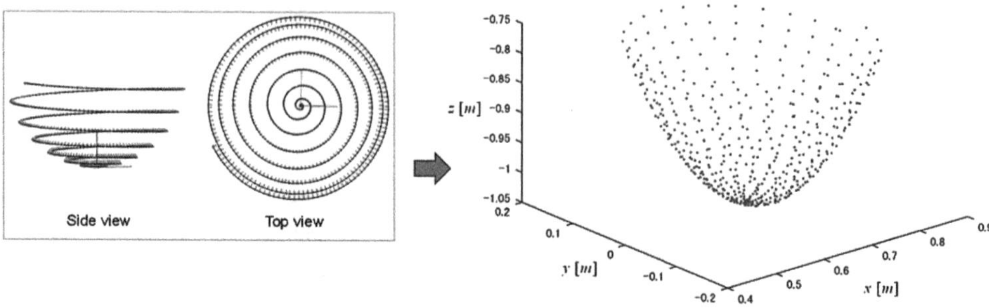

problem of industrial robots with a serial link structure becomes more complex with the increase of the DOFs.

NEURAL NETWORKED-BASED INVERSE KINEMATICS

Design of Neural Networks

In this section, the joint angle vector $\boldsymbol{\theta} = [\theta_1 \ldots \theta_6] \Rightarrow$ the homogeneous transform matrix ${}^0\boldsymbol{T}_6$ is called the forward kinematics and can be analytically calculated through ${}^0\boldsymbol{T}_6 = fkine(\boldsymbol{\theta})$ (Corke 1996). Reversely, ${}^0\boldsymbol{T}_6 \Rightarrow \boldsymbol{\theta}$ is to be the inverse kinematics. It is not easy but complex to obtain the analytical solutions of the inverse kinematics. In this section, it is tried that the mapping of inverse kinematics ${}^0\boldsymbol{T}_6 \Rightarrow \boldsymbol{\theta}$ is acquired in the neural network shown in Fig. 12, in which $\boldsymbol{x}_d = [x_d\ y_d\ z_d\ \alpha_d\ \beta_d\ ]^T$ is transformed from ${}^0\boldsymbol{T}_6$; α_d and β_d are the roll and pitch angles, respectively. The yaw angle can be always fixed to 0 because of the axis symmetry of the sixth axis. The number of the hidden layers and that of neurons are not important for researching the proposed weights update method, so that they were tentatively set to 2 and 20, respectively. The neurons in two adjacent layers are connected with weights randomly initialized within the range from −1 to 1. In the learning process shown in this section, the back propagation algorithm is applied for training of the weights.

The performance of the proposed efficient learning method is evaluated using the regularly prepared training data set consisting of multiple pairs of $\boldsymbol{x}_{dj} = [x_{dj}\ y_{dj}\ z_{dj}\ \alpha_{dj}\ \beta_{dj}\ ]^T$ and $\boldsymbol{\theta}_{dj}$ $(1 \le j \le 481)$, i.e., the sampled number in the training data set is 481. The desired points $[x_{dj}\ y_{dj}\ z_{dj}]^T$ included in $\boldsymbol{x}_{dj}$ generated along a spiral path can be viewed as shown in Fig. 13. Roll angle α_{dj} and pitch angle β_{dj} are converted from the normal vector at the respective points. The joint angle $\boldsymbol{\theta}_{dj}$ in the training data set was made by using an inverse kinematics function $\boldsymbol{\theta}_{dj} = ikine(\boldsymbol{x}_{dj})$ available on MATLAB (Corke 1996).

When the back propagation algorithm is used for adjusting weights in neural network, it is serious problem that much time is required for satisfactory convergence. Especially, undesirable stagnation of learning always depresses us. Here, an efficient training method is introduced by using the training data set shown in Fig. 13. In this method, when the learning falls into the situation of stagnation, a stimulating extraction from original training data set is applied and temporally creates another training data set with worse evaluation for more intensive training.

Conventional and Orthodox Learning Process

First of all, the conventional and orthodox learning process is described. Error E_i for criterion at the i-th learning process is defined by

$$E_i = \bar{e} = \frac{1}{481}\sum_{j=1}^{481} e_j \tag{40}$$

where e_j is the error in case that the j-th sample in training data set is given to input and output layers, which is obtained by

$$e_j = \sqrt{\sum_{k=1}^{6} (\theta_{djk} - \theta_{\mathrm{nn}jk})^2} \tag{41}$$

where $\boldsymbol{\theta}_{dj} \in \Re^6$ and $\boldsymbol{\theta}_{\mathrm{nn}j} \in \Re^6$ are the training vector for the output layer and the actual output from the NN, respectively. k is the number of each joint. j is the sampled number in training data set. The learning, i.e., the updating of weights, is continued until the following condition is satisfied.

$$E_i < E_d \tag{42}$$

where E_d is the maximum allowable error. The stagnation of training process can be judged by

$$E_{i-1} - E_i < E_s \tag{43}$$

where E_s is the stagnation index. In the conventional learning method used in this section, all samples in the training data set, i.e., 481 pairs, are sequentially given for updating 620 weights.

Figure 14 shows the learning result, in which the updating of weights by using a back propagation algorithm was conducted 481 times in one learning process on the horizontal axis. In other words, one learning process means 481 times iterative computations using 481 input and output pairs. Also, note that one computation for updating the weights consists of a pair of a forward propagation to calculate the error and a back propagation to update 620 weights based on the error. Figure 14 has the abscissa with a logarithmic plot to clearly show the progress of training at the early stage. The structure of the industrial robot with six-DOFs is more complex than that of the serial link with four-DOFs, so that it also takes more time for convergence. That is the reason why the logarithmic plot is used in the horizontal axis in Figs. 14 and 16.

It was confirmed that totally the updating of weights was conducted 81,702×481= 39,298,662 times. In this case, the desired maximum allowable error E_d was set to 0.03. The specification of CPU was Intel(R) Core(TM) i7 4790 3.6 GHz. In this research, the calculation time is not used for criterion because the value of the time depends on the performance of CPU built in PC.

Proposed Efficient Learning Process

On the other hand, in our proposed method, a new temporary training data set consisting of pairs of input $\boldsymbol{x}_l$ and output $\boldsymbol{\theta}_l$, that have not been well trained, is extracted from the original training data set. Three conditions of extraction are compared in the latter experiments, which are given by

Figure 14. Learning result using a conventional method

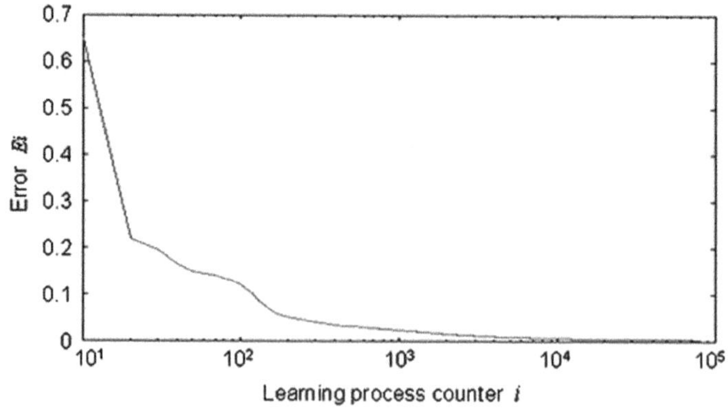

$$e_l > \bar{e} \tag{44}$$

$$e_l > \bar{e} + e_{sd} \tag{45}$$

$$e_l > \bar{e} + 2e_{sd} \tag{46}$$

Where

$$e_{sd} = \sqrt{\frac{1}{481}\sum_{j=1}^{481}(\bar{e} - e_j)^2} \tag{47}$$

where e_{sd} is the standard deviation of e_j ($1 \le j \le 481$). At the present time, some standards and criterions to systematically decide the condition have not been considered.

The pairs of $\boldsymbol{x}_l$ and $\boldsymbol{\theta}_l$ satisfying Eq. (44), (45) or (46) are extracted and form a new training data set with less number of pairs than the original training data set. When this new training data set is applied instead of the original set, the error $\tilde{E}_i$ at the i-th learning process is calculated by

$$\tilde{E}_i = \frac{1}{m}\sum_{l=1}^{m} e_l \tag{48}$$

where m is the number of the pairs in the training data set extracted from the original set based on Eq. (44), (45) or (46). E_i and $\tilde{E}_i$ are alternately used for the criterion in the back propagation algorithm in our proposed method. Figure 15 illustrates the conceptual comparison between the conventional and proposed learning methods. In the proposed method, the original training data set and the extracted one are alternately applied for the iterative learning 1000 and 4000 times, respectively. Note that the times of iterative learning for original and extracted sets can be arbitrarily changed according to actual problem settings. In this experiment, 1000 and 4000 times are found experientially. The vertical axis means the number of the pairs in the training data set.

Finally, the proposed method was applied to the same problem explained in the previous section to evaluate the effectiveness. As an example, Table 3 shows three results of the total times of weights updating in case that the numbers of iterative

Figure 15. Concept of the proposed learning method, where original training data set and extracted one are alternately applied

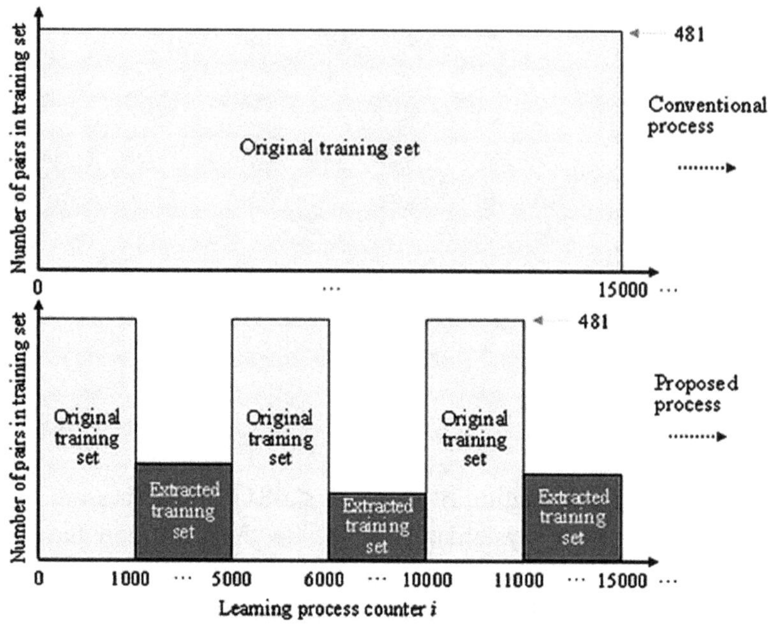

learning of original and extracted sets were set to 1,000 and 4,000, respectively. Figure 16 shows the learning result until E_i reached to 0.03 in the case of Eq. (44) in Table 3, in which sequential 1000 times of E_i and 4,000 times of $\tilde{E}_i$ are used alternately and repeatedly. It is observed that although the iterative learning counter is almost the same, total times of weights updating were improved by about 31%, i.e., from 39,298,662 to 27,050,036. It was experimentally observed that the numbers of sequential times for evaluating E_i and $\tilde{E}_i$ occurred different learning results, so that some method will be needed to systematically determine the numbers of times.

DISCUSSION

The designed NNs in this chapter have only 4 layers. The layers of the NN designed for the industrial manipulator has 5, 20, 20 and 6 neurons respectively so that total number of weights within this NN is 100+400+120 = 620 at most. In training, the number of pairs in teaching data is only 481. That is the reason why the calculation load to converge the error to be smaller than a desired value using the back propagation algorithm is significantly small compared to deep NNs such as convolutional neural

Table 2. Denavit-Hartenberg notation for Motoman HP20

k	a	α	d	θ
1	150	$-90°$	0	θ_1
2	760	0	0	θ_2
3	140	$-90°$	0	θ_3
4	0	90°	795	θ_4
5	0	90°	0	θ_5
6	0	0	-105	θ_6

Table 3. Three results of the total times of weights updating, in which numbers of iterative learning of original and extracted sets were 1,000 and 4,000, respectively

Extraction	Times of weights updating
Eq. (44)	27,050,036
Eq. (45)	29,130,464
Eq. (46)	31,901,186

Figure 16. Learning result in the case of using the extraction condition given by Eq. (44) in Table 3

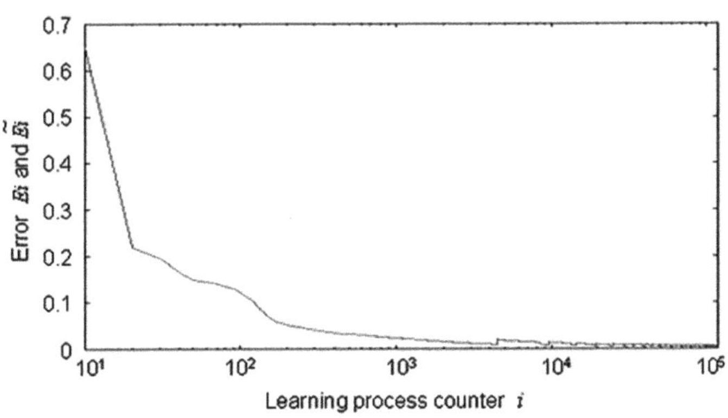

networks (CNNs). In the case of the NN for the inverse kinematic model of the industrial robot, the convergence time to a desired allowable error was took several hours. Note that no GPU was used in the training.

Incidentally, long short term memory (LSTM) approach was proposed in 1997 as an extended network system of recurrent NNs (Hochreiter, 1997). The LSTM is one of effective models or structures that can well deal with time series data such as generation of sentence and conversation, recognition of sound, voice and movie. The most promising characteristic is the ability to learn the so-called long term independencies. However, as for the problem addressed in this chapter that deal with robotic inverse kinematics, it seems that the LSTM technique reflects over specs and it is more complicated to be implemented compared to standard NN.

CONCLUSION

In this chapter, efficient learning technique of inverse kinematics using neural networks with effective ability to update NN weights has been developed for a serial link structure and industrial robot. Generally, in supervised learning a desired training data set should be prepared in advance to help neural networks learn a relation among multi inputs and outputs. The training data set consists of multiple pairs of input and output vectors. Each input vector is introduced to the input layer for forward computation and the desired output vector is compared with the yielded vector from the output layer. The time required for the learning process of the neural networks depends on the number of total weights in the neural networks and that of the input-output pairs in the training data set.

In the proposed learning process, input-output pairs, which have had the worse errors in the learning process, are extracted from the original training data set and form a new temporary training data set. From the following iteration of the learning, the temporary training data set is applied to the neural network instead of the original data set. That means only pairs with worse errors are used for updating weights until the mean value of errors decreases to the level of the desired value. Once the learning process is conducted using the temporary training data set, the original training data set is applied again instead of the temporary one. By repeatedly using these two training data sets alternately, the convergence time could be efficiently reduced. The effectiveness was proved through simulation experiments using a kinematic model of a leg with four-DOFs. In the conventional approach when the tip of a leg with a serial link structure is controlled kinematically in Cartesian space, an inverse Jacobian matrix is needed to generate velocities in joint space. However, the calculation of the inverse Jacobian is not easy. Hence, the use of neural networks based inverse kinematics was introduced to overcome the complexity.

In addition, neural networks with the effective weights update ability have been applied to the inverse kinematics of an industrial robot with six-DOFs. In the proposed learning process, after the learning is progressed, e.g., 1000 iterations, input-output pairs with worse errors are extracted from the original training data set and form a new temporary set. From the next iteration, the temporary set is applied instead of the original set. After the learning conducted using the temporary set, the original set is applied again instead of the temporary set. The effectiveness of the proposed method was also proved through simulation experiments using the kinematic model of an industrial robot MOTOMAN HP20, i.e., the convergence time could be efficiently reduced by alternately applying the two kinds of sets.

In future work, better conditions for extraction from original training data set will be investigated. Also, how to systematically determine the times of iterative learning for original training data set and the extracted one will be considered in order to reduce the load of weights updating.

REFERENCES

Aggarwal, L., Aggarwal, K., & Urbanic, R. J. (2014). Use of artificial neural networks for the development of an inverse kinematic solution and visual identification of singularity zone(s). *Procedia CIRP*, *17*, 812–817. doi:10.1016/j.procir.2014.01.107

Almusawi, A. R. J., Dülger, L. C., & Kapucu, S. (2016), A new artificial neural network approach in solving inverse kinematics of robotic arm (Denso VP6242). Computational Intelligence and Neuroscience, 2016(5720163).

Carlos, L.F., Jesus, H.B., & Alanis, A.Y. (2018). Inverse kinematics of mobile manipulators based on differential evolution. *International Journal of Advanced Robotic Systems*, *15*(1).

Corke, P. (1996). A robotics toolbox for MATLAB. *IEEE Robotics & Automation Magazine*, *3*(1), 24–32. doi:10.1109/100.486658

Duka, A. V. (2014). Neural network based inverse kinematics solution for trajectory tracking of a robotic arm. *Procedia Technology*, *12*, 20–27. doi:10.1016/j.protcy.2013.12.451

Duka, A. V. (2015). ANFIS Based Solution to the Inverse Kinematics of a 3DOF Planar Manipulator. *Procedia Technology*, *19*, 526–533. doi:10.1016/j.protcy.2015.02.075

Gan, Y., Dai, X., & Li, D. (2013). Off-line programming techniques for multi-robot Cooperation System. *International Journal of Advanced Robotic Systems*, *10*(7), 1–17. doi:10.5772/56506

Hoang, G., Min, J. H., Lee, G. M., Jun, B. H., Kim, H. K., & Kim, S. B. (2014, October), Omni-directional walking control for a six-legged robot using differential kinematics algorithm, In *Proceedings of 2014 14th International Conference on Control, Automation and Systems (ICCAS 2014)* (pp. 1163–1168). Academic Press. 10.1109/ICCAS.2014.6987735

Hochreiter, S., & Schmidhuber, J. (1997). Long short-term memory. *Neural Computation*, *9*(8), 1735–1780. doi:10.1162/neco.1997.9.8.1735 PMID:9377276

Koker, R. (2013, February). A genetic algorithm approach to a neural network-based inverse kinematics solution of robotic manipulators based on error minimization. *Information Sciences*, *222*, 528–543. doi:10.1016/j.ins.2012.07.051

Koker, R., Oz, C., Çakar, T., & Ekiz, H. (2004, December). A study of neural network based inverse kinematics solution for a three-joint robot. *Robotics and Autonomous Systems*, *49*(3), 227–234. doi:10.1016/j.robot.2004.09.010

Maeda, Y., Fujiwara, T., & Ito, H. (2014, September), Robot control using high dimensional neural networks. In *Proceedings of SICE Annual Conference 2014 -International Conference on Instrumentation, Control, Information Technology and System Integration-* (pp. 738–743). Academic Press. 10.1109/SICE.2014.6935220

Mao, Z., & Hsia, T. C. (1997, January). Obstacle avoidance inverse kinematics solution of redundant robots by neural networks. *Robotica*, *15*(1), 3–10. doi:10.1017/S0263574797000027

Nagata, F., Inoue, S., Fujii, S., Otsuka, A., Watanabe, K., & Habib, M. K. (2015, August). *Learning of inverse kinematics using a neural network and its application to kinematic control of position-based servo motor*. Paper presented at The 2015 World Congress on Advances in Aeronautics, Nano, Bio, Robotics, and Energy (ANBRE15).

Nagata, F., Otsuka, A., Sakakibara, K., Watanabe, K., & Habib, M. K. (2013, September), Experiment Systems Using Three Types of Motors for Biomimetic Machine Research. In *Proceedings of SICE Annual Conference 2013 -International Conference on Instrumentation, Control, Information Technology and System Integration* (pp. 2711–2717). Academic Press.

Nagata, F., & Watanabe, K. (2002, September), Learning of contact motion using a neural network and its application for force control. In *Proceedings of the 4th Asian Control Conference (ASCC2002)* (pp. 420–424). Academic Press.

Nagata, F., & Watanabe, K. (2016). Neural network-based inverse kinematics for Motoman HS20 and its efficient learning method. *Journal of the Institute of Industrial Applications Engineers*, *4*(4), 166–171. doi:10.12792/JIIAE.4.166

Tejomurtula, S., & Kak, S. (1999). Inverse kinematics in robotics using neural networks. *Information Sciences*, *116*(2-4), 147–164. doi:10.1016/S0020-0255(98)10098-1

Xia, Y., & Wang, J. (2001, February). A dual neural network for kinematic control of redundant robot manipulators. *IEEE Transactions on Systems, Man, and Cybernetics. Part B, Cybernetics*, *31*(1), 147–154. doi:10.1109/3477.907574 PMID:18244777

Xu, W., Mu, Z., Liu, T., & Liang, B. (2017). A modified modal method for solving the mission-oriented inverse kinematics of hyper-redundant space manipulators for on-orbit servicing. *Acta Astronautica*, *139*, 54–66. doi:10.1016/j.actaastro.2017.06.015

Zhou, Z.; Guo, H.; Wang, Y.; Zhu, Z.; Wu, J. & Liu, X. (2018, August). Inverse kinematics solution for robotic manipulator based on extreme learning machine and sequential mutation genetic algorithm. *International Journal of Advance Robotic Systems*.

KEY TERMS AND DEFINITIONS

Back Propagation Algorithm: Is the most famous and reliable algorithm based on the concept of the steepest decent method and can train weights between each neuron in neural networks based on errors. The errors are defined as the quantity between the output of neural network and teaching signal prepared in advance.

Industrial Robots: Have progressed remarkably as one of the powerful automation machines and been applied to many tasks such as welding, paining, handling, assembling and so on. General industrial robots have a serial link structure with five, six or seven DOFs pursuing the function of a human arm, so that more flexible motion than NC machine tools can be realized.

Kinematics: Quantitatively handles the relation between joint angles in joint coordinate system and the corresponding position/orientation elements of the arm tip in Cartesian coordinate system, without considering the masses of those parts nor the torques and forces that drive the joints.

Neural Networks: Are typically composed of multiple layers and the signal traverses from the input layer to the output layer of neurons. Trained neural networks can approximate an arbitrary nonlinear function.

Serial Link Structure: Is the typical configuration of articulated-type industrial robots. Each link of the industrial robot is connected with a servo motor-driven joint for rotational motion. Flexible position and orientation motion with multi-DOFs is characterized.

Weights Update Method: Multi-layered neural networks have to be trained to acquire the non-linear relation between input and output teaching signals. The training means weights update process to enable the neural networks to produce the output teaching signals according to the input teaching signals. In this chapter, a novel efficient weights update method is proposed.

Chapter 9
Hybrid Dynamic Modelling and Bioinspired Control Based on Central Pattern Generator of Biped Robotic Gait

Luis Miguel Izquierdo-Córdoba
https://orcid.org/0000-0003-4520-1482
University of Campinas, Brazil

João Maurício Rosário
University of Campinas, Brazil

Darío Amaya Hurtado
Nueva Granada Military University, Colombia

ABSTRACT

This chapter presents the theoretical foundations and methodology to develop a bioinspired hybrid control architecture for a biped robotic device that reproduces gait and human motor control strategies with the ability to adapt the trajectory to environmental conditions. The objective is to design robotic devices (such as exoskeletons), through the functional integration of hybrid dynamic system modeling (event-driven and continuous dynamics) with efficient and robust conventional control techniques and bioinspired control algorithms, with a near-natural human gait pattern. The human gait cycle is modeled as a hybrid dynamic using a finite state machine (FSM). The gait trajectories are to be generated in such a way that they will be capable of adapting to disturbances in the path followed by the robotic device; this will be achieved using a neuronal oscillator that simulates the behavior of a central pattern generator (CPG).

DOI: 10.4018/978-1-7998-1382-8.ch009

1. INTRODUCTION

Anthropomorphic robotics is an interdisciplinary field of study that integrates concepts of biomechanics, anatomy and human cognition, together with traditional robotic areas such as mechatronics and control theory. Current interest in this field presents attractive challenges, particularly in bipedal robotics, to design robots with a walking pattern that is as close to natural dynamics as possible.

In general, considering an application, for example, a lower limb exoskeleton, the device is required to operate in parallel to the human body or to mimic with high precision its movements, which generates the need to consider control techniques to achieve these goals.

Control in bipedal robots is challenging because they are performed interacting with dynamic environment and exhibit time variant dynamics. Bipedal robot locomotion requires the development of new technologies and more sophisticated techniques to overcome wide range of issues, such as motion control, torque control, stability, trajectory generation, generation of different gait patterns depending on the environment conditions, sensing and perception (Habib, Liu, Watanabe, & Izumi, 2007). The design of biped robots has traditionally been based on constant torque control of the joints with a local stability criterion, usually based on the Zero Moment Point (ZMP) approach, which lead to patterns of slow and jerky movement, and therefore do not reproduce the trajectories of joints exhibited in normal gait, the which makes the non-biomimetic gait (Batista, 2017).

Anthropomorphic robots can theoretically operate in many human environments, and to take advantage of their potential, robots must be able to operate and navigate in complex and unstructured environments. This requires a flexible walking system that can adapt quickly to new situations (Buschmann, Ewald, Ulbrich, & Buschges, 2012). The progress of humanoid robotics, therefore, requires the study and understanding of the human neuromusculoskeletal system (NMS) along with human motor control paradigms, and propose bioinspired control strategies to generate better performance rhythmic movements, and adaptable to the environmental conditions.

Findings from neurobiology suggest that human gait is a dynamic, partly self-stabilizing, process based on the interaction of the biomechanical structure with the neural control. The coordination of this process is a very complex problem and has been proposed in several kinds of research, involving a hierarchy of levels, where the lowest, for example, interactions between the muscles and the spinal cord, are mostly autonomous and where the top-level, cortical control, acts timely, as needed. This requires an architecture of several nested sensorimotor loops, where the gait process provides feedback signals to the person's sensory systems, which can be used to coordinate their movements (Manoonpong, Geng, Kulvicius, Porr, & Wörgötter, 2007).

Therefore, neurobiology plays a crucial role in hypothesizing engineering-inspired biological models. The integration between neuroscience and applications in robotics has developed innovations in rhythmic movement generation research by neural oscillators acting as Central Pattern Generator (CPG), enhancing temporal synchronization with the environment, and the online correction of spatial errors for robots. As an example, there is the Darpa Robotics Challenge, which was created to stimulate research on bipedal humanoid robots capable of intervening in dangerous areas to perform emergency maneuvers instead of humans (Suekichi, 2017). Control architectures based on neural oscillators, such as the Matsuoka oscillator, generate almost sinusoidal trajectories to perform stable rhythmic movements and adapted by sensorimotor couplings. Conventionally, the oscillator output is implemented in the framework of feedforward control, where it is directly regarded as any manipulated quantity, such as torque, rate of angle, angle, etc., for each active joint in a robot (Habib, Watanabe, & Izumi, 2009).

Incorporating these new bioinspired control laws with an event-based control system that triggers phase transitions in gait cycle based on the sensed walking state instead of only relying on time-based reference trajectories, increases the robustness of biped robots on uneven or unstructured terrain (Buschmann et al., 2012).

The goal of this chapter is to provide an overview and a comprehensive discussion about the integration between bioinspired and conventional control algorithms, and hybrid dynamic modeling to project a robotic mechanism able to reproduce the anthropomorphic biped gait as near as possible, adapting the kinematic and kinetic gait properties to environmental conditions.

2. BACKGROUND

The integration of robotics, biomechanics, and neuroscience has influenced the development of more sophisticated techniques in anthropomorphic and mainly assistive robotics, such as wearable robots, whose objective is to perform physical tasks that contribute to the well-being of the people (Kuteken, Batista, & Rosario, 2017). Research in this area, besides the solid knowledge of robotic techniques, also requires some background in anatomical behavior of the human body and the human neurological and cognitive systems (Pons, 2008). In this context, the bioinspired and biomimetic design is a particular issue for robotic bipedal gait. Principles relating to motion and motor control using neural oscillator networks have provided significant advantages. Biological systems can deal with unpredictable situations; they can adapt, learn and are robust to disturbances. It is therefore desirable to project robotic devices, e.g., with the same level of performance.

In concerning with bipedal gait using neural oscillators have been applied in several prototypes, as in the Honda ASIMO Humanoid (Chestnutt et al., 2005), the humanoid robot QRIO (Gen Endo, Jun Nakanishi, Jun Morimoto, & Cheng, 2005), HAL-5 full-body exoskeleton (Pons, 2008), US Defense Advanced Research Projects Agency (DARPA) (Pratt & Manzo, 2013), and other researches that proposed bioinspired control approaches based on nonlinear oscillator as in (Habib et al., 2007), (G. L. Liu, Habib, Watanabe, & Izumi, 2007), (G. L. Liu, Habib, Watanabe, & Izumi, 2008a), (Habib et al., 2009), (Gomes, Siqueira, & Gobbo, 2011), (Sugimoto & Morimoto, 2011), (C. J. Liu, Fan, Seo, Tan, & Goodman, 2012), (Abedi, Moghaddam, & Firoozabadi, 2014), (Lu & Tian, 2015), (Ikeda & Horie, 2016), (C. Liu, Wang, Goodman, & Chen, 2016), (Xia et al., 2017), (Santos, Alves, & Moreno, 2017), etc.

The Laboratory of Integrated Automation and Robotics (LAIR, UNICAMP, Brazil), of which the authors are part of, has developed several works in this field of biped robotic gait, with particular focus on kinematics and dynamics control (Puerta Barrera, 2017), (Buitrago Salazar, 2018), hybrid dynamic control (continuous and event-based) (Batista, 2017), and biomechatronic design of lower limb exoskeleton and artificial gait trajectories generation (Rosário, Suekichi Kuteken, & Izquierdo Cordoba, 2019). The author's current research is related to the integration of these works with biologically inspired control approaches and artificial intelligence algorithms to generate and control a walking pattern to biped robots.

3. DYNAMICS MODEL OF BIPED GAIT

Human gait refers to the sequence of movements performed by the lower limbs to move the Center of Gravity (CoG) of the body through the environment. It is an activity that requires coordination between the nervous and musculoskeletal systems, being composed of a succession of controlled imbalances providing locomotion while minimizing energy expenditure (Suekichi, 2017). Gait is distinguished from other forms of movement by always keeping at least one foot on the ground, reaching an average speed between 4 and 5 km/h.

According to (Vaughan, Davis, Christopher, & Connor, 2005) human locomotion programming occurs in supraspinal centers and involves the conversion of an idea into the pattern of muscle activity that is required for walking. The neural output resulting from this supraspinal programming can be thought of as a central locomotor command being transmitted to the brainstem and the spinal cord. The execution of this command involves two components:

- Activation of the inferior neural centers, which later determine the sequence of muscular activation patterns.

- Sensory feedback of muscles, joints and other receptors, which modifies the movements.

This interaction between the central nervous system, the peripheral nervous system, and the musculoskeletal effector system is illustrated in Figure 1.

3.1 Anthropomorphic and Biomechanics Concepts – Model in the Sagittal Plane

The first step in the dynamics characterization of human gait requires an understanding from biomechanics. Gait is the most common of all human movements and depends on the coordinated dynamic interaction between the motor system and external forces (Winter, 2009). The musculoskeletal system is the biological system to be analyzed for the synthesis of a biped robot mechanism. The body is approximated as the composition of anatomical segments (especially for the lower limb) considered like rigid bodies with properties like length, mass, inertial moment. The legs are approximated as a kinematic chain consisting of rigid links. This model is the one with the most significant degree of simplicity and attractive for dynamic and control studies.

Because biped dynamics is high order and nonlinear, it is difficult to understand unless some simplification is made. Therefore, the movement considered in this research is restricted to the sagittal plane because it is where the joints of the lower limbs develop a greater range of motion.

A fundamental part of the dynamic modeling of lower limbs during the gait cycle is the determination of the geometry of its components and the inertial properties. In the sagittal plane, the two-dimensional modeling of the segments is sufficient. For three-dimensional modeling, one has the option to make use of geometric approximations of the segments of the human body. Currently, with the introduction of body measurement technologies such as 3D scanners as well as the use of spectrometers and image processing technology, human segmentation tends to evolve in precision, allowing the acquisition of models with greater complexity and promoting a better adjustment of parameters for the same (Batista, 2017).

3.2 Event-Based Gait Cycle

The human gait constitutes a dynamic and cyclical sequence of trajectories performed by the lower limbs. The sequence of positions and orientations assumed by the leg segments that occur between the first contact of a foot with the ground to the same foot again touch the ground is called the gait cycle (Bruce, Hanrahan, Vaughan, Mackinnon, & Pandy, 1998). A gait cycle is traditionally divided into a

Figure 1. Interactions between the components responsible for generating the motor commands

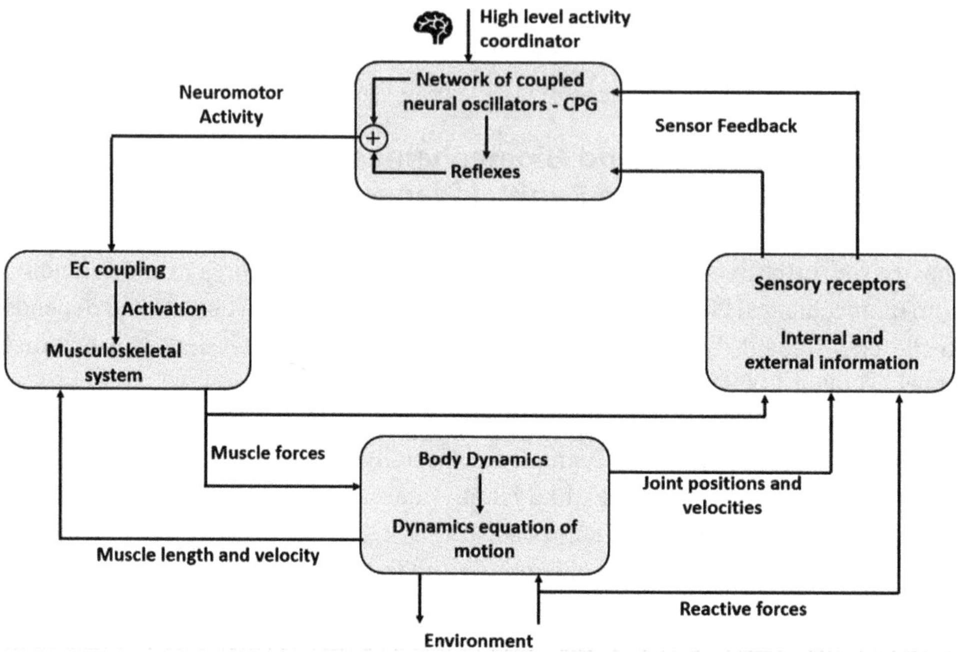

stance phase and a swing phase, where the lower limbs alternate in the function of body support and locomotion in space. According to (Vaughan et al., 2005) gait phases are conventionally described by the occurrence of events observed in foot movement as in Figure 2.

The stance phase is defined as the phase between the initial contact of the heel and the loss of contact of it with the ground so that the lower limb is supported on the ground. This phase lasts approximately 60% of the gait cycle and provides stable support. In the swing phase, the lower limb is flexed and in balance. This phase lasts about 40% of the cycle and provides the translation of the human body center of gravity (CoG).

3.3 Dynamic Model and Simulation

As defined above, the lower limbs are approximated as serial mechanisms consisting of rigid links connected by joints, each with three DOFs The geometry of a robotic mechanism is conveniently defined by attaching coordinate frames to each link. The model capable of meaningfully representing the leg during the swing phase is the triple pendulum, with joints in the hip, knee, and ankle, and with the referential

Figure 2. Stance and swing phases in the human gait cycle

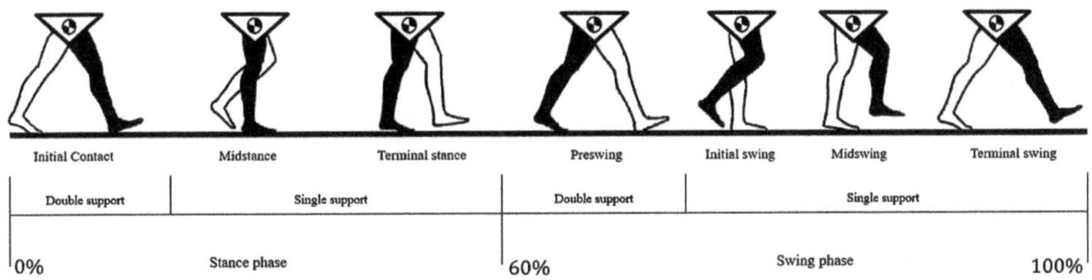

frame located at the hip joint. For the stance phase, the referential frame is located at the ankle joint, and the end-effector that moves through space is the CoG, and the model representing this kinematics chain becomes a triple inverted pendulum with the same three joints of the previous model. The geometric models of the triple pendulum for the swing phase and the inverted triple pendulum for the stance phase are described in Figure 3.

From the kinematic and dynamic model of the six-link biped robotic mechanism, each degree of freedom can be actuated by a torque that is proportional to the output of a Matsuoka oscillator. It is then necessary to know the trajectories that the hip, knee, and ankle joints must follow during a standard gait cycle.

Considering the geometric model for the swing phase, the hip joint position $\left(x_h, y_h\right)$ relative one absolute (inertial) frame is given by,

$$\begin{bmatrix} x_h \\ y_h \\ z_h \end{bmatrix} = \begin{bmatrix} x_0 \\ y_0 \\ 0 \end{bmatrix} \tag{1}$$

Knee position $\left(x_k, y_k\right)$, ankle position $\left(x_a, y_a\right)$ and toe position $\left(x_t, y_t\right)$ relative to the hip joint are computed as

$$\begin{bmatrix} x_k \\ y_k \\ z_k \end{bmatrix}_{swing} = \begin{bmatrix} x_h + L_1 \sin\theta_h \\ y_h - L_1 \cos\theta_h \\ 0 \end{bmatrix} \tag{2}$$

Figure 3. Forward geometric model (a) swing phase (b) stance phase

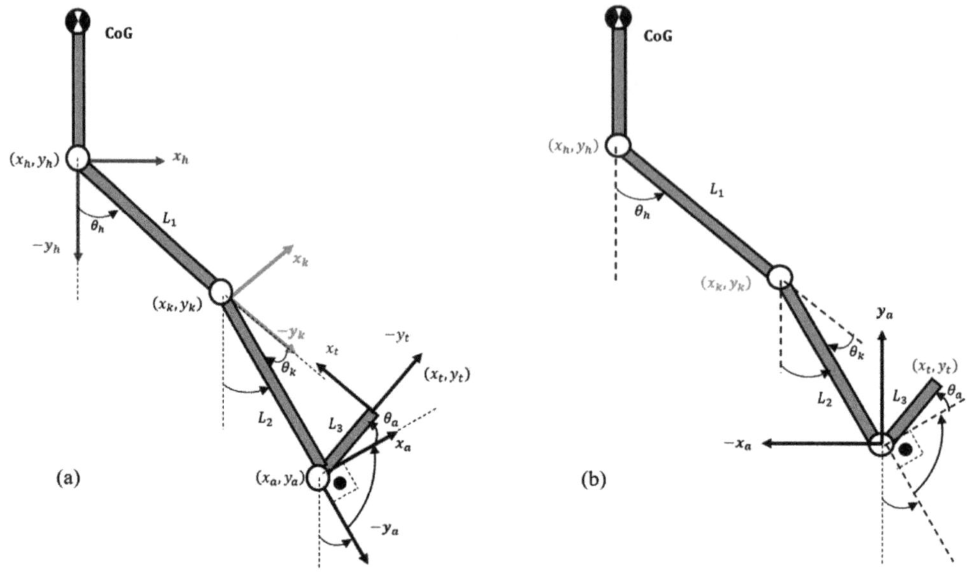

$$\begin{bmatrix} x_a \\ y_a \\ z_a \end{bmatrix}_{swing} = \begin{bmatrix} x_k + L_2 \sin\left(\theta_h - \theta_k\right) \\ y_k - L_2 \cos\left(\theta_h - \theta_k\right) \\ 0 \end{bmatrix} \tag{3}$$

$$\begin{bmatrix} x_t \\ y_t \\ z_t \end{bmatrix}_{swing} = \begin{bmatrix} x_a + L_3 \sin\left(\theta_h - \theta_k + \theta_t + 90^{\circ}\right) \\ y_a - L_3 \cos\left(\theta_h - \theta_k + \theta_t + 90^{\circ}\right) \\ 0 \end{bmatrix} \tag{4}$$

Angle θ_k is considered to have always a clockwise (negative) rotation. Angle θ_t of the foot is measured from a quadrature reference (90°) to the leg length L_2 as shown in Fig 3. As mentioned before, in the stance phase the reference system is moved to the ankle joint and its position in relation to the inertial frame is

$$\begin{bmatrix} x_a \\ y_a \\ z_a \end{bmatrix} = \begin{bmatrix} x_0 \\ y_0 \\ 0 \end{bmatrix} \tag{5}$$

Knee position (x_k, y_k), hip position (x_h, y_h) and toe position (x_t, y_t) relative to the ankle joint are computed as,

$$\begin{bmatrix} x_k \\ y_k \\ z_k \end{bmatrix}_{stance} = \begin{bmatrix} x_a - L_2 \sin(\theta_h - \theta_k) \\ y_a + L_2 \cos(\theta_h - \theta_k) \\ 0 \end{bmatrix} \tag{6}$$

$$\begin{bmatrix} x_h \\ y_h \\ z_h \end{bmatrix}_{stance} = \begin{bmatrix} x_k - L_1 \sin(\theta_h) \\ y_k + L_1 \cos(\theta_h) \\ 0 \end{bmatrix} \tag{7}$$

$$\begin{bmatrix} x_t \\ y_t \\ z_t \end{bmatrix}_{stance} = \begin{bmatrix} x_a + L_3 \sin(\theta_h - \theta_k + \theta_t + 90^{\circ}) \\ y_a - L_3 \cos(\theta_h - \theta_k + \theta_t + 90^{\circ}) \\ 0 \end{bmatrix} \tag{8}$$

3.4 Forward and Inverse Kinematics

It is clear that the inverse kinematics problem for serial-chain manipulators requires the solution of sets of nonlinear equations. Methods to solve the problem of the inverse geometric model are, in general, analytical and based on iterative numerical methods. However, since analytical methods may become extremely complicated, numerical methods are used to successively approximate the joint angles needed for a given endpoint desired in real time (Rosario, 2010).

There are several iterative numerical methods, including recursive methods, which use the forward geometric model and the inverse Jacobian matrix as denoted in Figure 4. This method considers the Jacobian as the representation of the infinitesimal change in the position vector $\delta \boldsymbol{r}$ as a result of an infinitesimal change in the generalized coordinates vector $\delta \boldsymbol{\theta}$. Linear approximation of the the Jacobian matrix can be defined as

$$\boldsymbol{J}(\boldsymbol{\theta}) = \frac{\delta \boldsymbol{r}}{\delta \boldsymbol{\theta}} \tag{10}$$

Figure 4. A recursive numerical method for solving the inverse kinematics problem

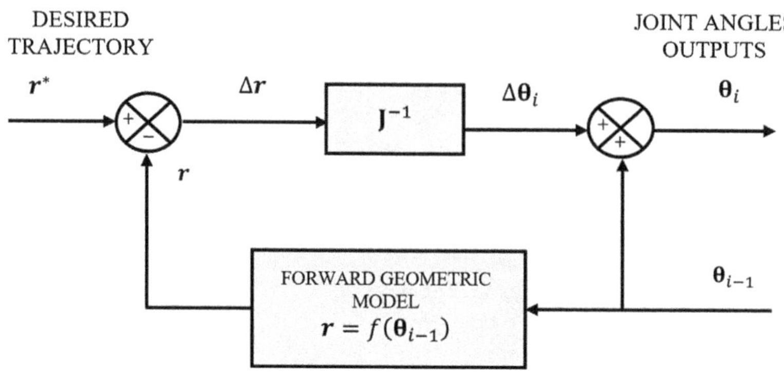

The recursive method using inverse Jacobian consists of a closed-loop algorithm which has as input the vector of desired positions (XY operational spatial data obtained in the laboratory of gait analysis laboratory), and as output the value of the joint angles as shown in Figure 5. The iterative sequence is defined in the following algorithm,

Algorithm to Solve the Inverse Kinematics Problem

Step 1. Load trajectory data $\boldsymbol{r}^* = \left[r_x^* \; r_y^*\right]^T$

Step 2. Set initial condition for the generalized coordinate vector $\boldsymbol{\theta}_0$

Step 3. Initialize iteration counter $i = 1$

Step 4. Set simulation time t_s

Step 5. for $i = 1$ to $i = \dfrac{t_s}{\Delta t}$

a. Calculate $\boldsymbol{r}$ with the forward geometric model using $\boldsymbol{\theta}_{i-1}$
b. Calculate inverse Jacobian matrix $\boldsymbol{J}^{-1}$ using $\boldsymbol{\theta}_{i-1}$
c. Calculate the infinitesimal change of position $\Delta\boldsymbol{r} = \boldsymbol{r}^* - \boldsymbol{r}$
d. Calculate the infinitesimal change of joint angles $\Delta\boldsymbol{\theta}_i = \boldsymbol{J}^{-1}\Delta\boldsymbol{r}$
e. Calculate joint angles for the goal position $\boldsymbol{\theta}_i = \Delta\boldsymbol{\theta}_i + \boldsymbol{\theta}_{i-1}$
f. Increase iteration counter $i = i + 1$

end.

In the simulations performed in this chapter, two DoF $\boldsymbol{\theta} = \left[\theta_h, \theta_k\right]^T$ are considered, and with the implementation of the recursive numerical method, the angles of the

Figure 5. Trajectories for hip, knee, and ankle joint angles in a gait cycle

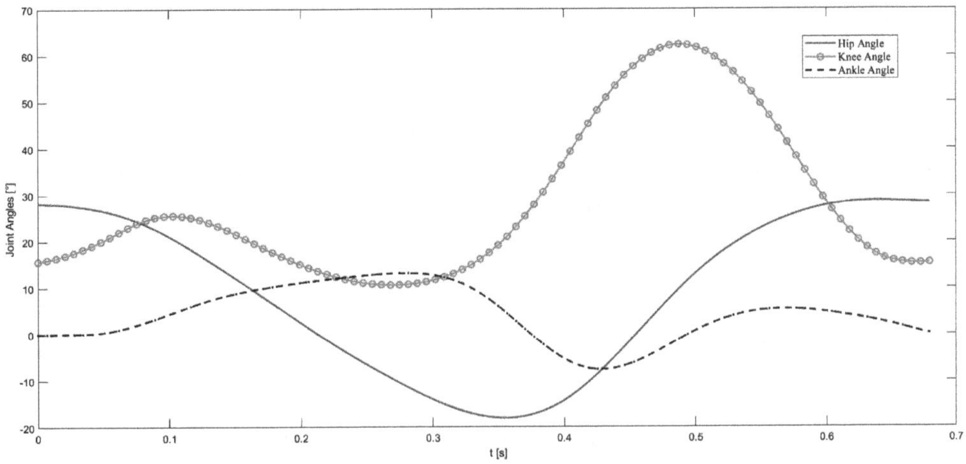

hip and knee joints are determined from the double pendulum model (swing phase), and the inverted double pendulum model (stance phase). The Jacobian matrix for the swing and support phases are given in equations 11 and 12, respectively.

$$\boldsymbol{J}\left(\boldsymbol{\theta}\right)_{swing} = \begin{bmatrix} L_1 \cos\theta_h + L_2 \cos\left(\theta_h - \theta_k\right) & -L_2 \cos\left(\theta_h - \theta_k\right) \\ L_1 \sin\theta_h + L_2 \sin\left(\theta_h - \theta_k\right) & -L_2 \sin\left(\theta_h - \theta_k\right) \end{bmatrix} \tag{11}$$

$$\boldsymbol{J}\left(\boldsymbol{\theta}\right)_{stance} = \begin{bmatrix} -\left(L_1 \cos\theta_h + L_2 \cos\left(\theta_h - \theta_k\right)\right) & L_2 \cos\left(\theta_h - \theta_k\right) \\ -\left(L_1 \sin\theta_h + L_2 \sin\left(\theta_h - \theta_k\right)\right) & L_2 \sin\left(\theta_h - \theta_k\right) \end{bmatrix} \tag{12}$$

Considering the lower limb as a robotic manipulator, a dynamic model can be obtained by the joint-space formulation derived using the Lagrangian approach. This formulation describes the relationship between the joint actuator forces and the motion of the mechanism. Lagrange formulation is given by,

$$\frac{d}{dt}\left(\frac{\partial L}{\partial \dot{q}_i}\right) - \frac{\partial L}{\partial q_i} = \tau_i, \tag{13}$$

where L is the difference between kinetic energy K and potential energy V, τ_i denotes the generalized torque at joint i, and q_i and $\dot{q}_i$ refer to the generalized coordinate and its time derivative, respectively.

According to (Rosario, 2010), the equation of the motor torque in a joint of a mechatronic system can be defined as a second order system, thus allowing its straightforward implementation in a control loop. Therefore, by organizing the terms of equation (13), we have that the torque at joint i is given by the approximate equation (14)

$$T_i = J_i \ddot{\theta}_i + B_i \dot{\theta}_i + T_{exti} \tag{14}$$

where J_i. is the effective inertia at the link i, B_i is the viscous friction and T_{exti} is the sum of the external forces and torques related to the system. Generally, T_{exti} is a nonlinear term composed of the torques produced by the centripetal force T_{p1}, the coupling torques between the links of the system T_{p2}, the torques produced by Coriolis force T_{p3}, and the gravitational effects T_{p4}, as expressed in equation (14) (Buitrago Salazar, 2018),

$$T_{exti} = \left(T_{p1} + T_{p2} + T_{p3} + T_{p4}\right)_i \tag{15}$$

The serial robotic mechanism shown in Figure 6 has the masses of the L_1 and L_2 links concentrated in their centers of mass (CoM), simplifying the dynamic modeling. The distances of the centers of mass of the links that constitute the mechanism are determined by L_{1CM} an L_{2CM}, respectively.

The definition of the Lagrangian determines the kinetic and potential energies for each link, and it is possible to calculate the approximate torque at the hip joint T_h and in the knee joint T_k in the form given in equation (14). Torque parameters of each joint of the serial mechanism are listed in Table 1.

The C_j $(j = 1,2\ldots,5)$ variables are defined as a function of geometric and mass parameters of the mechanism and are given by equation (16),

$$C_1 = M_1 L_{1CM}{}^2 + M_2 L_1^2 + I_1$$

$$C_2 = M_2 L_{2CM}{}^2 + I_2$$

Figure 6. A dynamic model for the 2 DoF serial mechanism

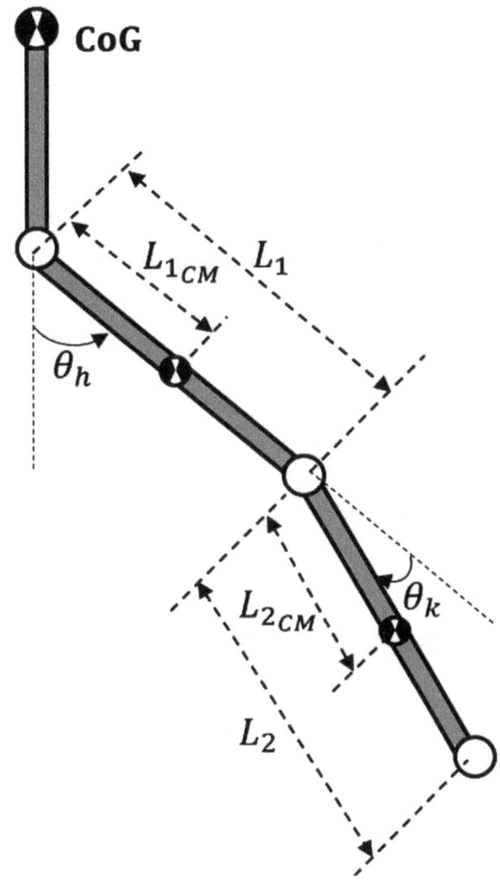

$$C_3 = M_2 L_1 L_{2CM}$$

$$C_4 = -M_1 g L_{1CM} - M_2 g L_1$$

$$C_5 = -M_2 g L_{2CM} \tag{16}$$

With the approximate dynamic model, it is possible to project a continuous control system to follow the angular trajectories of the joints of the serial robotic mechanism in both the stance and the swing phases. For this purpose, kinematic database of joint trajectories was obtained with marker and image processing at the Laboratory of Instrumentation for Biomechanics at the Faculty of Physical Education (Unicamp).

Table 1. Parameters for calculating the torques at hip joint T_h and knee joint T_k

Parameters	T_h	T_k
J_i	$C_1 + C_2 + 2C_3 \cos\theta_k$	C_2
B_i	0	0
T_{p1}	$-C_2 - C_3 \cos\theta_k$	$-C_2 - C_3 \cos\theta_k$
T_{p2}	$C_3 \sin\theta_k$	$C_3 \sin\theta_k$
T_{p3}	$-2C_3 \sin\theta_k$	0
T_{p4}	$C_4 \sin\theta_h + C_5 \sin\left(\theta_h - \theta_k\right)$	$-C_5 \sin\left(\theta_h - \theta_k\right)$

3.5 Equilibrium and Stability – ZMP Scope

Human gait has been defined as the translation of the body Center of Gravity (CoG) in the space with minimal expenditure of energy. Walking can be analyzed as a sequence of controlled falling periods, where the system recovers stability by placing its feet at appropriate positions. Thus, for biped robots, falling towards the current walking direction is necessary, since it has to place its feet further in the same direction to prevent instability (Buschmann et al., 2012).

The body has both the CoG and the center of pressure (CoP), and while the CoP can be directly measured with force plates or pressure mats, the CoG is indirectly estimated using multi marker-based analysis with motion tracking software. The CoP is related to the CoG, in that the CoP works to keep the CoG within the base of support. A widely approach used in simplified robot model is to generate CoG trajectories by solving the equations of motion for a given zero moment point (ZMP). ZMP is a mathematical formulation to find a stable point that causes equilibrium of action and reaction momentum (momentum equal to zero). It can be approached using a Single Linear Inverted Pendulum Model (SLIPM). In the SLIPM mathematical equation, there are two primary components which are the position vector of CoG and Acceleration (Linear) of CoG (Pristovani, Henfri, Sanggar, & Dadet, 2018), as illustrated in Figure 7.

Current locomotion control strategies for bipedal robots use the Zero Moment Point Tracking (ZMP) algorithm. Even though this control strategy is a simple and

Figure 7. Schematic biped model of ZMP

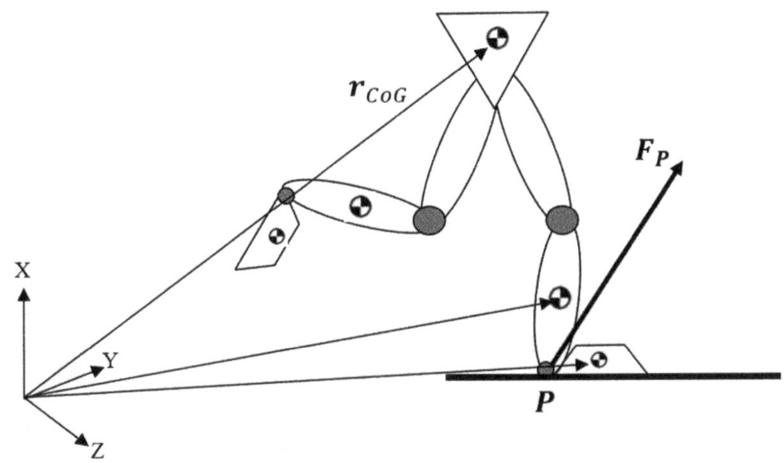

straightforward methody to perform bipedal robots, it requires a precise modeling and joint performance with high gain of joint control for successful locomotion (Liu et al., 2007).

3.6 Continuous Control System

For hardware design and detailed dynamics analysis, a model that includes motor and gear dynamics is needed for projecting a closed-loop continuous control system. Local feedback at each joint of the biped robot is projected by considering the dynamic model of each joint according to equations (14), (15) and (16). Control system with PID controller and PID in the RST form (continuous and discrete) are tuned for the dynamic model (transfer function) of the hip and knee robot joints.

Time-based trajectories for the lower limb joints, standard anthropomorphic parameters (mass, lengths of links, and CoM distances of each link) and DC motor parameters found in the literature, are used in such process of parameter controller syntonization. Figure 8 shows the closed-loop controls for hip and knee joints.

Current locomotion control strategies are based mainly on trajectory plans that prescribe the desired state of movement as a function of time [1]. Most humanoids developed use the target zero moment point (ZMP) tracking algorithm for bipedal locomotion to describe system stability and to control a system by following a target ZMP trajectory [2]. While this control strategy is a simple, straightforward way to realize biped robots, it requires precise modeling and precise joint actuation with high joint control gain to achieve successful locomotion.

Figure 8. Continuous control system for the dynamic model of hip and knee joint

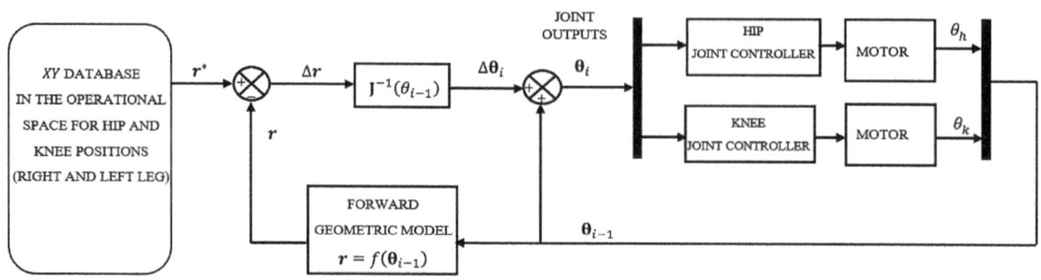

4. HYBRID DYNAMIC CONTROL STRATEGY FOR HUMAN GAIT

The most evident division of the gait cycle is between the swing and the stance phase, in which not only the system adopts a different trajectory for each phase, but the lower limb adopts a distinct dynamic behavior; it perform like a double pendulum in the swing phase, and as an inverted double pendulum in the support phase. Two different physical models generate a discontinuity. This functional discontinuity between two continuous dynamic models raises the need to use, along with continuous modeling, the theory of discrete event system which qualifies the modeling task as a hybrid.

The hybrid control system consists of event-based control and continuous control. The event-based controller acts as a supervisor for a continuous controller, detecting the gait stage and determining the trajectory that the continuous controller must be tracking (Batista, 2017). Robotics has a great interest in this kind of control strategies since they allow discontinuities and nonlinearities of systems commonly described employing differential equations, to be modeled with DEVS (Discrete-Event System) formalism (van der Schaft & Schumacher, 2000).

4.1 Gait Event Detection

In gait analysis, the entire cycle is generally studied as a cyclical occurrence of events. The gait modeling requires initially determining the occurrence of the main events, and, from these events, define the transitions that govern the behavior of the model. Event detection can be performed from kinetic and kinematic data, offline and online. In general, the gait cycle is divided by the support and balance phases. The timing of heel strike (HS) and toe-off (TO), are the events that determined the transitions between stance and swing, and the accurate identification of these phases is essential when analyzing bipedal gait.

Force plate recordings are routinely used to identify these events. A simple threshold on the force level can accurately provide the timing of HS and TO, determining the sample at which the vertical force rises above some threshold F_{min} level for heel-strike, and falls below that F_{min} level again for toe-off (TO). This method is known as the golden method and is generally restricted to a gait laboratory setting and the number of available force platforms (O'Connor, Thorpe, O'Malley, & Vaughan, 2007).

Pressure sensitive switches placed at the heel and toe, have often been used as an alternative to determining the onset of the stance and swing phases (Hreljac & Marshall, 2000). Assuming that the foot switches are appropriately positioned, and synchronized with other data collection methods, accurate data regarding heel-strike and toe-off timing could be obtained.

Events identification depends on the instrumentation available. Sensors such as accelerometers, gyroscopes, and force sensor resistor are used to estimate the HS and TO events in an online manner. Algorithms for determining heel-strike and toe-off times during normal walking using only available kinematic data are useful when no advanced instrumentation or equipped laboratories are available. Its accuracy can be evaluated by comparing the results to the "gold standard" method (Hreljac & Marshall, 2000).

In gait event detection using sampling kinematics data, a large variety of kinematic algorithms exist, which use marker position data and derivatives of this data to identify gait events. In this case, it is implemented an algorithm developed by (O'Connor et al., 2007) called foot velocity algorithm (FVA). The FVA uses data from the heel and toe markers and identifies features in the vertical velocity of the foot which corresponds to the gait events.

By convention, initial foot contact is generally taken as the starting point of a complete gait cycle and marks the beginning of the stance phase. Termination of foot contact marks the beginning of the swing phase. In normal gait, these events correspond to heel strike (HS) and toe-off (TO), respectively. The timing information of these events is used in the analysis of temporal gait parameters such as stride time and periods of single and double support and allows for the time normalization of data per gait cycle.

The FVA algorithm only requires the vertical information of the foot center, which is obtained by calculating the midpoint of the heel and toe marker positions. With the vertical position data of the foot center, its vertical velocity is computed by taking the first derivative using finite difference equations (O'Connor et al., 2007). It can be noted that the cyclic movement of the gait is described exclusively on the y-axis, while in the x-axis certainly demonstrates that the advancement in space happens almost linearly during the swing phase followed by the stance phase.

This vertical velocity of the foot center has a simple characteristic shape repeated for each gait cycle during normal gait. In Figure 9 it can be noted that the vertical

Figure 9. Vertical velocity of the foot center for the right and left leg in a gait cycle

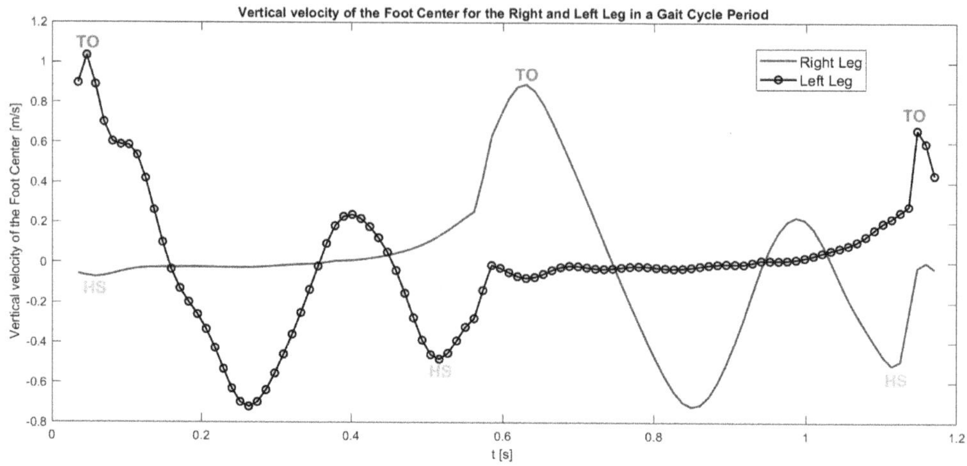

velocity curve has easily identifiable features – maximum and minimum values– which allow to identify the correct HS and TO times automatically for each leg.

The timing of the TO events are identified by searching for the maximum values in the vertical curve within a window of gait cycle. The minimum values in the signal are identified using a smaller window size, 10% of the gait cycle period, which provides a set of possible HS times. The correct minimum for the HS event is identified by imposing a constraint on the possible HS times: that the heel marker must be close to the ground at the time of heel strike. Data were processed offline, and the FVA algorithm was implemented in MATLAB®.

Even though the above solution is simpler, more sophisticated and accurate algorithms have been developed to identify gait events, based on Support Vector Machines, Artificial Neural network, intelligent fuzzy computational algorithms, etc.

4.2 Event-Based Control Strategy for Gait Using Finite State Machine (FSM)

As stated before and illustrated in Figure 10, the two leading states in the gait cycle correspond to:

- Right Leg Stance - Left Leg Swing: RL_ST – LL_SW.
- Right Leg Swing - Left Leg Stance: RL_SW – LL_ST.

Also considered are the auxiliary states, which correspond to the states of double support, as shown in Figure 11.

Figure 10. Leading states in the human gait cycle

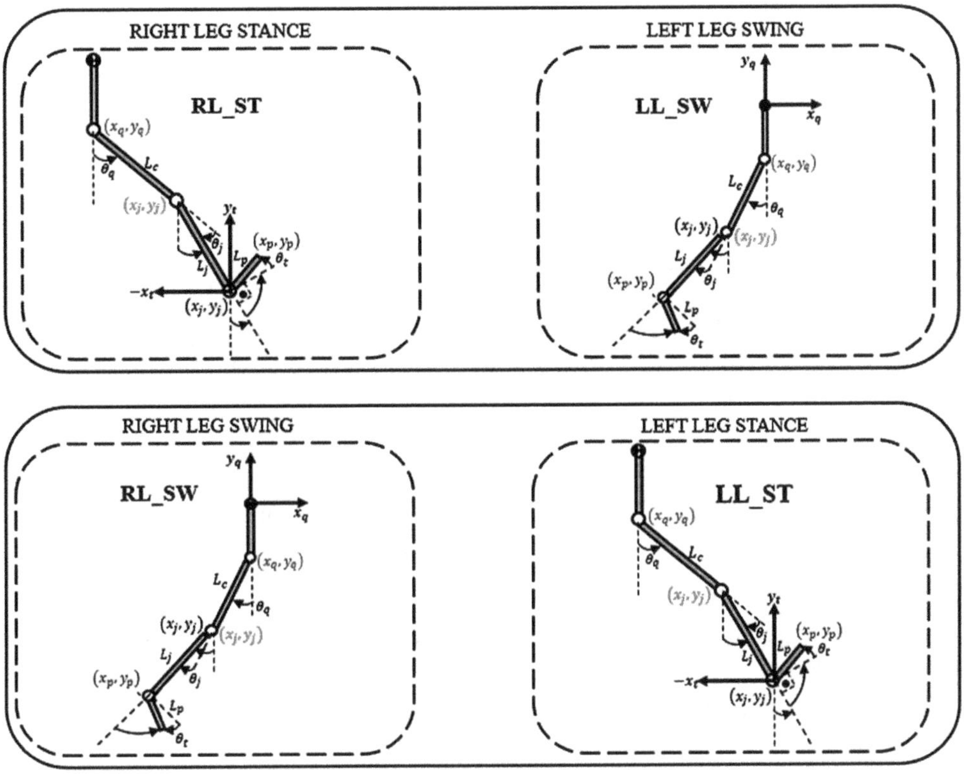

- Double Stance Right: DST_R - right leg is ahead.
- Double Stance Left: DST__L – left leg is ahead.

Table 2 defines the states considered in the event-based model of the human gait cycle.

The FSM gait model in this work is implemented in Simulink/Stateflow® A finite state machine (FSM) is a model that represents the behavior of a reactive system also known as an event-driven system. A reactive system makes a transition from one state (mode) to another, if the condition defining the change is “true” (MathWorks, 2018) FSM for gait cycle considered four states, as illustrated in Figure 11, has the transition conditions defined in Table 3. The Heel Strike (HS) and the toe-off (TO) events determine the transition conditions as stated in figure 9.

The HS and TO detection in each leg during a gait cycle is performed with the FVA algorithm explained above, which uses the vertical velocity of the center foot. Input signals for the algorithm are the displacement–time data for heel and

Figure 11. FSM with two main states and two auxiliary states for gait cycle

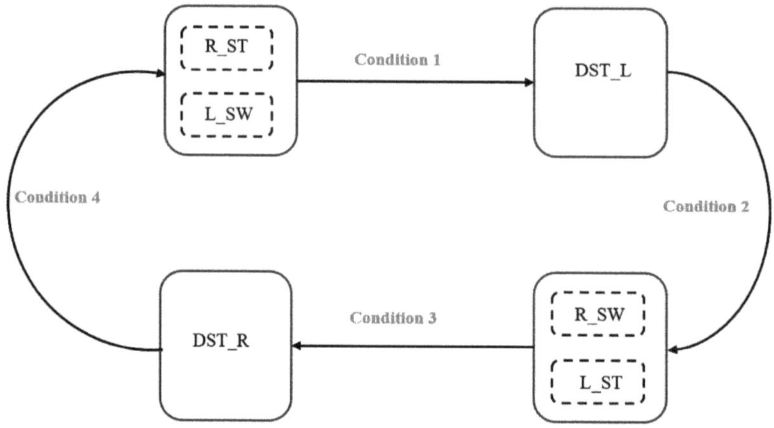

toe joints; the FVA algorithm identifies minimum/maximum values values in the velocity curve and outputs the time of TO and HS occurrence.

Each time instant t_i will depend on the standard gait speed considered. Thus, using event-triggered transitions between stance and swing phases implemented in an FSM instead of merely following a continuous time-based trajectory may give more robustness and maneuverability to the biped robot.

Table 2. Phases for right and left lower limb according to the considered state in the gait cycle

State	Right Leg	Left Leg
1. R_ST – L_SW	Stance	Swing
2. DST_L	Anterior Stance	Posterior Stance
3. R_SW – L_ST	Swing	Stance
4. DST__R	Posterior Stance	Anterior Stance

Table 3. Event - define state transitions

Condition 1	TO_L	Toe -Off Left
Condition 2	HS_L	Heel Strike Left
Condition 3	TO_R	Toe- Off Right
Condition 4	HS_R	Heel Strike Right

Figure 12. Sub-states for stance and swing phases in a standard gait cycle period

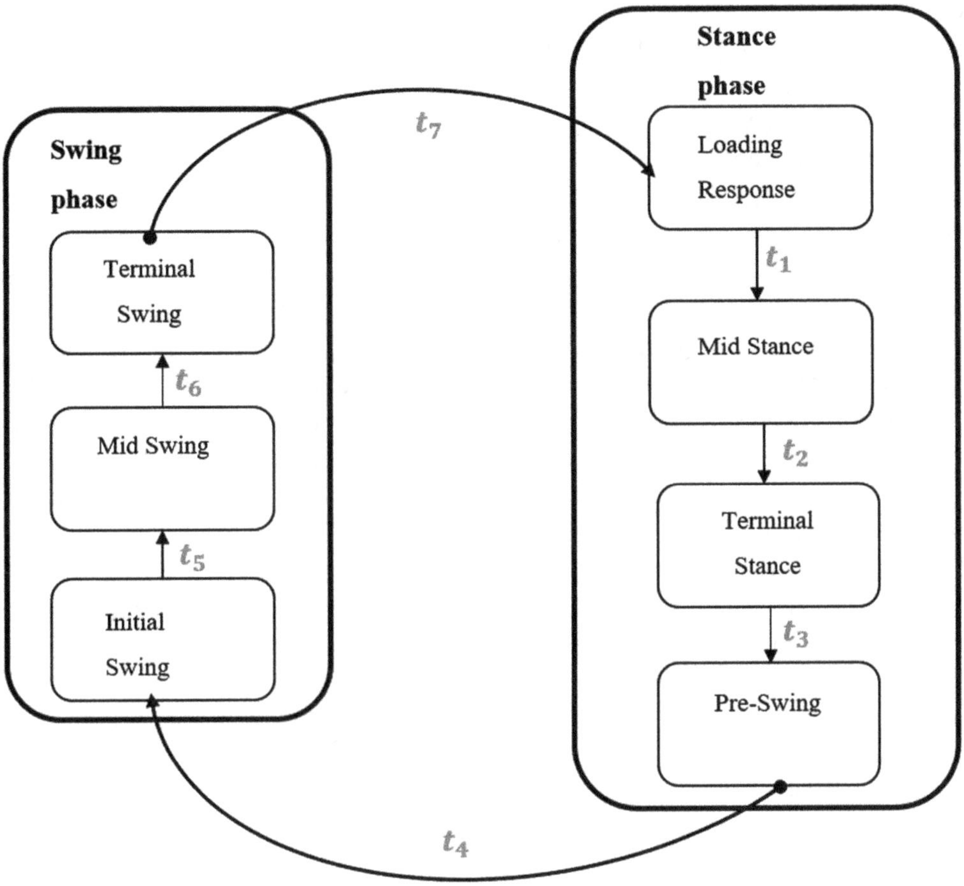

5. NEURAL OSCILLATOR APPLIED TO BYPED ROBOTIC GAIT CONTROL

The control architecture proposed in this study considers the generation of periodic rhythmic movements in the articulations of the 2 DoF robotic mechanism, based on a Matsuoka neural oscillator performing the role of a Central Pattern Generator (CPG). The neural oscillator generates endogenously rhythmic trajectories adapted by discrete sensorimotor couplings. The dynamic oscillator operation is modulated in real time by a higher level compensator based on information-motion laws. This two-stage architecture is similar to the organizational logic of human motor control (Siegler, Bardy, & Warren, 2010).

The human movement shows a level of robustness and dexterity still unmatched by the robotic mechanisms. A deep understanding of the origin of such robustness for replication in robotics remains a field of study nowadays, showing much prospecting. The integration between neuroscience and robotics has generated innovations in rhythmic movement generation using neural networks (neural oscillators), along with temporal synchronization with the environment, and online correction of spatial errors for robots (Avrin, 2017).

Rhythmic movements, such as gait, breathing, etc., are generated through spinal neural networks called Central Pattern Generators (CPGs). These networks provide robust and efficient control of rhythmic tasks without the need for complex dynamic internal models. CPGs are represented by different mathematical models such as oscillators, vector fields, artificial neurons, etc, and modulated by sensory signals (proprioceptive and exteroceptive sensors) and descending signals of the cerebral cortex to adapt behavior to the changing environment (Habib et al., 2007). Traditionally in robotics have been used for bipedal locomotion control strategies based on Zero Moment Point (ZMP), which generate slow and non-biomimetic motion patterns (Li, Vanderborght, Tsagarakis, & Caldwell, 2013), creating the need to implement bioinspired motion control strategies.

CPGs can produce patterns of rhythmic activity even when isolated from the rest of the nervous system, without the need for peripheral sensory feedback or commands from the supraspinal nervous system. Despite this, both sensory feedback stimuli and the hierarchically superior commands of the nervous system play essential roles in the functioning of motor activities coordinated by CPGs (Suekichi, 2017).

Models based on CPG are increasingly applied in the control of biped locomotion in humanoid robotic systems. According to (Ijspeert, 2008) and (Heliot, Azevedo, & Espiau, 2012), CPG-based control structures have some advantages for robots performing cyclic trajectories as,

- Robustness to small disturbances, due to the CPG's limit cycle behavior.
- Easy process of modulating amplitude and frequency of the trajectories, since CPGs have few control parameters. When properly implemented, a CPG-based control system reduces the dimensionality of the problem.
- CPG-based controllers are suitable for distributed architectures.

CPGs constitute a motor control structure with a low level of abstraction that, in isolation, would not take into account factors such as the intentionality of the actions and the environmental information. The highest level motor control depends on the descending commands generated in a deliberative way in the supraspinal nervous system. For the CPG-based control system to have the ability to make decisions, it is necessary a coordinate knowledge in the form of perceptual mental schemas (Suekichi, 2017).

5.1 Matsuoka Neural Oscillator

The CPG models can be designed with variable complexity. For this reason, the neural oscillator proposed by Matsuoka is widely used as a CPG model in neuroscience and robotics. It is a half-center oscillator constitute by two mutually inhibitory neurons described by a set of coupled nonlinear differential equations. The Matsuoka oscillator is considered a nonlinear oscillator with real biological inspiration and a level of abstraction sufficient to allow its use in computational models and to study the sensorimotor couplings involved in the control of rhythmic tasks. One essential feature of CPGs is their ability to synchronize their intrinsic oscillations with input signals (entrainment), thus allowing adaptability to change environments (Buschmann et al., 2012). However, it often requires a costly parameter tuning.

Nonlinear oscillators have crucial properties compared to linear systems or sinusoidal generators. In particular, they can produce oscillations in the form of limit cycles, which are stable orbits in phase space, and which are more robust than those of a linear system. The Matsuoka neural oscillator consists of two mutually inhibitory identical neurons that activate the flexor and extensor muscle groups of the limb in motion, as shown in Figure 13. Each Matsuoka's neural model consisted of two first-order coupled differential equations, one representing the membrane potential and the other the degree of fatigue of the neuron (Liu, Habib, Watanabe, & Izumi, 2008b),

$$\text{Neuron 1}\ \tau_r\dot{x}_1 = -x_1 - \beta v_1 - \rho y_2 - h_0\left[m\right]^+ + u,\ \tau_a\dot{v}_1 = -v_1 + y_1, \tag{17}$$

$$\text{Neuron 2}\ \tau_r\dot{x}_2 = -x_2 - \beta v_2 - \rho y_1 - h_0\left[m\right]^- + u,\ \tau_a\dot{v}_2 = -v_2 + y_2, \tag{18}$$

The mathematical neuron model has two state variables and constant parameters whose values must be selected appropriately The states $x_i\left(t\right)$ and $v_i\left(t\right)$ $\left(i{=}1{,}2\right)$. are, respectively, the membrane potential and the self-inhibition responsible for the fatigue phenomenon of the $i-ésimo$ neuron. Neurons are coupled by the terms $y_i\left(t\right) = \max\left(x_i\left(t\right),0\right)$.

The oscillator output y_{out} is given by equation (19), and represents the mean activation rate (or average pulse rates) of the CPG, shown in Figure 14. $m\left(t\right)$ represents the sensory input, with $\left[m\left(t\right)\right]^+$ and $\left[m\left(t\right)\right]^-$ defined by equations (20) and (21). The parameters that define the oscillator dynamics are the intensity of the

Figure 13. Matsuoka neural oscillator model

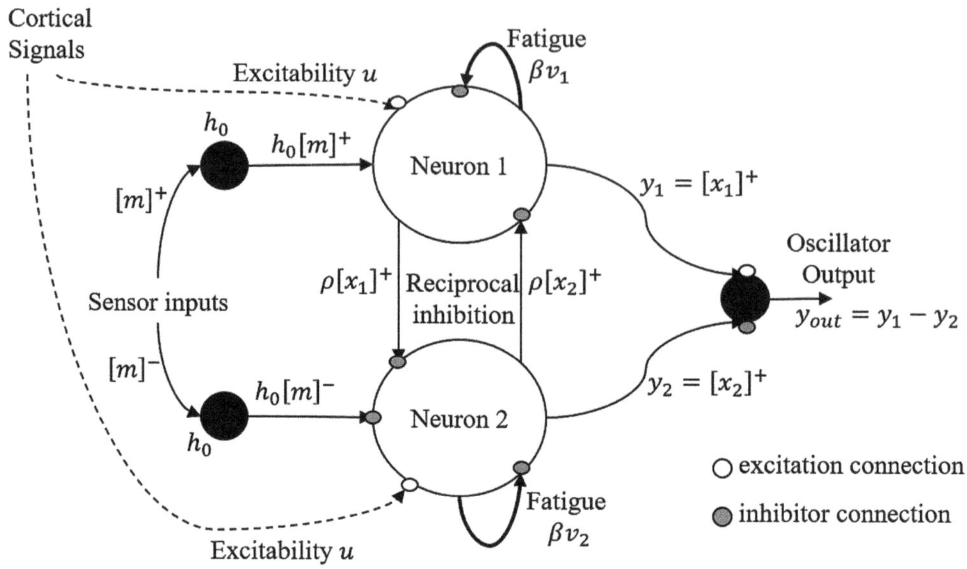

mutual inhibition ρ, and the intensity of the autoinhibition β (which determines the intensity of the fatigue effect). u is the CPG excitability parameter, which determines the output amplitude of the oscillator. h_0 is a constant gain in the input $m(t)$.Thtime constants τ_r and τ_a determine the response times of the states $x_i(t)$ e $v_i(t)$.

$$y_{out}(t) = \max(x_1(t),0) - \max(x_2(t),0) \tag{19}$$

$$[m(t)]^+ = \max(m(t),0). \tag{20}$$

$$[m(t)]^- = \max(-m(t),0) \tag{21}$$

The two neurons linked reciprocally, alternately inhibit and excite each other to produce oscillation as output. Such activity represents the alternating and mutually inhibition of the flexor and extensor muscles at joints during walking. Matsuoka neural oscillator has two modes of operation. In the first mode, called forced

Figure 14. Rhythmic output signals of matsuoka neural oscillator

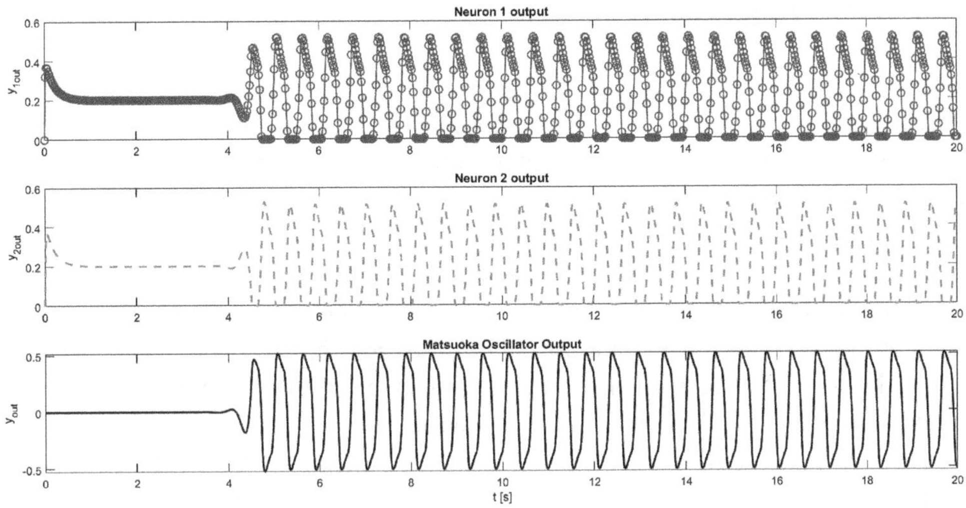

oscillation, the oscillator is driven by an external signal coupled as input. In the second mode, called autonomously, the oscillator independently produces (without stimulus input $m(t)$) a periodic pulse signal equato the oscillator pulse itself,

$$\omega_n = \frac{1}{\tau_a}\sqrt{\frac{(\tau_r + \tau_a)\beta}{\tau_r \rho} - 1} \tag{22}$$

Where ω_n is the output frequency of the oscillator in autonomous mode

A network of muttually coupled nonlinear oscillators to generate a rythmic gait pattern, can be structured with several configurations, assigning symmetrical and asymmetrical coupling mechanism between oscillators within the network structure under different possibilities of inhibitions and excitations, as in (Habib et al., 2007), (Liu et al., 2007) and (Liu et al., 2008a). The parameters, as well as the necessary interconnection coupling coefficients within the CPG network, in a mutually inhibited and coupled matsuoka oscillator, directly influence the generation of the rythmic signals in a bipedal robotic mechanism with four—link interconnection and 4 DoF.

Since the Matsuoka oscillator is nonlinear, its parameters are difficult to adjust to a specific task. Generally, optimization heuristic techniques are implemented to adjust the oscillator parameters, such as Particle Swarm Optimization (PSO), Genetic Algorithms (GA), reinforcement learning, etc (Avrin, Makarov, Rodriguez-Ayerbe, & Siegler, 2016).

5.2 Behavioral Dynamics Approach and Perception-Action Cycle

In neuroscience, behavioral dynamics suggests that states of the nervous system, from which cognition emerges, can be characterized by dynamic laws, and then human behavior is formalized using differential equations. This approach considers the motion to be the result of an emergent attractor of the agent-environment coupling, that is, a solution to which it always converges, from specific initial conditions of the set of coupled differential equations, which describe the dynamics of the individual and the environment interacting mutually (Warren & Fajen, 2008). For example, for the bipedal locomotion (Thelen & Smith, 2009) point out that the cyclic alternation between the movements of the two legs during a normal gait (with opposing movements) is a preferential behavior.

The notion of stability is essential in dynamic systems theory. System response to disturbances allows estimating stability. An equilibrium point will be considered stable if the system tends to return to its equilibrium position after a disturbance. The attractor corresponding to a biped step in phase opposition thus demonstrates excellent stability as it is maintained even after slight modifications in the pattern. It is possible to keep this pattern in phase opposition to walking with jumps, or in a different kind of surface. However, if these changes in the individual or the environment exceed a specific limit, it is possible that the behavior can no longer be maintained. Thus, if the slope of the walking surface exceeds a threshold, we must proceed from a biped to a quadruped gait (Avrin, 2017).

Among human movements, rhythmic actions seem to rely more on low-level action-perception couplings than on preprogrammed motor planes (Warren, 2006). Knowledge of such couplings can help simplify the control problem faced by robots.

CPGs receive proprioceptive and exteroceptive sensory information, which allows their activity to be adapted. CPGs generally receive two kinds of input that modify the dynamics of their neurons (in particular the amplitude and frequency of the generated action potentials). The first input, a signal from the descending cortex, can provide general instructions such as starting, stopping, or changing the speed. The second one is sensory feedback, usually proprioceptive, that modulates CPG behavior in response to environmental changes. Both types of signals are retransmitted to the CPG in the form of synaptic or neuromodulatory inputs.

Visual information is also coupled to neural oscillator networks, and these couplings are responsible for the temporal and spatial synchronizations observed during the performance of rhythmic visuomotor tasks.

This proposal is confronted with experimental results on the cyclical impact of balls, a benchmark well known to neuroscientists and dynamists due to their intrinsic dynamic properties. The concept of self-organization proposes that movement results from the interaction between the neural system, the musculoskeletal system, and the environment. The kinematics of motion is not only determined internally but is an emergent property of the agent-environment coupled system (Avrin, 2017).

The perception-action cycle, therefore, assumes the following form: the agent perceives the environment $\left(e\right)$ and extracts information $\left(i\right)$ from it used to act $\left(a\right)$. This action is expressed by the application of forces $\left(f\right)$ on the environment $\left(e\right)$. The scheme proposed by (Warren & Fajen, 2008) to summarize this cycle is given in Figure 15.

The cycle of perception-action described above will perform an essential role in understanding human movement. The goal is to identify and express control laws in the form of mathematical equations by coupling the agent and the environment dynamics. The term sensorimotor control or coupling law will be used concerning the relationships that bind control variables (control parameters, state variables) of a dynamic system and sensory information. Thus, the term sensorimotor control describes the sensory feedback that allows controlling and adapting the gait coupled with the environment dynamics.

5.3 Bioinspired Control Applied to a Hybrid Dynamic Simulation Using MATLAB/Simulink

CPGs are related to motor commands that define the kinematics of the human gait (position, velocity, and acceleration) as reference trajectory generators. These reference trajectories can be tracked by closed-loop controllers. By coupling the dynamics of the neural oscillator with proprioceptive and exteroceptive sensory information, the parameters of the discrete event dynamics in the FSM can be modulated to generate a stable and adapted joint trajectory. Thus, the frequency and amplitude of joint

Figure 15. Representation of the perception-action cycle
(adapted from Warren & Fajen, 2008)

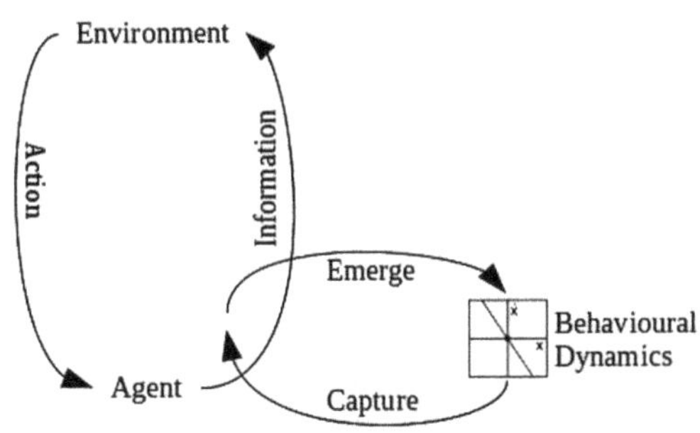

trajectories can be modified by readjusting of the state variables or the parameters of the Matsuoka Oscillator differential equations.

The hierarchy of a hybrid control system, consisting of a Matsuoka neural oscillator, an event-based control system and continuous controller to guarantee the execution of a biomimetic biped gait in a robot is proposed in Figure 16.

The asymptotic stability analysis of the CPG-based bipedal gait control can be performed through Poincaré maps, by calculating the eigenvalues of the Jacobian matrix of the nonlinear differential equations that define the Matsuoka Oscillator dynamics.

6. SIMULATION AND RESULTS

The event-based control of the gait cycle was implemented in Simulink/ Stateflow®, as shown in Figure 17. Kinematic data used in simulations were obtained in the gait laboratory tracking of trajectories of the body joints employing markers and image treatment.

The joints trajectories are the references of the continuous controllers. It was considered a PID and a PID in the RST form (continuous and discrete). In Figure 18 and Figure 19 is shown the simulation results obtained.

Figure 16. Hierarchical control architecture based on CPG, FSM and PID (RST) controller

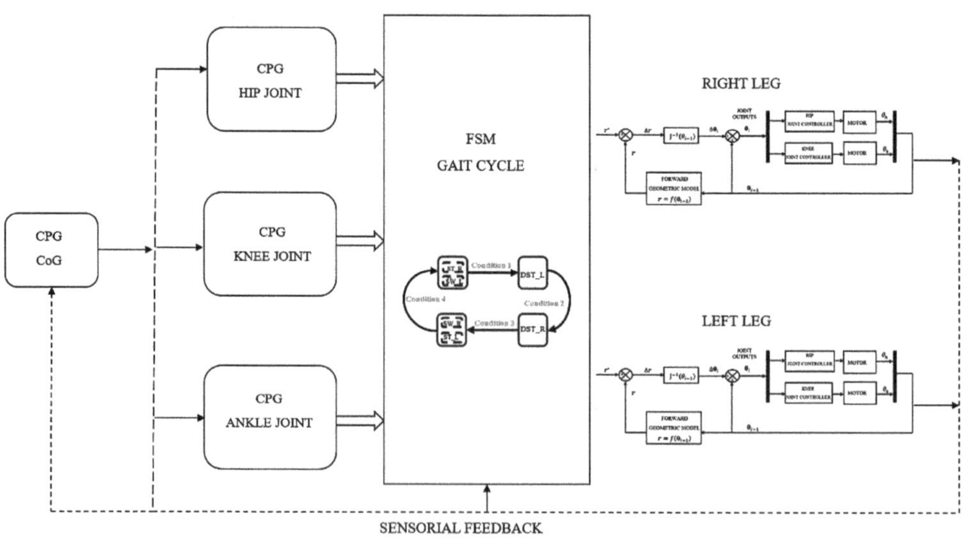

Figure 17. Control architecture for hybrid dynamic model for bipedal gait

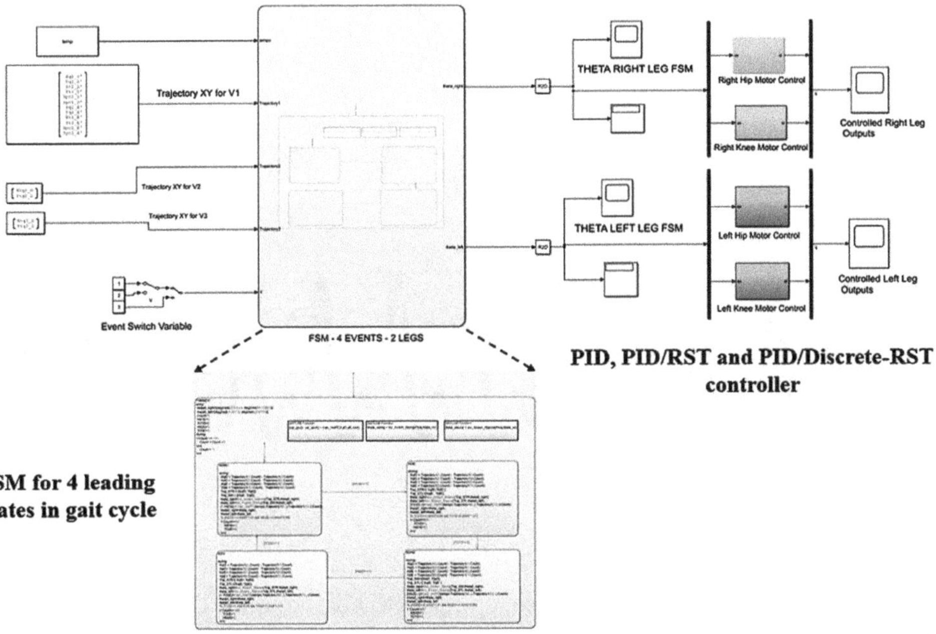

Figure 18. Angle trajectories of the right leg joints generated by the FSM

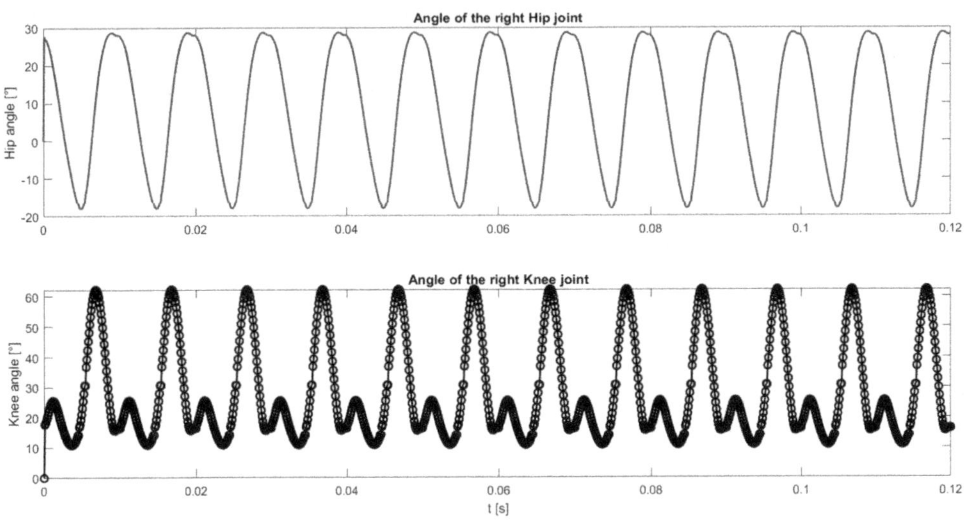

The control inputs are obtained from the error signal between the controller output and reference signals. This error signal is compensated by the controller, which after power amplification address it to the DC motors, as illustrated in Figure 20. The PID

Figure 19. Angle trajectories of the left leg joints generated by the FSM

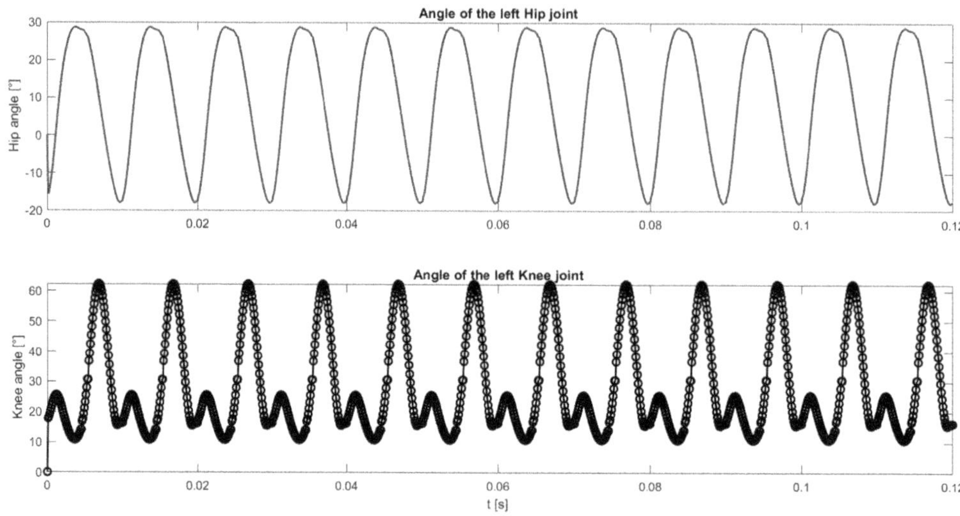

controller showed the worst response, and the PID/discrete RST controller showed the best performance in terms of steady-state error analysis.

In the event-based control approach implemented in this work, The Foot Velocity Algorithm (FVA) was used for detecting heel strike and toe-off events inside the Finite State Machine gait model. Better performance of the FSM model depends on the accurate determination of the To and HC events. Thus, if higher precision is required, more robust techniques are suggested, such as support-vector machines (SVM) or Artificial Neural Networks (ANN).

Figure 20. Responses of PID, RST and discrete RST controller for right hip and knee joints

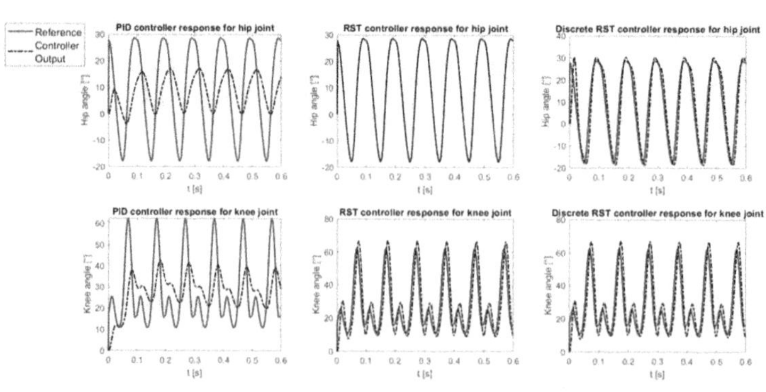

The cyclic movement of the gait is described exclusively on the y-axis, while the x-axis clearly demonstrates that the advancement in space occurs almost linearly during the swing phase followed by the support phase. Therefore, the authors our current work is to implement Matsuoka neural oscillators for the degrees of freedom of the biped robot mechanism analyzed, to generate the cyclic vertical trajectories. Stability analysis of the Matsuoka neural oscillator trajectories will be performed using Poincaré maps.

7. FINAL CONSIDERATIONS AND FURTHER DEVELOPMENTS

In this paper proposes a control strategy integrating event-based and bioinspired laws for robotic biped gait. Hybrid dynamic modeling and theory of nonlinear dynamical systems are essential sources for the study of human motor control and, therefore, for the synthesis of bioinspired control laws, which help to generate a more natural and robust movement in robots. Controllers based on neural oscillators that perform the functions of a CPG, incorporating the behavioral dynamics approach, allow developing a coupling between robot and environment through sensorial information, which generates motor commands that enable a stable and adapted gait to environmental conditions.

As future research, we seek the integration of control strategies based on neural oscillators, with recent approaches of Artificial Intelligence (AI) for robotics, allowing more sophisticated real-time behavior of robotic biped systems. However, these techniques are still limited by their computational cost and the number of degrees of freedom of the robot.

REFERENCES

Abedi, M., Moghaddam, M. M., & Firoozabadi, S. M. P. (2014). A full bio-inspired bipedal gait locomotion system. *2014 2nd RSI/ISM International Conference on Robotics and Mechatronics, ICRoM 2014*. 10.1109/ICRoM.2014.6990897

Avrin, G. (2017). *Modélisation du contrôle moteur humain lors de tâches rythmiques hybrides et application à la commande de robots anthropomorph*. Université Paris-Saclay.

Avrin, G., Makarov, M., Rodriguez-Ayerbe, P., & Siegler, I. A. (2016). Particle Swarm Optimization of Matsuoka's oscillator parameters in human-like control of rhythmic movements. *Proceedings of the American Control Conference*. 10.1109/ACC.2016.7524938

Batista, R. F. (2017). *Sistema dinâmico híbrido aplicado a modelo mecânico bípede antropomórfico*. University of Campinas.

Bruce, A., Hanrahan, S., Vaughan, K., Mackinnon, L., & Pandy, M. G. (1998). The Biophysical Foundations of Human Movement. *Medicine &amp Science in Sports & Exercise*. doi:10.1097/00005768-199801000-00025

Buitrago Salazar, G. D. (2018). *Concepção de Diferentes Estruturas para Exoesqueletos de Membro Inferior baseado no Estudo Dinâmico e Utilização de Biomateriais Concepção de Diferentes Estruturas para Exoesqueletos de Membro Inferior baseado no Estudo Dinâmico e Utilização de Biomateria*. University of Campinas.

Buschmann, T., Ewald, A., Ulbrich, H., & Buschges, A. (2012). Event-based walking control - From neurobiology to biped robots. In *2012 IEEE/RSJ International Conference on Intelligent Robots and Systems* (pp. 1793–1800). IEEE. 10.1109/IROS.2012.6385783

Chestnutt, J., Lau, M., Cheung, G., Kuffner, J., Hodgins, J., & Kanade, T. (2005). Footstep Planning for the Honda ASIMO Humanoid. In *Proceedings of the 2005 IEEE International Conference on Robotics and Automation* (pp. 629–634). IEEE. 10.1109/ROBOT.2005.1570188

Endo, G., Nakanishi, J., Morimoto, J., & Cheng, G. (2005). Experimental Studies of a Neural Oscillator for Biped Locomotion with QRIO. In *Proceedings of the 2005 IEEE International Conference on Robotics and Automation* (pp. 596–602). IEEE. 10.1109/ROBOT.2005.1570183

Gomes, M. A., Siqueira, A. A. G., & Gobbo, R. G. (2011). Improving the parameters of neural oscillators to generate joint trajectories of an exoskeleton for lower limbs. In *2011 9th IEEE International Conference on Control and Automation (ICCA)* (pp. 286–291). IEEE. 10.1109/ICCA.2011.6138064

Habib, M. K., Liu, G. L., Watanabe, K., & Izumi, K. (2007). Bipedal Locomotion Control via CPGs with Coupled Nonlinear Oscillators. In *Mechatronics, 2007 IEEE International Conference on* (pp. 1–6). IEEE. 10.1109/ICMECH.2007.4280021

Habib, M. K., Watanabe, K., & Izumi, K. (2009). Biped locomotion using CPG with sensory interaction. *IEEE International Symposium on Industrial Electronics*, 1452–1457. 10.1109/ISIE.2009.5219063

Heliot, R., Azevedo, C., & Espiau, B. (2012). Functional Rehabilitation: Coordination of Artificial and Natural Controllers. Rehabilitation Robotics. doi:10.5772/5160

Hreljac, A., & Marshall, R. N. (2000). Algorithms to determine event timing during normal walking using kinematic data. *Journal of Biomechanics*, *33*(6), 783–786. doi:10.1016/S0021-9290(00)00014-2 PMID:10808002

Ijspeert, A. J. (2008). Central pattern generators for locomotion control in animals and robots: A review. *Neural Networks*, *21*(4), 642–653. doi:10.1016/j.neunet.2008.03.014 PMID:18555958

Ikeda, R., & Horie, R. (2016). Control of bipedal locomotion with a neural oscillator-based brain-computer interface. *Journal of Information and Communication Technology*, *15*(2), 19–37.

Kuteken, R. S., Batista, R. F., & Rosario, J. M. (2017). Cognitively inspired artificial bipedal humanoid gait generation. *2017 International Work Conference on Bio-Inspired Intelligence: Intelligent Systems for Biodiversity Conservation, IWOBI 2017 - Proceedings*. 10.1109/IWOBI.2017.7985518

Li, Z., Vanderborght, B., Tsagarakis, N. G., & Caldwell, D. G. (2013). Quasi-straightened knee walking for the humanoid robot. Cognitive Systems Monographs. doi:10.1007/978-3-642-36368-9_9

Liu, C., Wang, D., Goodman, E. D., & Chen, Q. (2016). Adaptive walking control of biped robots using online trajectory generation method based on neural oscillators. *Journal of Bionics Engineering*, *13*(4), 572–584. doi:10.1016/S1672-6529(16)60329-3

Liu, C. J., Fan, Z., Seo, K., Tan, X. B., & Goodman, E. D. (2012). Synthesis of Matsuoka-Based Neuron Oscillator Models in Locomotion Control of Robots. In *2012 Third Global Congress on Intelligent Systems* (pp. 342–347). IEEE. 10.1109/GCIS.2012.99

Liu, G. L., Habib, M. K., Watanabe, K., & Izumi, K. (2007). The Design of Central Pattern Generators Based on the Matsuoka Oscillator to Generate Rhythmic Human-Like Movement for Biped Robots. *Journal of Advanced Computational Intelligence and Intelligent Informatics*, *11*(8), 946–955. doi:10.20965/jaciii.2007.p0946

Liu, G. L., Habib, M. K., Watanabe, K., & Izumi, K. (2008a). Central pattern generators based on Matsuoka oscillators for the locomotion of biped robots. *Artificial Life and Robotics*, *12*(1–2), 264–269. doi:10.100710015-007-0479-z

Lu, Q., & Tian, J. (2015). Research on Walking Gait of Biped Robot Based on a Modified CPG Model. *Mathematical Problems in Engineering*, *2015*, 1–9. doi:10.1155/2015/584954

Manoonpong, P., Geng, T., Kulvicius, T., Porr, B., & Wörgötter, F. (2007). Adaptive, Fast Walking in a Biped Robot under Neuronal Control and Learning. *PLoS Computational Biology*, *3*(7), 1305–1320. doi:10.1371/journal.pcbi.0030134 PMID:17630828

MathWorks. (2018). *Stateflow - MATLAB & Simulink*. Author.

O'Connor, C. M., Thorpe, S. K., O'Malley, M. J., & Vaughan, C. L. (2007). Automatic detection of gait events using kinematic data. *Gait & Posture*, *25*(3), 469–474. doi:10.1016/j.gaitpost.2006.05.016 PMID:16876414

Pons, J. L. (2008). *Wearable Robots: Biomechatronic Exoskeletons*. doi:10.1002/9780470987667

Pratt, G., & Manzo, J. (2013). The DARPA robotics challenge. *IEEE Robotics & Automation Magazine*, *20*(2), 10–12. doi:10.1109/MRA.2013.2255424

Pristovani, R. D., Henfri, B. E., Sanggar, D., & Dadet, P. (2018). Walking strategy model based on zero moment point with single inverted pendulum approach in "T-FLoW" humanoid robot. In *Proceedings - 2017 2nd International Conferences on Information Technology, Information Systems and Electrical Engineering, ICITISEE 2017*. 10.1109/ICITISEE.2017.8285498

Puerta Barrera, J. F. (2017). *Proposta de arquitetura de controle para prótese robótica de membro inferior*. University of Campinas; doi:10.1901/jaba.2012.45-839

Rosario, J. M. (2010). *Robótica Industrial I Modelagem, Utilização e Programação*. Sao Paulo, Brazil: Editora Baraúna.

Rosário, J. M., Suekichi Kuteken, R., & Izquierdo Cordoba, L. M. (2019). *Proposal of Methodology of a Bipedal Humanoid Gait Generation Based on Cognitive Algorithm*. doi:10.1007/978-3-030-13835-6_13

Santos, C. P., Alves, N., & Moreno, J. C. (2017). Biped Locomotion Control through a Biomimetic CPG-based Controller. *Journal of Intelligent & Robotic Systems*, *85*(1), 47–70. doi:10.100710846-016-0407-3

Siegler, I. A., Bardy, B. G., & Warren, W. H. (2010). Passive vs. active control of rhythmic ball bouncing: The role of visual information. *Journal of Experimental Psychology. Human Perception and Performance*, *36*(3), 729–750. doi:10.1037/a0016462 PMID:20515200

Suekichi, R. (2017). *Geração de Padrões de Marcha para Sistemas Robóticos Bípedes Humanoides*. University of Campinas.

Sugimoto, N., & Morimoto, J. (2011). Phase-dependent trajectory optimization for CPG-based biped walking using path integral reinforcement learning. In *2011 11th IEEE-RAS International Conference on Humanoid Robots* (pp. 255–260). IEEE. 10.1109/Humanoids.2011.6100908

Thelen, E., & Smith, L. B. (2009). Dynamic Systems Theories. In Handbook of Child Psychology. Academic Press. doi:10.1002/9780470147658.chpsy0106

van der Schaft, A., & Schumacher, H. (2000). *An introduction to hybrid dynamical systems* (Vol. 251). London: Springer London; doi:10.1007/BFb0109998

Vaughan, C. L., Davis, B. L., Christopher, L., & Connor, J. C. O. (2005). Dynamics of human gait. *European Journal of Physics.*

Warren, W. H. (2006). The dynamics of perception and action. *Psychological Review, 113*(2), 358–389. doi:10.1037/0033-295X.113.2.358 PMID:16637765

Warren, W. H., & Fajen, B. R. (2008). Behavioral dynamics of visually guided locomotion. *Understanding Complex Systems*, 45–75. doi:10.1007/978-3-540-74479-5_3

Winter, D. A. (2009). *Biomechanics and Motor Control of Human Movement.* Hoboken, NJ: John Wiley & Sons; doi:10.1002/9780470549148

Xia, Z., Deng, H., Zhang, X., Weng, S., Gan, Y., & Xiong, J. (2017). A central pattern generator approach to footstep transition for biped navigation. *International Journal of Advanced Robotic Systems, 14*(1), 1–9. doi:10.1177/1729881416682708

KEY TERMS AND DEFINITIONS

Biped Gait: Dynamic and cyclical sequence of trajectories performed by the two lower limbs over a period of time.

Event-Based: Graphical description of system operation through events, where an event is a mode of system operation.

Exoskeleton: Passive or active mechatronics device integrated into the human body, or a part of it. An exoskeleton is intended to map and/or enhance motor functions.

Finite State Machine: Graphical description of a reactive system, in which the system makes a transition from one state to another, if the condition defining the change is true.

Neural Oscillator: Nonlinear oscillator capable of autonomously generating a periodic signal in the form of a limit cycle.

PID/RST Controller: A proportional integral derivative control system, which can be transformed into a Reference Signal Tracking control system.

Zero Moment Point: Dynamic stability analysis criterion of the biped locomotion, achieved by calculating the point on the surface of the foot where the total of horizontal inertia and gravity forces equals zero.

Chapter 10
Model-Based Multi-Objective Reinforcement Learning by a Reward Occurrence Probability Vector

Tomohiro Yamaguchi
Nara College, National Institute of Technology (KOSEN), Japan

Shota Nagahama
Nara College, National Institute of Technology (KOSEN), Japan

Yoshihiro Ichikawa
Nara College, National Institute of Technology (KOSEN), Japan

Yoshimichi Honma
Nara College, National Institute of Technology (KOSEN), Japan

Keiki Takadama
The University of Electro-Communications, Japan

ABSTRACT

This chapter describes solving multi-objective reinforcement learning (MORL) problems where there are multiple conflicting objectives with unknown weights. Previous model-free MORL methods take large number of calculations to collect a Pareto optimal set for each V/Q-value vector. In contrast, model-based MORL can reduce such a calculation cost than model-free MORLs. However, previous model-based MORL method is for only deterministic environments. To solve them, this chapter proposes a novel model-based MORL method by a reward occurrence probability (ROP) vector with unknown weights. The experimental results are reported under

DOI: 10.4018/978-1-7998-1382-8.ch010

the stochastic learning environments with up to 10 states, 3 actions, and 3 reward rules. The experimental results show that the proposed method collects all Pareto optimal policies, and it took about 214 seconds (10 states, 3 actions, 3 rewards) for total learning time. In future research directions, the ways to speed up methods and how to use non-optimal policies are discussed.

INTRODUCTION

Reinforcement learning (RL) is a popular algorithm for automatically solving sequential decision problems such as robot behavior learning and most of them are focused on single-objective settings to decide a single solution. A single objective RL can solve a simple learning task under a simple situation. However, in real world robotics, a robot often faces that the optimal condition on its own objective changes such as an automated driving car in a public road where many human driving cars move. So the real world learner has to treat multi-objective which may conflict by subsumption architecture (Tajmajer 2017)or the weights of these objectives may depend on the situations around the learner. Therefore, it is important to study multi-objective optimization problems in both research fields for robotics and reinforcement learning.

In multi-objective reinforcement learning (MORL), the reward function emits a reward vector instead of a scalar reward. A scalarization function with a vector of n weights (weight vector) is a commonly used to decide a single solution. The simple scalarization function is linear scalarization such as weighted sum. The main problem of previous MORL methods is a huge learning cost required to collect all Pareto optimal policies. Hence, it is hard to learn the high dimensional Pareto optimal policies. To solve this, this chapter proposes the novel model-based MORL method by reward occurrence probability (ROP) with unknown weights. There are two main features. The first feature is that the average reward of a policy is defined by inner product of the ROP vector and the weight vector. The second feature is that it learns ROP in each policy instead of Q-values. Pareto optimal deterministic policies directly form the vertices of a convex hull in the ROP vector space. Therefore, Pareto optimal policies are calculated independently with weights and just once. The experimental results show that the authors' proposed method collected all Pareto optimal policies under three dimensional stochastic environments, and it takes a small computation time though previous MORL methods learn at most two or three dimensions deterministic environments.

The objectives of this chapter are as follows:

1. Solving multi-objective reinforcement learning problems where there are multiple conflicting objectives with unknown weights.
2. To learn all Pareto optimal solutions which maximize the average reward defined by the reward occurrence probability (ROP) vector of a solution and unknown weights.
3. Visualizing the distribution of all Pareto optimal solutions in the ROP vector space.

BACKGROUND

Reinforcement learning (RL) is a popular algorithm for a learning agent to automatically solve sequential decision problems which are commonly modeled as Markov decision processes (MDPs). A MDP is a discrete time stochastic control process where outcomes are partly random and partly under the control of a decision maker. At each time step, the process is in some state s, and the decision maker may choose any action a that is available in state s. The process responds at the next time step by randomly moving into a new state s', and giving the decision maker a corresponding reward $R_a(s, s')$. In almost reinforcement learning methods, the reward is usually simplified as $R_a(s) = R(a, s)$. The probability that the process moves into its new state s' is influenced by the chosen action. Specifically, it is given by the state transition function $P_a(s, s')$. When a next state s' only depends on the current state s and the decision maker's action a, (it is independent of all previous states and actions), this property is called *simple Markov property*. A discrete MDP model is represented as both the state action transition matrix $P_a(s, s')$ and reward matrix $R_a(s, s')$ for all triple among state s, action a and a new state s' in the environment.

Despite numerous reinforcement learning methods, most of them focused on single objective settings where the goal of a decision maker (an agent) decides a single solution by the optimality criterion. This reinforcement learning methods are classified according to the learning algorithm and the optimality criterion. The former, there are two kinds of learning algorithms whether directly estimating the MDP model or not, one is the *model-based* approach such as real-time dynamic programming (RTDP) (Barto, 1995) and H-Learning (Tadepalli, 1998) which takes a small time complexity but a large space complexity, and another one is the *model-free* approach (Yang 2016) such as Q-learning. The *model-based* approach starts with directly estimating the MDP model statistically, in which both the state action transition matrix $P_a(s, s')$ and reward matrix $R_a(s, s')$ for all triple (s, a, s') are estimated by observing a new state s' and received reward r when performing an action *a* on state *s*, This process is called the model identification. Using the estimated MDP model by model identification process, the value of each state as $V(s)$ or the quality of each state action $Q(s, a)$ are calculated to search the optimal solution that maximizes $V(s)$ of each state. In contrast, the *model-free* approach directly learns $V(s)$ or $Q(s, a)$ without estimating $P_a(s, s')$ and $R_a(s, s')$ of the MDP model. The latter, there are two kinds of optimality criteria whether using a discount factor or not, one is maximizing the sum of the *discounted rewards*, and another one is maximizing the *average reward* without any discount factor (Mahadevan, 1996)(Tadepalli, 1998)(Gao, 2006)(Yang, 2016). Most previous RL methods are model-free approach with a discount factor since the model-based approach takes the large space complexity.

A multi-objective MDP (MOMDP) is an MDP in which the reward function emits a *reward vector* instead of a scalar reward. A scalarization function with a vector of *n* weights (*weight vector*) is a commonly used to decide a single solution. The simple scalarization function is linear scalarization such as weighted sum. In this chapter, the authors mainly target the weighted sum function for the scalarization function. However, the authors' method can be applied to other scalarization function such as Tchebycheff norm method.

Multi-objective reinforcement learning (MORL) (Roijers, 2013)(Roijers, 2015) (Liu, 2016)(Pinder, 2016) has several methods which can be divided into two main approaches, the scalar combination and *Pareto optimization* (Herrmann, 2015). In the former case, the scalar combination finds a single policy that optimizes a combination of the rewards. MOMDP and known weights are input to the learning algorithm, then it output a single solution. In the latter case, *Pareto optimization* finds multiple policies that cover the Pareto front, which requires collective search for sampling the Pareto set (Natarajan, 2005). MOMDP is input to the learning algorithm, then it output a solution set. Note that there are two ways to select a single solution in the set, one is the scalarization with known weight, another is a user selection.

Most of the state-of-the-art MORL are model-free value-based reinforcement learning algorithms (Barrett, 2008)(Hiraoka, 2009)(Moffaert, 2014)(Lizotte, 2012) (Roijers, 2013) with a main problem that is MORL previous methods incorporate a huge learning cost to collect all Pareto optimal policies. First, they need sufficient number of executions for each state-action pair to satisfy the convergence condition toward optimality as with *Q*-learning. Secondly, Pareto optimal set is calculated for each $V(s)$ or $Q(s, a)$ since these methods are value-based. Thirdly, when updating *V*-values or *Q*-values, Pareto candidates are added or updated as *V*/*Q*-value vector, it must keep a large number of candidates until each Pareto optimal set is converged. Therefore, previous MORL methods take large number of calculations to collect Pareto optimal set for each *V*/*Q*-value vector. In contrast, model-based MORL can reduce such a calculation cost (Wiering, 2014) than model-free MORLs. However, second and third problems as described above are still remained, and the method proposed by Wiering et al is for only deterministic environments. Thus, it is hard to learn high dimensional Pareto optimal policies by previous methods.

To solve these problems, this chapter proposes the novel model-based MORL method by reward occurrence probability with unknown weights. Our approach is one of the average reward model-based reinforcement learning (Mahadevan, 1996) and is based on not *V/Q*-value but *reward occurrence probability* (ROP) for each reward rule. The essential point is that it learns *reward occurrence probability* in each policy instead of *V/Q*-values for each state. A *reward occurrence probability* (ROP) of a reward is the probability to acquire the reward per step when executing some policy constantly. For example, when a reward is acquired once for each five steps, the ROP is 0.2 Note that a ROP is only depend on the occurrence of a reward, not depend on the value of the reward. When the environment has multiple rewards, *ROP vector* is introduced in which each element of the ROP vector corresponds to the ROP of each reward. After the MDP model is identified, all reward acquisition policies are collected by tree search, then the *ROP vector* for each reward acquisition policy is calculated to decide Pareto optimal deterministic policies in the ROP vector space. *The ROP vector space* is the n-dimensional space, in which the dimension is the number of rewards, and each axis is the ROP of each reward. Note that the ROP vector is index of the policy, so several policies may have same ROP vector. All ROP vectors are located in the ROP vector space,

Thus, there are two main features to it. The first feature is that the average reward of a policy is defined by the inner product of the ROP vector and a weight vector. Note that a weight vector represents a trade-off among multi objective, and each element of the vector represents a weight of each objective. The second feature is that Pareto optimal ROP vectors corresponding to deterministic policies directly form the vertices of a convex hull in the ROP vector space. Therefore, Pareto optimal policies are calculated independently with weights and just one time.

The key points of our approach are as follows:

1. Each objective is defined by a reward rule and its unknown weight.
2. Multi-objective is defined by the fixed reward rule set and the unknown weight vector.
3. Each policy is assigned to the reward occurrence probability (ROP) vector where n^{th} value is the occurrence probability of nth reward rule of the policy.
4. In the ROP vector space (rectangular coordinate system), any stationary policy is mapped to a point where coordinates are indicated by its ROP vector.
5. Optimal deterministic stationary policies with unknown weights form the vertex of convex hull in the ROP vector space.

Figure 1. It is an overview of the model-based reinforcement learning system

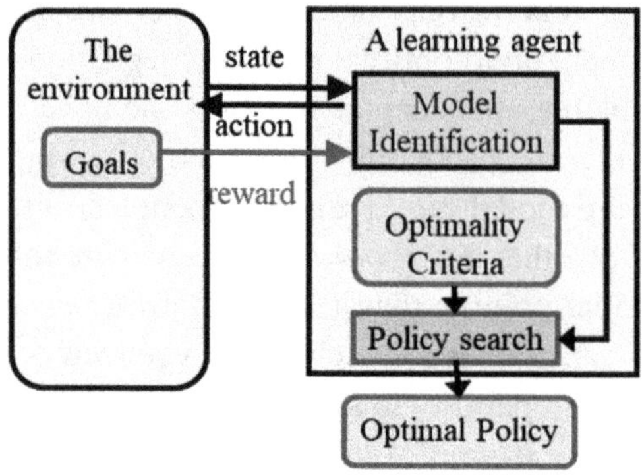

6. Average reward of a policy is defined by the inner product of the ROP vector of the policy and the weight vector.
7. The range of the weight vector of each optimal deterministic stationary policy can be calculated geometrically.
8. The stochastic learning environment can be learned by standard MDP model identification method and our proposed search method collects all Pareto optimal policies.

MODEL-BASED REINFORCEMENT LEARNING

This section describes the framework of model-based RL methods that estimate a model of the environment while interacting with it and search for the best policy of its current estimated model under some optimality criterion. Some comparisons have shown model-based RL are much more effective than model-free methods such as Q-learning (Barto, 1995). Figure 1 shows an overview of the model-based reinforcement learning system. A learning agent is a decision maker who is in the environment. The environment has some goal which is modeled as a MDP model with a reward function.

At each time step, the agent as a decision maker is in some state which can be observed as s, and the agent may choose any action a that is available in state s. At the next time step, the environment responds by partially randomly moving into a new state s', sometimes giving the agent a corresponding reward $\boldsymbol{R}(s, a)$.

Next, the agent as a learner is described. In Figure 1, the learning agent consists of three blocks which are model identification block, optimality of policies block and policy search block. The details of these blocks are described in following subsections. The novelty of our method lies in policy search block which collects all *reward acquisition policies* according to average reward optimality. The detail of this block is described in next section.

MDP Model and Markov Chains

A *Markov decision process* (MDP) (Puterman, 1994) in this chapter is a discrete time and a discrete state space stochastic control process. It provides a mathematical framework for modeling decision making in situations where outcomes are partly random and partly under the control of a decision maker. Note that for a decision maker, actions are the allowing choice, and rewards are giving motivation. A MDP model is defined by of following four elements:

1. **Set of States**: $\boldsymbol{S} = \{s_0, s_1, s_2, , s_n\}$
2. **Set of Actions**: $\boldsymbol{A} = \{a_0, a_1, a_2, , a_n\}$
3. **State Transition Probabilities** $\boldsymbol{P}(s'| s, a)$: A probability of occurring state s' when execute action a at state s.
4. **Reward Function** $\boldsymbol{R}(s, a)$: Return a scalar value as an acquired reward when execute action a at state s.

In this paper, state action pair (s, a) is called a *rule*. Reward rule (s, a) is defined when $\boldsymbol{R}(s, a)$ is not zero. $\boldsymbol{R}(s, a)$ means that a reward is assigned to a rule (s, a). $\boldsymbol{P}(s'| s, a)$ means that the probability that the process moves into its new state s' is influenced by the chosen action. Thus, the next state s' depends on the executed rule, that is the current state s and the decision maker's action a, and is independent of all previous executed rules as described as *simple Markov property*. A *stochastic policy* π.is a probability distribution over actions for every possible state. A *deterministic policy* is defined by a function that selects an action for every possible state. This chapter mainly deals with *deterministic policies.*

A *Markov chain* is a stochastic model describing a sequence of possible states in which the probability of each state depends only on the previous state. It's an intension of Markov decision processes, the difference is that there is neither actions nor rewards in a *Markov chain*. This chapter focuses on the property that a *policy* of a MDP forms a *Markov chain* of the MDP, it is described later in next section.

The authors assume a MDP model is *ergodic* where the model satisfies these conditions as follows:

1. **Irreducible:** All states can be reached from all others.
2. **Aperiodic:** Each state is visited without any systematic period.
3. **Finite States:** The number of states is finite.

Model Identification

In the model identification block, the state transition probabilities $\boldsymbol{P}(s'| s, a)$ and the reward function $\boldsymbol{R}(s, a)$ are estimated incrementally by observing a sequence of (s, a, r). This estimated model is generally assumed by MDP. Model-based RL methods learn the transition and reward models of the environment by making use of counters that are used in a Maximum-Likelihood Estimation (MLE) to compute approximate transition probabilities and average rewards (Wiering 2014).

Note that the MLE probability is same as the occurrence based probability. Each time the agent selects rule (s, a) and makes a transition to state s', the transition model's counter values $C(s, a)$ and $C(s, a, s')$ are increased by one. In a similar fashion, the obtained reward r is added to the value $R_t(s, a)$ which computes the sum of all rewards obtained by selecting rule (s, a). Finally, the maximum likelihood model of the MDP is computed as Eq. (1).

$$P(s'| s, a) = C(s, a, s') / C(s, a) \text{ and } R(s, a) = R_t(s, a) / C(s, a) \tag{1}$$

Average Reward Optimality Criterion

The optimality of policies block defines the optimality criterion of the learning policy. In this research, a policy which maximizes average reward is defined as an optimal policy. There are two kinds of optimality criteria on average reward RL, one is gain-optimal which considers acquired rewards only in a stationary cycle, the other is bias-optimal which considers acquired rewards both on a temporally path and the stationary cycle (Mahadevan 1996). This research is based on gain-optimal average reward. Eq. (2) shows the definition of gain optimal average reward.

$$g^{\pi}\left(s\right) \equiv \lim_{N\to\infty} E\left(\frac{1}{N}\sum_{t=0}^{N-1} r_t^{\pi}\left(s\right)\right) \tag{2}$$

where N is the number of step, $r_T^{\pi}\left(s\right)$ is the expected value of reward that an agent acquired at step t where policy is π and initial state is s and $E()$ denotes the expected value.

LC-Learning

This subsection summarizes *LC-Learning* (Konda 2002)(Satoh 2006) which is our basic method of policies search. LC-Learning is one of the average reward model-based reinforcement learning methods. It collects all reward acquisition *deterministic* policies under *unichain* condition. The *unichain* condition requires that every policy in an MDP result in a single *ergodic* class, and guarantees that the optimal average cost is independent of the initial state (Tsitsiklis 2006). The features of LC-Learning are as follows:

1. Collecting all reward acquisition policies by breadth search started by each *reward rule*.
2. Calculating gain-optimal average reward using reward acquisition cycle of each policy.

Next section, ROP based LC-learning is described.

MODEL-BASED MORL BY REWARD OCCURRENCE PROBABILITY

This section starts with describing the definition of a weighted reward vector to apply reinforcement learning to multiple objective. Then, *reward occurrence probability (ROP) vector* is introduced to define average reward with a weighted vector.

Figure 2 shows the flow of the policy search as shown in Figure 1. After collecting all reward acquisition policies by LC-learning, *ROP vector* for each collected policy is calculated. The detail of this process is described later and is shown in Figure 6. Then, these ROP vectors are mapped in the ROP vector space, Pareto optimal policies are decided by calculating a convex hull from the ROP vector set. Note that the ROP vector for each vertex of the convex hull is associated with Pareto optimal policies.

Weighted Reward Vector

To represent multi objective, the reward is divided into *d* reward types one for each objective, and a weight which represents the importance or preference of that reward is associated with each reward type (Natarajan 2005)(Hiraoka 2009). In this chapter, the reward function is defined by a vector of *d* rewards (reward vector) $\vec{r} = \left(r_1, r_2, \ldots, r_d\right)$ where each r_i represents a position of reward rule and the weight vector $\vec{w} = \left(w_1, w_2, \ldots, w_d\right)$ which represents a trade-off among multi objective. A scalarization function with a weight vector is called a *weighted sum of rewards*. Eq. (3) shows a weighted sum of rewards defined by inner product of the reward vector $\vec{r} = \left(r_1, r_2, \ldots, r_d\right)$ and the weight vector $\vec{w} = \left(w_1, w_2, \ldots, w_d\right)$. Note that $\left|\vec{w}\right| = 1$.

Figure 2. It is the flow of the policy search

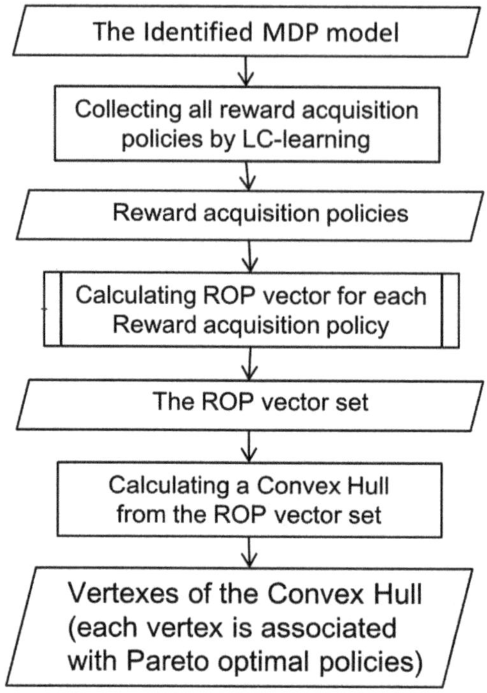

$$r\left(\vec{w}\right)=\sum_{i=1}^{d}w_i r_i=\vec{w}\cdot\vec{r} \tag{3}$$

Average Reward by Reward Occurrence Probability

Average reward is the expected received rewards per step when an agent performs state transitions routinely according to a policy. A step is a time cost to execute an action. Under an unichain policy π and reward vector $\vec{r}=\left(r_1,r_2,\ldots,r_d\right)$. *Reward Occurrence Probability* (ROP) vector $\vec{\boldsymbol{P}}_\pi$ is defined as shown in Eq. (4). In that, $\boldsymbol{P}_i$ is the expected occurrence probability per step for reward r_i.

$$\vec{\boldsymbol{P}}_\pi\equiv\left(\boldsymbol{P}_1,\boldsymbol{P}_2,\ldots,\boldsymbol{P}_d\right) \tag{4}$$

Average reward ρ_π under a policy π is defined by the inner product of the ROP vector $\vec{\boldsymbol{P}}_\pi$ and the weight vector $\vec{w}$ as shown in Eq. (5).

Figure 3. It shows the geometrical meaning of average reward by the inner product of the ROP vector $\vec{P}_\pi$ and the weight vector $\vec{w}$

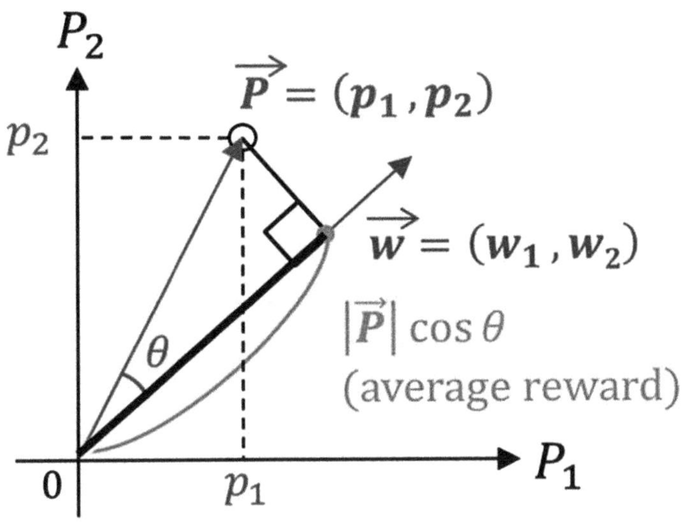

$$\rho_\pi(\vec{w}) = \sum_{i=1}^{d} w_i P_i = \vec{w} \cdot \vec{P}_\pi \tag{5}$$

Figure 3 shows the geometrical meaning of Eq. (5) under d = 2. There are two reward rules and their occurrence probabilities are $\boldsymbol{P}_1$, and $\boldsymbol{P}_2$. An ROP vector $\vec{\boldsymbol{P}}_\pi$ = (p_1, p_2) which is in two dimensional ROP vector space, where the horizontal axis is $\boldsymbol{P}_1$ and the vertical axis is $\boldsymbol{P}_2$. The average reward of inner product for ROP vector $\vec{\boldsymbol{P}}_\pi$ is $\left|\vec{\boldsymbol{P}}_\pi\right|$ **cos θ** since $\left|\vec{w}\right| = 1$. It is equal to the distance between (0, 0) and the foot of a perpendicular on the $\vec{w}$ from the point (p_1, p_2). Note that the average reward for any ROP vector which point (p_1, p_2) is on the $\vec{w}$ is the same. This research assumes that $\vec{\boldsymbol{P}}_{\pi 0}$ is zero vector where π_0 is any policy receiving no reward.

Collecting All Reward Acquisition Policies (Satoh 2006)

Our searching method for reward acquisition policies is based on LC-learning as the authors described before. Deterministic policies which acquire some rewards are searched by converting a MDP into the tree structures where reward acquisition rules are root rule. Figure 4 shows an illustrated example. Figure 4 (a) shows an example of the identified MDP model which consists of four states, six rules and two rewards is converted into the two tree structures. Figure 4 (b) shows the two kinds of search

Figure 4. It is an illustrated example collecting all reward acquisition cycles from an identified MDP model. (a) is an example of the identified MDP model, (b) is the two kinds of search trees searching reward acquiring cycles, and (c) is the collected Three kinds of reward acquiring cycles

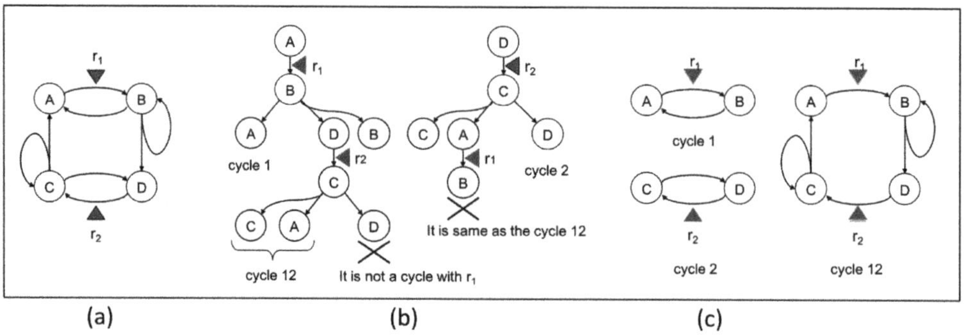

trees searching reward acquiring cycles. In each search tree, the root node and its transition starts from each reward rule. In this tree structure, the path from the root node to the state which is same to the initial node is the reward acquisition cycle. In stochastic environment, some of the rule is deliquesce stochastically. In such case, path from parent node of stochastic rule to the state that is already extracted is part of a policy that contains the stochastic rule. Figure 4 (c) shows all reward acquisition cycles in Figure 4 (a). After collecting all reward acquisition cycles, for each cycle, it is expanded to policies by searching temporary path from states which is not in the cycle to any states in the cycle. For example of cycle 1 in Figure 4 (c), there is one policy where the temporary path is D to C and C to A.

After collecting all reward acquiring policies, for each policy, the state transition probability matrix $\boldsymbol{P}_{\pi}$ of the policy π is prepared as *Markov chain* as the authors described in previous section. Figure 5 shows an illustrated example of calculating *Markov chains of the reward acquisition policies*. Figure 5 (a) shows the identified MDP model, Figure 5 (b) shows examples of the two kinds of Markov chains for reward acquisition policies with two rewards r_1 and r_2. Figure 5 (c) is the two kinds of state transition probability matrices of each policy. Note that any element p_{ij} in the matrix is the state transition probability from state i to state j.

Calculating ROP vector for each Reward Acquisition Policy

When the occurrence probability of state i is α_i, it is equivalent to the occurrence probability of the reward rule (i, a) under a deterministic policy π. Calculating

Figure 5. It is an illustrated example of calculating Markov chains. (a) is an example of the identified MDP model, (b) is the two kinds of Markov chains for reward acquisition policies, and (c) is the two kinds of state transition probability matrices of each policy.

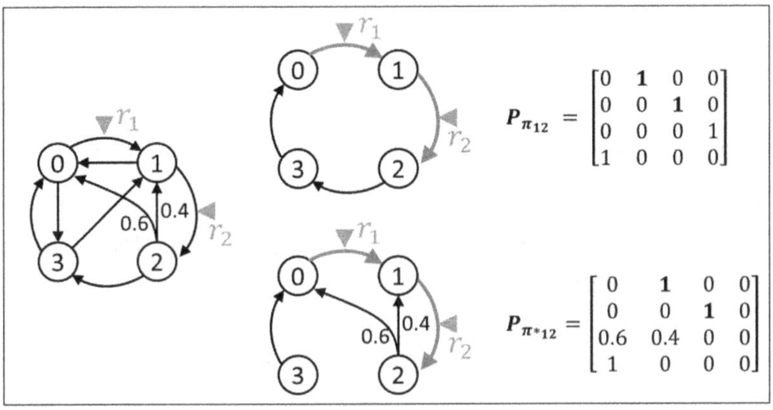

method of each ROP vector $\vec{P}_\pi$ for each reward acquisition policy π associated with P_π from the reward acquisition policy set is as follows:

Step 1: Set up simultaneous linear equations for each P_π.

Under a deterministic policy π, the occurrence probability vector for all states $\vec{\alpha}_\pi$ defined as Eq. (6) is the solution of simultaneous linear equations Eq. (7).

$$\vec{\alpha}_\pi \equiv \left(\alpha_1, \alpha_2, \ldots, \alpha_{|S|}\right) \tag{6}$$

$$\vec{\alpha}_\pi P_\pi = \vec{\alpha}_\pi \tag{7}$$

Step 2: For each P_π, solve Eq.(7) by Gaussian elimination.

Step 3: For each $\vec{\alpha}_\pi$ derived at step 2, forms ROP vector $\vec{P}_\pi$ by picking up the occurrence probability of each reward rule.

Step 4: Make a ROP vector set from the ROP vectors derived at step 3.

Step 5: Make a mapping from the reward acquisition policy set to the ROP vector set.

Note that in Eq. (6), α_i is the occurrence probability of state i, the sum of all α_i is 1. In Eq. (7), $\boldsymbol{P}_\pi$ is the state transition probability matrix of the policy π since the occurrence probability of each state under a policy forms the *Markov chain* as the authors described in previous section. Figure 6 shows the detail of these processes from step1 to step 5.

For example, two Markov chains as shown in Figure 5 (c) are used. Now solving Eq. (7) by $\boldsymbol{P}_{\pi^*12}$, $\vec{\alpha}_{\pi^*12}$ = (3/13, 5/13, 0, 5/13) where occurrence probability of r_1 is 3/13 and occurrence probability of r_2 is 5/13, therefore, ROP vector $\vec{\boldsymbol{P}}_{\pi^*12}$ = (3/13, 5/13). Solving Eq. (7) by $\boldsymbol{P}_{\pi12}$, $\vec{\alpha}_{\pi12}$ = (1/4, 1/4, 1/4, 1/4) where occurrence probability of r_1 and r_2 is 1/4, therefore, ROP vector $\vec{\boldsymbol{P}}_{\pi12}$ = (1/4, 1/4). Note that these ROP vectors will be shown in Figure 7 (a).

Calculating a Convex Hull From a ROP Vector Set

After collecting all ROP vectors as a set, each ROP vector is located **at** a point in the reward occurrence probability (ROP) vector space. Figure 7 shows an illustrated example of the convex hull in two dimensional ROP vector space of the identified

Figure 6. It shows the detail of calculating ROP vector for each collected policy

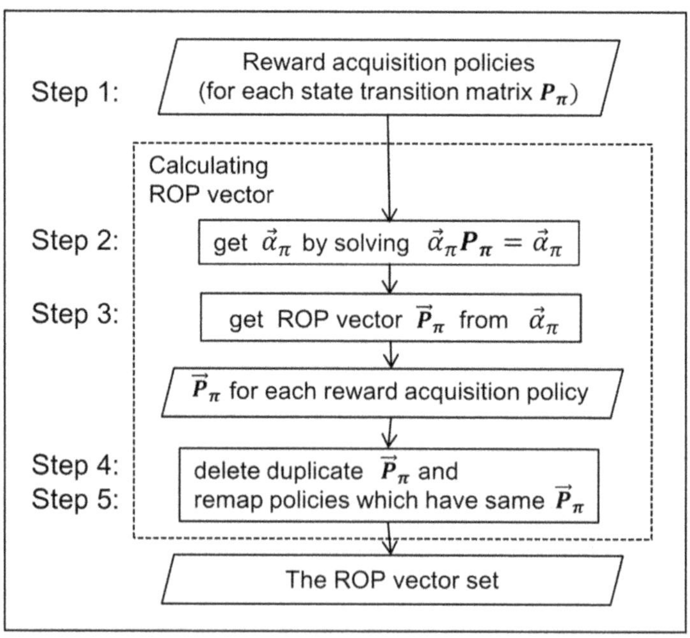

MDP model as shown in Figure 5 (a). In Figure 7 (a) and (b), there are two axis, where the horizontal axis is ROP $\boldsymbol{P}_1$, and the vertical axis is ROP $\boldsymbol{P}_2$. Figure 7 (a) shows the covex hull of six ROP vectors in the P_1 x P_2 ROP vector space. There are five kinds of reward acquisition policies and five kinds of different ROP vectors and one zero vector. In the ROP vector space, there are six ROP vectors including zero vector (0,0). Among them, four ROP vectors are Pareto optimal which are the vertices of the convex hull, and rest two ROP vectors are non-Pareto optimal which are inside of the convex hull, Note that the authors use *n* dimensional Quick Hull algorithm (Barber 1996) of Python library.

Figure 7 (b) shows four divisions of weight vectors among four vertices of the convex hull as Pareto optimal policies. Each vertex of ROP vector has the optimal zone of weighted vectors. The dotted arrows are the limits of optimal zones of weighted vectors. In Figure 7 (b), there are four limits on weighted vectors, $\vec{w}_0$, $\vec{w}_1$, $\vec{w}_2$ and $\vec{w}_3$. For example, $\vec{w}_1$ is the limit weight vector between $\vec{\boldsymbol{P}}_{\pi 1}$ and $\vec{\boldsymbol{P}}_{\pi 12}$ where average reward of them is the same under $\vec{w}_1$. The optimal zone for each ROP vector is as follows;

1. $\vec{\boldsymbol{P}}_{\pi 1}$ is from $\vec{w}_0$ to $\vec{w}_1$
2. $\vec{\boldsymbol{P}}_{\pi 12}$ is from $\vec{w}_1$ to $\vec{w}_2$
3. $\vec{\boldsymbol{P}}_{\pi 3}$ is from $\vec{w}_2$ to $\vec{w}_3$
4. $\vec{\boldsymbol{P}}_{\pi 0}$ is from $\vec{w}_3$ to $\vec{w}_0$

The advantage of the proposed method is that Pareto optimal deterministic policies directly form the vertices of a convex hull in the ROP vector space. Therefore, Pareto optimal policies are calculated independently with weights and just one time. Besides, our method can be easily applied **to** other scalarization function such as Tchebycheff norm method in the ROP vector space.

EXPERIMENTS

Experimental Setup

The authors conduct the experiment to analyze the bottle neck of the proposed method by evaluating the computation cost of our major processes described in previous section. Experimental conditions on stochastic and ergodic MDP model as the learning environment are as follows:

Figure 7. It is an illustrated example of the convex hull in two dimensional ROP vector space of Figure 5. (a) is the covex hull of six ROP vectors in the P_1 x P_2 ROP vector space. (b) is four divisions of weight vectors among four vertices of the convex hull as Pareto optimal policies

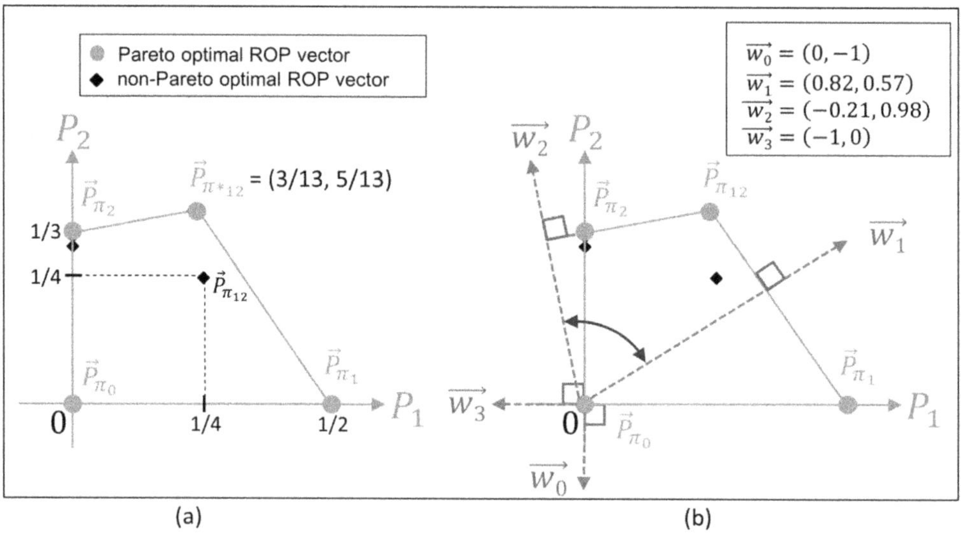

1. The number of states is seven cases, 4,5,6,7, 8, 9, and 10 states.
2. The number of actions is three.
3. The number of rewards is three.
4. The transition probability of each rule is setup randomly under the condition that the number of branches of transitions is randomly setup between 1, 2 or 3.

Experimental results are averaged one hundred experiments with different MDP models. Note that for each measurement item, average and standard deviation (SD) of one hundred data are calculated. Measurement items on memory cost and time cost are as follows:

1. The number of reward acquisition policies.
2. The number of reward acquisition cycles.
3. The number of ROP vectors.
4. The number of vertexes of convex hull.
5. Time cost of model identification [sec].
6. Time cost of search for reward acquisition cycles [sec].
7. Time cost of calculating ROP vectors [sec].

8. Time cost of calculating convex hull [msec].
9. The total performance time [sec].

Experimental Results

The authors conducted the learning experiment for one hundred times and each measurement item is an averaged. In stochastic environment, our proposed method successfully collected all reward acquisition deterministic policies including all Pareto optimal policies under a weighted sum scalarization function where weights are unknown. Figure 8 shows the number of vertexes of convex hull according to the number of states. It suggests that the increase of it is linear according to the number of states.

To analyze the bottle neck of the proposed method, Figure 9 shows the overview of the analysis of the experimental results. Figure 9 (a) shows the number of reward acquisition policies and cycles according to the number of states as the memory costs, and Figure 9 (b) shows the total performance time according to the number of states as the time cost. It suggests that the increase of them is exponential according to the number of states.

Table 1 shows the detailed analysis of the experimental results on memory cost according to the number of states. It includes the number of reward acquisition policies, the number of reward acquisition cycles, the number of ROP vectors and the number of vertexes of convex hull. First, the number of reward acquisition policies increases triple which is equal to the number of actions. The reason is that the number of deterministic policies are $|A|^{|S|}$ where $|A|$ is the number of actions and $|S|$ is the number of states. Second, the number of reward acquisition policies, the number of reward acquisition cycles and the number of ROP vectors increase exponentially according to the number of states. As the experimental result, the bottle neck of the memory cost is the number of reward acquisition policies. This memory cost problem is discussed in next subsection.

Table 2 shows the detailed analysis of the experimental results on time cost according to the number of states. It includes the time costs of model identification, search for reward acquisition cycles, calculating ROP vectors, calculating convex hull and the total performance time. Note that the measure of time of calculating convex hull is milli-second, rest of the measure of time costs is second. As the experimental result, proportion of the time costs of calculating ROP vectors to the total performance time becomes largest according to the number of states. It becomes more than 90% to the total performance time under 9 and 10 states. Therefore, the bottle neck of the time cost is calculating ROP vectors. This time cost problem is discussed in next subsection.

Figure 8. The number of vertexes of convex hull is shown according to the number of states

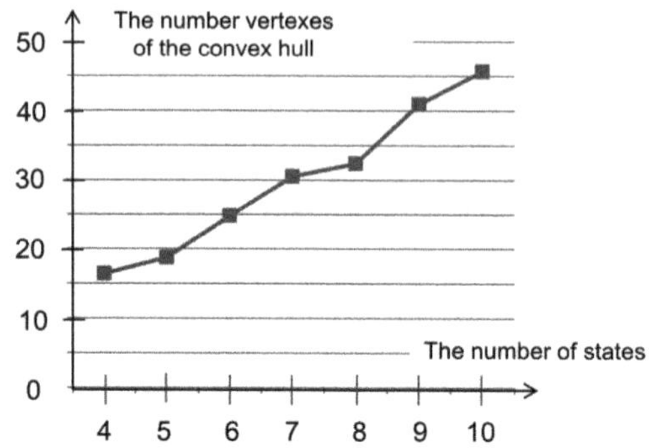

SOLUTIONS AND RECOMMENDATIONS

This section discusses the memory cost and time cost problems of the proposed method. First, the way to reduce the bottle neck of the memory cost is discussed. The authors start to discuss the partial computation for searching reward acquisition policies. Main idea is to put off collecting reward acquisition policies after calculating convex hull. It is possible to calculate ROP vectors from reward acquisition cycles which are shown in figure 4 (c). Main issue is to treat the state transition possibility matrix for the states out of the cycle since their possibilities are not determined since these states form some temporary path toward the cycle. To solve this, the

Figure 9. The overview of the analysis of the experimental results is shown. (a) shows the number of reward acquisition policies and cycles according to the number of states as the memory costs. (b) shows the total performance time according to the number of states as the time cost

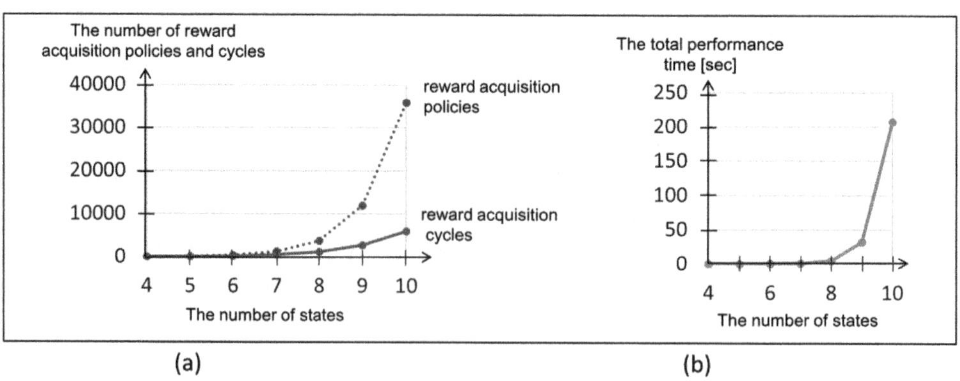

Table 1. It is the detailed analysis of the experimental results on memory cost according to the number of states

The Number of States	The Number of Reward Acquisition Policies	The Number of Reward Acquisition Cycles	The Number of ROP Vectors	The Number of Vertexes of Convex Hull
	Average (SD)	Average (SD)	Average (SD)	Average (SD)
4	53.3 (13.1)	46.6 (12.6)	34.2 (14.0)	16.4 (5.8)
5	148.2 (41.5)	106.6 (36.4)	71.2 (31.6)	18.8 (7.1)
6	455.0 (123.5)	253.8 (89.7)	168.4 (84.2)	24.8 (11.4)
7	1327.5 (435.4)	599.9 (249.4)	400.1 (203.3)	30.5 (14.0)
8	3849.1 (1255.5)	1276.0 (499.4)	844.6 (415.4)	32.4 (16.7)
9	12003.8 (3545.4)	2846.6 (1330.1)	2016.9 (1086.0)	40.9 (19.4)
10	35947.1 (12007.5)	5999.2 (2988.0)	4318.4 (2604.0)	45.6 (21.0)

Table 2. It is the detailed analysis of the experimental results on time cost according to the number of states

The Number of States	Model Identification [sec]	Search for Reward Acquisition Cycles [sec]	Calculating ROP Vectors [sec]	Calculating Convex Hull [msec]	The Total Performance Time [sec]
	Average (SD)	Average (SD)	Average (SD)	Average (SD)	Average (SD)
4	0.10 (0.06)	0.01 (0.01)	0.01 (0.01)	0.47 (2.66)	0.12 (0.06)
5	0.23 (0.20)	0.03 (0.01)	0.03 (0.01)	1.25 (4.24)	0.29 (0.20)
6	0.54 (0.74)	0.07 (0.02)	0.15 (0.08)	1.42 (4.47)	0.76 (0.72)
7	1.46 (3.78)	0.18 (0.06)	0.90 (0.58)	2.19 (5.42)	2.56 (3.68)
8	1.25 (1.18)	0.47 (0.17)	4.93 (3.27)	1.87 (5.08)	6.67 (3.38)
9	2.10 (2.33)	1.09 (0.39)	32.53 (24.36)	3.75 (6.67)	35.78 (24.33)
10	4.31 (9.38)	2.83 (1.08)	206.96 (167.36)	6.06 (6.69)	214.15 (168.54)

authors proposes to assume the states out of the cycle as isolated states since the occurrence probability of a state out of the cycle which in on some temporary path will converge to zero when solving Eq. (7). Figure 10 shows an illustrated example of the way to calculate ROP vectors from reward acquisition cycles. Figure 10 (a) shows an example of the identified MDP model as shown in figure 5 (a). Figure 10 (b) shows an example of reward acquisition cycle where states 2 and 3 are out of cycle. Figure 10 (c) shows the assumed state transition possibility matrix of Figure 10 (b) where the state transition possibilities from state 2 and 3 are all zero.

Second, the time cost problem of the proposed method is discussed. To reduce the bottle neck of the time cost, the authors make a study on analysis of calculating ROP vectors in the latest preliminary experiments as shown in Figure 6. Then, most of time cost of calculating ROP vectors is concerned with two sub-processes. One is to delete duplicate ROP vectors in the ROP vector set. As shown in Table 1, the ratio of the number of reward acquisition policies to the number of ROP vectors is about eight times under the number of states is ten. In other words, about eight different or similar reward acquisition policies have the same ROP vector on average. To solve these problems, the authors use to sort a ROP vector set using multiple keys to delete duplicate ROP vectors in review. Note that the program is implemented by Python list. Another sub-process to be reduced is mapping ROP vector to associated policies. This problem will be solved after solving memory cost problem as previously discussed. Alternate approach will be discussed in Future research directions.

FUTURE RESEARCH DIRECTIONS

This section discusses the possibilities of this research focusing on the aspect of the distribution of ROP vector set in the convex hull. Figure 11 shows the aspect of the distribution of ROP vector set in the convex hull.

First, alternate approach to solve the memory cost and time cost problems of the proposed method is discussed. Figure 11 (a) shows the density distributed ROP vectors on the axes. The authors discovered that most of ROP vectors acquire only one reward which are located on any of axes in the ROP vector space by sorting collected ROP vectors by the number of acquired rewards.

This property is useful to reduce the computation cost since optimizing ROP vectors with one reward can be preprocessed with low time cost. The main idea is that after classified ROP vectors with one reward by acquired reward, pick up two kinds of ROP vectors with maximum reward occurrence probability and minimum reward occurrence probability without zero. Figure 11 (b) shows the two kinds of non Pareto-optimal ROP vector subset. In each axis, there are three ROP vectors,

Figure 10. An illustrated example of the way to calculate ROP vectors from reward acquisition cycles is shown. (a) shows an example of the identified MDP model as shown in figure 5 (a). (b) shows an example of reward acquisition cycle where states 2 and 3 are out of cycle. (c) shows the assumed state transition possibility matrix of Figure 10 (b) where the state transition possibilities from state 2 and 3 are all zero

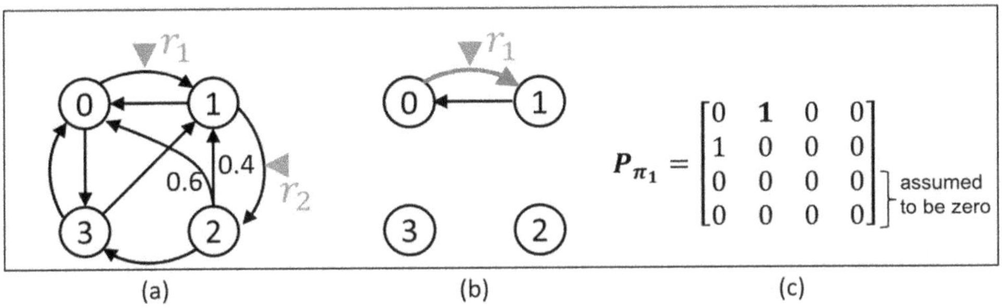

one is zero vector and rest of them are maximum and minimum reward occurrence probability without zero. Note that the procedure of this preprocess ii before sorting the ROP vector set using multiple keys to delete duplicate ROP vectors described as previous section.

Second, how non Pareto-optimal ROP vectors work is discussed. Figure 11 (b) shows the two kinds of non Pareto-optimal ROP vector subset. One is minimum and positive ROP vector set, the other is non-Pareto optimal ROP vectors inside of the convex hull. Now the meaning of Pareto optimal ROP vectors in the proposed method is described. In this research, Pareto optimal policies has a shorter length of cycle than non-Pareto optimal ones since the Reward occurrence probability per step is larger. As figure 4 (b) shows the policy search process by breadth search, policy acquisition cycles with short length is found at first. This suggests that the mastery task of human or robot behavior learning, the length of the reward acquisition cycle means the complexity of the learned output. Therefore, considering the ROP vector space as the mastery space as shown in Figure 10 (b), minimum and positive ROP vector set suggest the mastery goals, and ROP vectors in the convex hull are sub-goals toward the mastery goals.

CONCLUSION

This chapter proposed the novel model-based MORL method by reward occurrence probability (ROP) with unknown weights. There are two main features. First one is that average reward of a policy is defined by inner product of the ROP vector and the weight vector. Second feature is that it learns ROP in each policy instead

Figure 11. It shows the aspect of the distribution of ROP vector set in the convex hull. (a) shows the density distributed ROP vectors on the axes. (b) shows the two kinds of non Pareto-optimal ROP vector subset

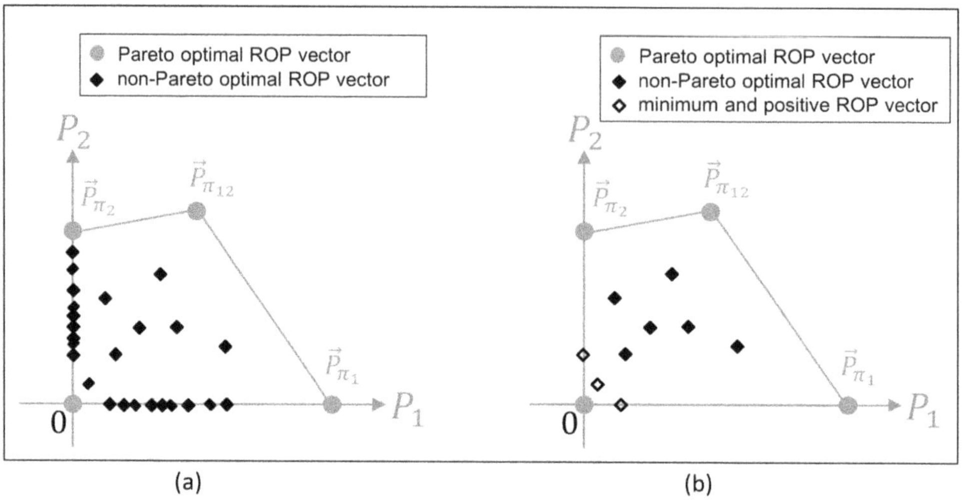

of Q-values. Pareto optimal deterministic policies directly form the vertices of a convex hull in the ROP vector space. Therefore, Pareto optimal policies are calculated independently with weights and just once. The experimental results show that the authors' proposed method collected all Pareto optimal policies under three dimensional stochastic environments, and it takes a small computation time though previous MORL methods learn at most two or three dimensions deterministic environments.

Future works is to reduce calculation cost for collecting Pareto optimal policies. The authors are planning two ways, one is pruning in the policy search on collecting all reward acquisition policies by estimating the length of the reward acquisition cycle. Another approach is applying parallel computing techniques, multi-processing by Multi-core CPU for policy search, and GPGPU for calculating ROP vector.

ACKNOWLEDGMENT

The authors would like to thank Prof. Shimohara and Prof. Habib for offering a good opportunity to present this research. This work was supported by JSPS KAKENHI(Grant-in-Aid for Scientific Research ©) [Grant Number 16K00317].

REFERENCES

Barber, C. B., David, P., Dobkin, D. P., & Huhdanpaa, H. (1996). The Quickhull Algorithm for Convex Hulls. *ACM Transactions on Mathematical Software*, *22*(4), 469–483. doi:10.1145/235815.235821

Barrett, L., & Narayanan, S. (2008). Learning All Optimal Policies with Multiple Criteria. In *Proceedings of the 25th International Conference on Machine Learning (ICML-2008)* (pp. 41-47). ACM. 10.1145/1390156.1390162

Barto, A. G., Steven, J., Bradtke, S. J., & Singh, S. P. (1995). Learning to act using real-time dynamic programming. *Artificial Intelligence*, *72*(1-2), 81–138. doi:10.1016/0004-3702(94)00011-O

Gao, Y. (2006, November). *Research on Average Reward. Reinforcement Learning*. Algorithms, National Laboratory for Novel Software Technology, Nanjing University. Retrieved from http://lamda.nju.edu.cn/conf/MLA06/files/Gao.Y.pdf

Handa, H. (2009). Solving multi-objective reinforcement learning problems by EDA-RL -, acquisition of various strategies, In *Proceedings of the Ninth International Conference on Intelligent Systems Design and Applications* (pp. 426-431). Pisa, Italy: IEEE 10.1109/ISDA.2009.92

Herrmann, M. (2015). *RL 16: Model-based RL and Multi-Objective Reinforcement Learning*. University of Edinburgh, School of Informatics. Retrieved from http://www.inf.ed.ac.uk/teaching/courses/rl/slides15/rl16.pdf

Hiraoka, K., Yoshida, M., & Mishima, T. (2009). Parallel reinforcement learning for weighted multi-criteria model with adaptive margin. *Cognitive Neurodynamics*, *3*(1), 17–24. doi:10.100711571-008-9066-9 PMID:19003453

Konda, T., Tensyo, S., & Yamaguchi, T. (2002). LC-Learning: Phased Method for Average Reward Reinforcement Learning - Preliminary Results. In *Proceedings of 7th Pacific Rim International Conference on Artificial Intelligence (PRICAI2002): PRICAI2002 Trends in Artificial Intelligence, Lecture notes in Artificial Intelligence 2417* (pp. 208-217). Berlin, Germany: Springer. 10.1007/3-540-45683-X_24

Liu, C., Xu, X., & Hu, D. (2015). Multiobjective Reinforcement Learning: A Comprehensive Overview. *IEEE Transactions on Systems, Man, and Cybernetics. Systems*, *45*(3), 385–398. doi:10.1109/TSMC.2014.2358639

Lizotte, D. J., Bowling, M., & Murphy, S. A. (2012). Linear fitted-q iteration with multiplereward functions. *Journal of Machine Learning Research*, *13*, 3253–3295. PMID:23741197

Mahadevan, S. (1996). Average Reward Reinforcement learning: Foundations, Algorithms, and Empirical Results. *Machine Learning*, *22*(1-3), 159–196. doi:10.1007/BF00114727

Moffaert, K. V., & Nowe, A. (2014). Multi-Objective Reinforcement Learning using Sets of Pareto Dominating Policies. *Journal of Machine Learning Research*, *15*, 3663–3692.

Natarajan, S., & Tadepalli, P. (2005). Dynamic preferences in multi-criteria reinforcement learning. In *Proceedings of the 22nd international conference on machine learning (ICML-2005)* (pp. 601-60). ACM

Pinder, J. M. (2016). *Multi-Objective Reinforcement Learning Framework for Unknown Stochastic & Uncertain Environments* (Unpublished doctoral dissertation). Retrieved from http://usir.salford.ac.uk/id/eprint/39978/ 2/John%20Pinder%20 PhD%20Thesis%20Complete.pdf

Puterman, M. L. (1994). *Markov Decision Processes: Discrete Stochastic Dynamic Programming*. Hoboken, NJ: John Wiley & Sons, Inc. doi:10.1002/9780470316887

Roijers, D. M., Vamplew, P., Whiteson, S., & Dazeley, R. (2013). A Survey of Multi-Objective Sequential Decision-Making. *Journal of Artificial Intelligence Research*, *48*, 67–113. doi:10.1613/jair.3987

Roijers, D. M., Whiteson, S., Vamplew, P., & Dazeley, R. (2015). Why Multi-objective Reinforcement Learning? In *Proceedings of the 12th European Workshop on Reinforcement Learning (EWRL 2015)*, (pp. 1-2). Academic Press.

Satoh, K., & Yamaguchi, T. (2006). *Preparing various policies for interactive reinforcement learning*. Paper presented at the meeting of SICE-ICASE International Joint Conference 2006.

Tadepalli, P., & Ok, D. (1998). Model-Based Average Reward Reinforcement Learning. *Artificial Intelligence*, *100*(1-2), 177–224. doi:10.1016/S0004-3702(98)00002-2

Tajmajer, T. (2017). *Multi-objective deep Q-learning with subsumption architecture*. Retrieved from https://arxiv.org/pdf/1704.06676v1.pdf

Tsitsiklis, J. N. (2007). NP-Hardness of checking the unichain condition in average cost MDPs. *Operations Research Letters*, *35*(3), 319–323. doi:10.1016/j.orl.2006.06.005

Wiering, M. A., Withagen, M., & Drugan, M. M. (2014). Model-based multiobjective reinforcement learning. In *Proceedings of the IEEE Symposium on Adaptive Dynamic Programming and Reinforcement Learning* (pp. 1-6). IEEE.

Yang, S., Gao, Y., Bo, A., Wang, H., & Chen, X. (2016). Efficient Average Reward Reinforcement Learning Using Constant Shifting Values. *Proceedings of the Thirtieth AAAI Conference on Artificial Intelligence (AAAI-16)*, 2258-2264

ADDITIONAL READING

Drugan, M., Wiering, M., Vamplew, P., & Chetty, M. (2017). Special issue on multi-objective reinforcement learning. *Neurocomputing*, *263*(8), 1–2. doi:10.1016/j.neucom.2017.06.020

Drugan, M., Wiering, M., Vamplew, P., & Chetty, M. (2017). Multiobjective Reinforcement Learning: Theory and Applications. *Neurocomputing*, *263*(8), 1–86. doi:10.1016/j.neucom.2017.06.020

Gábor, Z., Kalmár, Z., & Szepesvári, C. (1998). Multi-criteria reinforcement learning. In *Proceedings of the Fifteenth International Conference on Machine Learning* (pp. 197-205). Morgan Kaufmann Publishers Inc.

Mnih, V., Kavukcuoglu, K., Silver, D., Rusu, A. A., Veness, J., Bellemare, M. G., & Petersen, S. (2015). Human-level control through deep reinforcement learning. *Nature*, *518*(7540), 529–533. doi:10.1038/nature14236 PMID:25719670

Moffaert, K. V., Drugan, M., & Nowe, A. (2013). Scalarized Multi-Objective Reinforcement Learning: Novel Design Techniques, In *Proceedings of IEEE Symposium on Adaptive Dynamic Programming and Reinforcement Learning (ADPRL)*, IEEE 10.1109/ADPRL.2013.6615007

Mossalam, H., Assael, Y. M., Roijers, D. M., & Whiteson, S. (2016). Multi-objective deep reinforcement learning. arXiv preprint arXiv:1610.02707. Retrieved from https://arxiv.org/pdf/1610.02707.pdf

Nguyen, T. T. (2018). A Multi-Objective Deep Reinforcement Learning Framework, Retrieved from http:// arxiv.org/ftp/arxiv/papers/1803/1803.02965.pdf

Tanner, B., & White, A. (2009). RL-Glue: Language-independent software for reinforcementlearning experiments. *Journal of Machine Learning Research*, *10*(Sep), 2133–2136.

Uchibe, E., & Doya, K. (2008). Finding intrinsic rewards by embodied evolution and constrained reinforcement learning. *Neural Networks*, *21*(10), 1447–1455. doi:10.1016/j.neunet.2008.09.013 PMID:19013054

Vamplew, P., Dazeley, R., Berry, A., Issabekov, R., & Dekker, E. (2011). Empirical evaluation methods for multiobjective reinforcement learning algorithms. *Machine Learning*, *84*(1-2), 51–80. doi:10.100710994-010-5232-5

Vamplew, P., Webb, D., Zintgraf, L. M., Roijers, D. M., Dazeley, R., Issabekov, R., & Dekker, E. (2017). MORL-Glue: a benchmark suite for multi-objective reinforcement learning. In *Proceedings of the 29th Benelux Conference on Artificial Intelligence (BNAIC 2017)*, 389, Groningen, The Netherlands. Retrieved from http://roijers.info/pub/morl-glueBnaic17.pdf

Vamplew, P., Yearwood, J., Dazeley, R., & Berry, A. (2008, December). On the limitations of scalarisation for multi-objective reinforcement learning of Pareto fronts. In *Proceedings of Australasian Joint Conference on Artificial Intelligence* (pp. 372-378). Springer, Berlin, Heidelberg. 10.1007/978-3-540-89378-3_37

KEY TERMS AND DEFINITIONS

Average Reward: The expected received rewards per step when an agent performs state transitions routinely according to a policy.

LC-Learning: One of the average reward model-based reinforcement learning methods. It collects all reward acquisition deterministic policies under the unichain condition.

Markov Chain: A stochastic model describing a sequence of possible states in which the probability of each state depends only on the previous state. It is an intension of Markov decision processes, the difference is the subtraction of actions and rewards.

Markov Decision Process (MDP): It is a discrete time and a discrete state space stochastic control process. It provides a mathematical framework for modeling decision making in situations where outcomes are partly random and partly under the control of a decision maker.

Model-Based Approach: The reinforcement learning algorithm which starts with directly estimating the MDP model statistically, then calculates the value of each state as V(s) or the quality of each state action pair Q(s, a) using the estimated MDP to search the optimal solution that maximizes V(s) of each state.

Multi-Objective MDP (MOMDP): An MDP in which the reward function describes a vector of n rewards (reward vector), one for each objective, instead of a scalar.

Pareto Optimization: It is to find multiple policies that cover the Pareto front, which requires collective search for sampling the Pareto set.

Reinforcement Learning: The popular learning algorithm for automatically solving sequential decision problems. It is commonly modeled as Markov decision processes (MDPs).

Reward Acquisition Probability (ROP): The expected occurrence probability per step for the reward.

Weight Vector: A trade-off among multi objective, and each element of the vector represents a weight of each objective.

Chapter 11
Autonomous Surgical Robotics at Task and Subtask Levels

Tamás Dániel Nagy
https://orcid.org/0000-0002-5485-9588
Óbuda University, Hungary

Tamás Haidegger
https://orcid.org/0000-0003-1402-1139
Óbuda University, Hungary

ABSTRACT

The revolution of minimally invasive procedures had a significant influence on surgical practice, opening the way to laparoscopic surgery, then evolving into robotics surgery. Teleoperated master-slave robots, such as the da Vinci Surgical System, has become a standard of care during the last few decades, performing over a million procedures per year worldwide. Many believe that the next big step in the evolution of surgery is partial automation, which would ease the cognitive load on the surgeon, making them possible to pay more attention on the critical parts of the intervention. Partial and sequential introduction and increase of autonomous capabilities could provide a safe way towards Surgery 4.0. Unfortunately, autonomy in the given environment, consisting mostly of soft organs, suffers from grave difficulties. In this chapter, the current research directions of subtask automation in surgery are to be presented, introducing the recent advances in motion planning, perception, and human-machine interaction, along with the limitations of the task-level autonomy.

DOI: 10.4018/978-1-7998-1382-8.ch011

INTRODUCTION

Minimally Invasive Surgery (MIS) reshaped surgical practice significantly during the last decades. Contrary to the traditional technique operating through large incisions, MIS is performed through few-centimeter wide ports—so-called keyholes—using laparoscopic instruments, the area of operation is observed on endoscopic camera stream. Smaller incisions offers benefits both for the patient and the hospital, like lower risk of complications, rapid recovery and thus shorter hospital stay. On the other hand, MIS presents new challenges to the surgeons, such as the limited range of motion, different forms of motion, with less Degrees of Freedom (DoF) and also fatigue from weary postures.

Robot-Assisted Minimally Invasive Surgery (RAMIS) was introduced to ease these difficulties. The idea of teleoperated master–slave surgical systems originates from space research: the intervention was to be performed on the patient—in this case an astronaut—by a slave device, controlled by a human surgeon through a master device on Earth (Márton, Szántó, Haidegger, Galambos & Kövecses, 2017; Takács, Nagy, Rudas & Haidegger, 2016). The slave side robot arms are equipped by laparoscopic instruments and an endoscopic camera, and copy the movement of the surgeon next to the remote master console, who is able to observe the operation on the endoscopic camera stream.

In the past years, there has been a strong uptake of Computer-Integrated Surgery (CIS) systems, and their current global annual market potential is estimated at $11.4 bn, with an expected annual growth of 7%, according to a recent study by LG Electronics. Also, more and more people accept the advancement of surgical robotics, although, the willingness was measured lower in the developed countries in a recent survey conducted by PricewaterhouseCoopers (PwC) (Fig. 1). A recent Eurobarometer poll found that 26% of responders would be comfortable with a robot surgeon operating on them (2% up from 2014)[1], while IEEE found

Figure 1. International acceptance of robotic surgery. A 2017 study by PwC found that there is growing number of people who would embrace these technologies (PricewaterhouseCoopers, 2017) Image credit: PwC

that 62% of people would be willing to allow an AI driven robot to operate on their children if need be[2].

However, real remote teleoperation has not become a daily practice, and stalled at the state of research mainly due to the issues caused by time delay, it turned out that teleoperation itself can present a number of benefits. The communication latency—being the biggest issue—can be reduced to a level that is insignificant for the surgeon by placing the master and the slave devices close to each other; in the case of commercial RAMIS, the master and the slave side devices are in the same room. This technique can reduce the fatigue of the surgeon, being able to operate in a more ergonomic, seated position. Furthermore, the movement of the surgeon can be scaled on the slave side—the most delicate movements can be controlled by relatively large hand movements, and hand tremors can also be filtered.

There is no doubt, that the most successful RAMIS device on the market is the da Vinci Surgical System (Intuitive Surgical Inc., Sunnyvale, CA). The 1st generation da Vinci was cleared by the U. S. Food and Drug Administration (FDA) in 2000, and soon became widely used. In 2019, its 4th generation is available (Fig. 2.), with more than 5300 units installed worldwide, which performed over 1 million procedures last year (Takács et al., 2016).

BACKGROUND

Many believe that the next necessary step in the evolution of surgery is automation. Already, as of today, the partial automation of surgery is feasible. The workflow of RAMIS procedures frequently contains monotonous and time-consuming elements, automation of such subtasks would ease the cognitive load on the surgeon, making them possible to pay more attention on the critical parts of the intervention (Elek et al., 2017, 2016).

Unfortunately, autonomy in the given environment, consisting mostly of soft tissues, suffers from grave difficulties. Unlike working on rigid-tissues, where exact registration to the instrument is possible, soft tissues are permanently in motion, and highly deformable, thus no pre-computed tool trajectories can be used. A key enabling approach to reduce complexity is dividing the surgical workflow into atomic elements. Identifying surgical subtasks makes it possible to guide the robotic instruments during soft tissue operations, following the decomposed surgical motions. This can be performed on different levels of granularity, and can be used to compile a parameterizable motion library (D. Á. Nagy, Nagy, Elek, Rudas & Haidegger, 2018; T. D. Nagy & Haidegger, 2018).

Another challenge of surgical automation is undoubtedly the implementation of perception algorithms usable in the complex human environment. Computer

Figure 2. The 4 generations of the da Vincy Surgical System; a) da Vinci Classic, b) da Vinci S, c) da Vinci Si and d) da Vinci Xi
Image credit: Intuitive Surgical Inc.

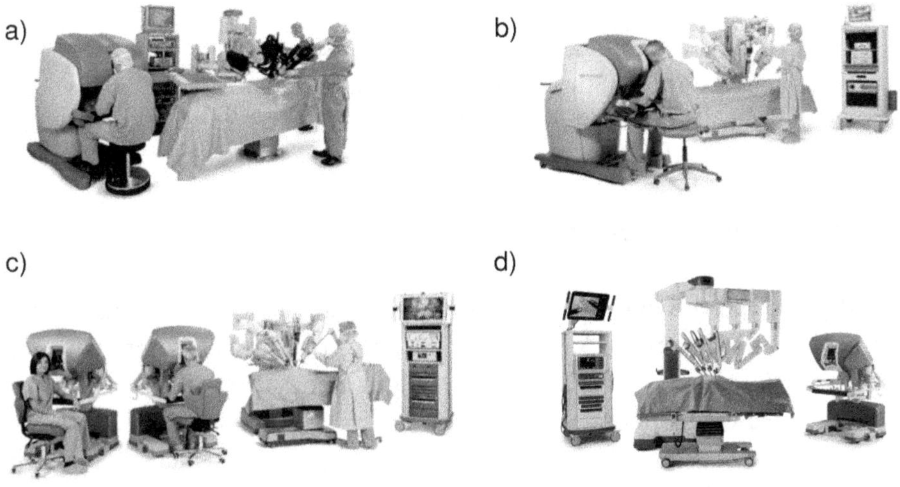

vision suffers from reflections and the fact that the visual features of different organs being very similar, yet it is still the gold standard. New methods relying on palpation emerged in the last years, where force sensors can be utilized e.g., to find the location of tumors or other anatomies/pathologies (Garg et al., 2016; Wei, Arbter & Hirzinger, 1997; Yamamoto, Vagvolgyi, Balaji, Whitcomb & Okamura, 2009).

Finally, irrespective of the nature of automation, the supervision of the surgeon is crucial and safety critical during the execution, that requires effective human—machine interaction. The surgeon should be able to parameterize and launch autonomous execution. They also must have the ability observe the area of operation during autonomous execution, and to gain back manual control anytime necessary, or the autonomous algorithm encounters events it cannot solve (Chen, 2016), similarly to the domain of self-driving cars (Takács, Drexler, Galambos, Rudas & Haidegger, 2018).

In this chapter, the main challenges associated with the development of autonomous surgical robotics are discussed, starting with the objective assessment of capabilities and functions. The above-mentioned aspects of subtask automation in surgery are presented, introducing the recent advances in surgical robot motion planning, perception and human–machine interaction, along with the limitations of the task-level autonomy.

STANDARDIZATION AND SAFETY CONCERNS

One of the benefits of international standards is that they may facilitate the safe implementation of new technologies into advanced prototypes and products. When it comes to the field of autonomous surgical robotics, safety is probably the biggest concerns, therefore the major Standards Development Organizations (SDOs) have long started to take action – working towards a safety standard framework for medical and surgical robots (Haidegger, 2019).

Degree of Autonomy of Surgical Equipment

Roughly 8 years ago the joint International Organization for Standardization and International Electrotechnical Commission (ISO/IEC) Technical Committee (TC) group analyzed the status of surgical robot standardization. Only one major gap was found: the Degree of Autonomy (DoA)—introduced in ISO 8373:2012 - Robots and robotic devices -- Vocabulary—was not defined properly. Understanding the fact that the proper definition of autonomy and its conjugated forms "autonomous", "automation", or related definitions can be unambiguous, the ISO/IEC joint working group decided to extend the scope of their work to all Medical Electrical Equipment (MEE) or Medical Electrical System (MES) with a DoA (other than zero). The discussion on the topic was concluded in a new Technical Report (TR) IEC/TR 60601-4-1: Medical electrical equipment – Part 4-1: Guidance and interpretation – Medical electrical equipment and medical electrical systems employing a degree of autonomy. The TR recommended to omit such words like "automation" or "automatic" within this robotic standard; DoA was defined instead as "taxonomy based on the properties and capabilities of the MEE or MES related to autonomy".

Derived from the field of industrial automation (Kaber & Endsley, 2004) and service robotics (Huang, 2008), the TR recommends the parameterization of DoA along four cognition-related functions of a system, which are affecting options of an MES:

- Generate an option: to formulate possible options, based on the result of the monitoring task for achieving predefined goals.
- Execute an option: to carry out the selected option. Robots can typically be active or passive supporters of a surgical task execution.
- Monitor an option: to collect necessary information to perceive the status of MEE or MES, patient, operator or environment. Therefore, signals beyond the internal (proprioceptive) control signals of the robot.
- Select an option: to decide on a particular option from the pool of generated.

$$DoA = \mathcal{F}\{G\big|E\big|M \mid S\},$$

where the overall DoA metric is normed sum of the four function of the system assessed on a linear scale, 0 meaning fully manual and 1 fully autonomous.

The objective assessment of the DoA of a system can be utilized along the four described functions, each can be performed by a human or by a computer. The DoA of the system is defined on a scale from 0 to 1; DoA = 0 means "no autonomy", and the highest DoA represents "full autonomy". The low-level electronic and computational functions of MEE or MES, like communication or motor control, are excluded from this definition, as the term "no autonomy" is meant on the system level. Also, a classification of DoA can be given at different levels of granularity, depending on the level where are those functions implemented.

Autonomous execution is already present is surgery (especially in image-guided surgery), however decision making (selecting) is mostly done by human experts (Haidegger, 2019). On the other hand, computer systems are more capable in monitoring task compared to humans, hence most critical processes happen at a faster scale humans can perceive; this is the safety concept of Situation Awareness (SA).

The TR offers various alternatives for DoA assessment; the most applicable one, based on industrial automation, defines 10 levels of DoA (Table 1). Practically, during the risk management of a surgical robot (most commonly according to ISO 14971 – Application of risk management to medical devices), DoA should also be taken into account; DoA does not necessarily correlates with the level of risk, but it may impact risk management gravely. In case of an error or malfunction, the take-over of the human operator may be necessary. Due to loss of SA the operator may not be able to control the situation properly, thus the DoA determines the handling of the hazards. At lower DoA, the responsibility can be shared between the human and the robot; at higher DoA malfunctions presents critical hazards.

Basic Safety and Essential Performance of Surgical Robots

From the aspect of the user (and also of the manufacturer), avoiding and managing any kind of failure (software, hardware, communication, system-level) is critical. In the past three decades, 33 documented casualties were caused by industrial robots, which is still only giving 0.0005% of all work-related deaths (Delvaux, 2016) None of the surgical robotic cases are included in this statistics. In surgical robotics—especially in RAMIS—the human surgeon takes full liability for the outcome of the intervention (as ruled in all juridical cases up to now).

With the increase of DoA, risk mitigation and management becomes crucial. Methodologies to support the safety of design and development of robotic devices

Table 1. Descriptive classification of degree of autonomy adapted from (Kaber & Endsley, 2004) based on IEC/TR 60601-4-1. H: the human operator performs the given function. C: the computer-driven system performs the given function

DoA	Description	Monitor	Generate	Select	Execute
1. 0	**Full Manual (FM):** No autonomy involved. The operator performs all tasks including monitoring the state of the system, generating performance options, selecting the option to perform (decision making) and executing the decision made, i.e, physically implementing it.	H	H	H	H
2. 1	**Teleoperation (TO):** The equipment assists the operator with the execution of the selected action, although continuous operator control is required. The operator performs all tasks, including monitoring the state of the equipment, generating options, selecting the desired option and execution of it. (Master–Slave teleoperation.) Note: traditional robotics standards consider teleoperation as zero DoA.	H/C	H	H	H/C
3. 2	**Pre-Programmed Execution (PE):** The operator generates and selects the options to be performed without any analysis or selection by the equipment. Note: traditional robotic standards considered this as "autonomous" or "automatic" operation.	H/C	H	H	C
4. 3	**Shared Decision (SD):** Both the operator and the equipment generate possible decision options. The operator retains full control over the selection of which option to execute. Both the operator and the equipment participate in the execution.	H/C	H/C	H	H/C
5. 4	**Decision Support (DS):** Me equipment generates a list of decision options, which the operator can select from, or the operator may generate alternative options. Once the human has selected an option, it is turned over to the equipment to execute it.	H/C	H/C	H	C
6. 5	**Blended Decision (BD):** The equipment generates a list of decision options, which it selects from and executes if the operator consents. The operator may also generate and select an alternative option; the equipment will then execute the selected action. BD represents a high-level decision support system that is capable of selecting among alternatives as well as executing the selected option.	H/C	H/C	H/C	C
7. 6	**Guided Decision (GD):** The equipment presents a set of actions to the operator. The operator's role is to select from this set, he/she cannot generate any other additional options. The equipment will fully executes the selected action.	H/C	C	H	C
8. 7	**Autonomous Decision (AD):** The equipment selects the best option and executes it, based upon a list of alternatives it generates (this list can be augmented by alternatives suggested by the operator).	H/C	H/C	C	C
9. 8	**Operator Monitoring (OM):** The equipment generates options, selects the option to implement and executes it. The operator monitors the equipment and intervenes if necessary. Intervention places the human in the role of making a different option selection. During the procedure there may be decision making points that will be decided by the equipment.	H/C	C	C	C
10. 9	**Full Autonomy (FA):** The equipment carries out all actions. The operator does not intervene except to e-stop the equipment (which is a general requirement).	C	C	C	C

were published by various groups, like the generic Hazard Identification and Safety Insurance Control (HISIC) policy that has been applied to several robotic systems (Fei, Ng, Chauhan & Kwoh, 2001).

In 2015 the ISO/IEC TC 62/SC 62D joint committee started a discussion on the minimum requirements for a practical degree of safety for surgical robots; the results are planned to be published in 2019 as IEC 80601-2-77: Particular requirements for the basic safety and essential performance of robotically assisted surgical equipment. The standard collects all relevant thermal and mechanical hazards, along with the required usability trials. It also defines the basic types of surgical robots and their relevant components. In terms of RAMIS, the focal points of the standard are:

- Robotic surgical instruments.
- The patient-side part of the robot.
- The operator-side part of the robot.
- The endoscope holder (if any).

The same committee also presented a standard, focusing on the hazards related to loss of SA, namely IEC/CD 80601-2-78: Particular requirements for the basic safety and essential performance of medical robots for rehabilitation, compensation or alleviation of disease, injury or disability. This standard defines SA as "the operator's perception, comprehension and prediction of a robot's behavior in its environment". SA is a key factor in tasks where human supervision or interaction with the robot is necessary to reduce risk. The standard states that the manufacturers will have to include necessary SA information for their future medical robotic systems. The quality of the Human–Robot Interface (HRI) is essential to keep SA; if the critical information is forwarded to the human operator through an adequate way, e.g., sound alerts, that may highly increase SA. Furthermore, using AI methods, the system may also offer decision support to handle complex situations (Chen, 2016).

It is believed by the community, that upcoming standards would focus more on the safety of the patients and the improvement of the treatments rather than technical metrics, and willing to continue work aiming that goal.

Level of Autonomy for Surgical Robots

Although the standards, mentioned above are fundamental for the assessment of the capabilities of surgical robotic systems, they are not definite enough to present taxonomy to generally assess the development phases of surgical robotics, or to perform benchmarking. Surgical robotic systems need to be categorized based on their advancement, relative in the field. A gradual mapping was presented in (Yang et al., 2017), to classify the autonomous capabilities of surgical robots. Some earlier work suggested to put the HRI into the center of the classification, defining a 0–7 scale (Rosen, 2013). Similar concepts are also presented in the field of self-driving; in (Takacs, Rudas, Bosl & Haidegger, 2018) a 6-grade scale was introduced for autonomous vehicles. At higher levels of autonomy, the role of the robot is still argued; the robot is not only a medical device anymore, but it also practices medicine, which is entirely different from the viewpoint of regulatory. FDA, for example, regulates medical devices, but not the practice of medicine.

The mapping of (Yang et al., 2017) has one fundamental problem in the middle ranges Level of Autonomy (LoA), where the most of the current autonomous capabilities would fall into: this mapping offers no metric to determine the level of human supervision required. The role of SA may be crucial to distinguish the cognitive level up to which the human may be able and shall be allowed to perform take-over; described as human-on-the-loop control (Mohamed, Hossny, Nahavandi, Dalvand & Asadi, 2018).

It is evident human supervision dependent scale in any mapping of autonomy. The sensory and processing bandwidth capabilities of humans are limited, thus the loss of SA would mean that the conditional enabling of a robot functionality, or a supervising its actions require SA, and only the cognitive time horizon is different, in which the human can react.

Coherent to the current standardization efforts, yet fitting to the commonly used terms, the following scale of LoA is suggested (Haidegger, 2019) (Fig. 3.):

- **LoA 0 – No Autonomy:** All system-level functions (generating, selecting, executing and monitoring actions) are performed by the human operator. Technically it means that during the surgery no active robotic equipment is used, thus it may be considered identical to a non-robotic case.
- **LoA 1 – Robot Assistance:** The surgical robot performs specific, low level functions only. E.g., teleoperated systems, tremor filtering, minor safety features.
- **LoA 2 – Task-level Autonomy:** The system is trusted to complete certain tasks or sub-tasks in an autonomous manner. E.g., image-guided bone drilling, wound closure. It may only happen for a short instance.
- **LoA 3 – Supervised Autonomy:** The system can autonomously complete large section of a surgical procedure, while making low-level cognitive decisions. All actions are performed under human supervision, assuming the operator's situation awareness.
- **LoA 4 – High-Level Autonomy:** The robotic system executes complete procedures based on human-approved surgical plans, while the human only has the capability to emergency stop (e-stop) the procedure. The robot shall be able to complete the task even if the human fails to respond appropriately to a request to intervene.
- **LoA 5 – Full Autonomy:** A full-time performance of the robotic system, handling all environmental and adverse conditions. The system succeeds in

scenarios where even the best human operator would fail, therefore there will be no need for a human fallback option.

Unlike DoA, this LoA definition is empirical, focusing on the key enabling robot capabilities of a system. Full autonomy of surgical robots is still belongs to science fiction, however, several techniques based on AI are being under intensive research by various research groups (Richter, Orosco & Yip, 2019). Many believe, that, similarly to domain of self-driving vehicles, the market will suddenly get interested in autonomy, as LoA 3 (Supervised Autonomy) becomes reliable and gain currency.

THE GRANULARITY LEVELS OF SURGERY

To gain a better understanding of surgery, including the workflow or the movement of the surgeon, hierarchic decomposition may be an expediential instrument (Gao et al., 2014; MacKenzie, Ibbotson, Cao & Lomax, 2001; Nagyné Elek & Haidegger, 2019; Vedula et al., 2016). This deeper knowledge can fuel automation, and also the development of surgical ontologies—the machine-readable representation of surgical knowledge (D. Á. Nagy, Nagy, et al., 2018).

The literature of the field presents various definitions on some levels of granularity, however, no consistent definitions are yet compiled for the whole domain. For the decomposition of the surgeon's motion, and also for the purpose of partial automation, the precise formulation of such definitions is essential. Thus, according to the current state of the art, the granularity levels of surgery is defined by the authors as follows (Fig. 4.):

1. **Operation:** The entire invasive part of the surgical procedure.
2. **Task:** Well delimited surgical activity with a given high-level target/goal to achieve.
3. **Subtask:** Circumscribed activity segments that accomplish specific minor landmarks in completing the surgical task.
4. **Surgeme:** An atomic unit of intentional surgical activity resulting in a perceivable and meaningful outcome.
5. **Motion primitive:** General elements of motion patterns, that can be directly translated into robot commands.

In the current studies, which target surgical automation, subtask level activities are chosen most frequently. The completion of subtasks usually results in the accomplishment of a specific milestone, that fits the scope of partial automation quite well. The next level downwards—the level of surgemes—contains elements,

Figure 3. The proposed 6-stage classification for assessing the autonomous capabilities of surgical robots. The concept of Level of Autonomy follows the ISO/IEC standardization framework, determining LoA based on the human versus robotic functions of the system
(Haidegger, 2019)

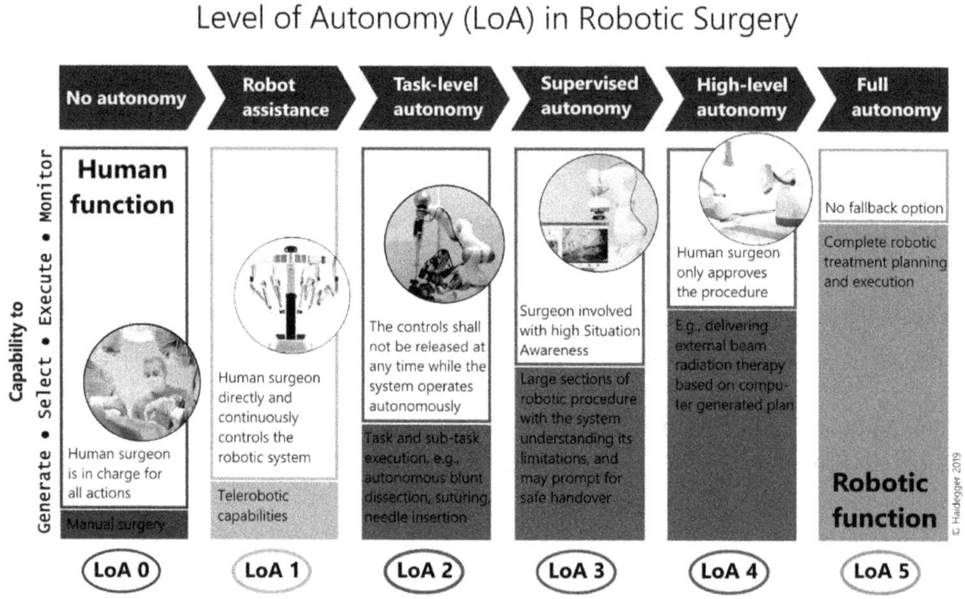

which are universal to various subtask, making possible the implementation of an universal motion library of parameterizable surgemes for automation.

Additionally, for the separation of subsequent surgemes from one another, certain events and features are defined. One, and probably the most obvious one is the shape of motion, which distinguishes e.g., grasping and releasing of an object. Tissue interaction during the surgeme execution is also crucial; the tool can be moving freely in the patient's body, or can grasp a free or and anchored piece of tissue. The change in environmental interaction usually means transitioning between different surgemes. The final feature listed here is the instrumentation—e.g., cutting can be executed using scissors, but scissors must not be used for grasping—, as the change of instruments always occurs at surgeme transitions (McKinley et al., 2016).

RESEARCH PLATFORMS

To realize the data collection and processing for the above, complete research platforms had to be built and constructed, bringing computer technology to the

Figure 4. Overview of surgical motion's granularity levels. Mapping of an example, Laparoscopic Cholecystectomy procedure onto different granularity levels (Elek et al., 2017; T. D. Nagy & Haidegger, 2018)

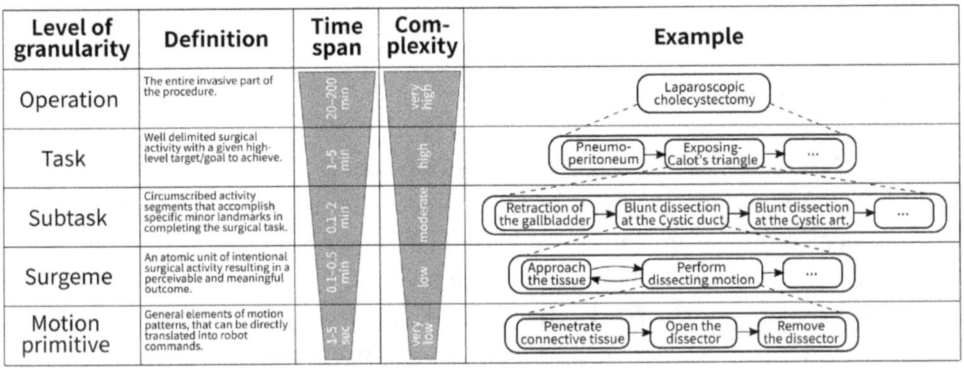

operating room. On the academic research side, the Robot Operating System (ROS) platform is widely used in the research of robotics, and also preferred by many in the medical robotics domain; most of the research centers, working on the two dominant research platforms presented below, rely on ROS. ROS is undoubtedly a powerful, modular tool with already implemented solutions for most of the frequently occurring problems of the field, such as stereo-camera calibration or acquisition of sensory data. Its more robust version, *ROS-Industrial* is also available (https://rosindustrial.org).

Da Vinci Research Kit

Approximately 5 years ago, when the 1st generation da Vinci robots (da Vinci classic) was sent to retirement due to the discontinued service and supply, the old systems found another purpose. Those systems were still functional and could be utilized in applications more tolerant to malfunctions. At the Johns Hopkins University the development of a research platform for those robots—the Da Vinci Research Kit (DVRK)—was concluded, and only in a few years, an active community was built with more than 35 setups worldwide (Kazanzides et al., 2014).

DVRK is a fully open-source platform, consists of custom hardware and software elements, in order to open the possibility of programming the da Vinci arms. The controllers—developed to operate the arms—built on custom boards: an IEEE-1394 FPGA board for computational power and low latency communication, and a Quad Linear Amplifier (QLA) for high-frequency low-level robot control. The controllers are connected to PC using IEEE 1394a (FireWire). On the PC side, the open-source *cisst* libraries (Kazanzides, 2005) are reliable for the handling of FireWire communication and the mid-level control of the robot. The *cisst* libraries offer the functionality to program the arms themselves. Additionally, *cisst* is also interfaced with ROS, which interface is currently used to program the da Vinci arms at more than half of the DVRK locations (Z. Chen, Deguet, Taylor & Kazanzides, 2017).

The RAVEN Platform

The RAVEN-I platform was originally developed at the University of Washington in the mid-2000s, aiming for space use, and other specific application areas. Hence most surgical robots were bulky, and used dominantly in the operating room. RAVEN-I was to be a new, lighter, portable and still durable surgical robot, with possibility to be used on the field. After it has proved its versatility and durability in a number of experiments, e.g., in a trial on an underwater research station, as a part of the NASA's NEEMO program, its research potential was also soon discovered. In the beginning of 2010s, the University of Washington Biorobotics Lab and the University of California Santa Cruz Bionics Lab developed an updated design of the system, named RAVEN-II. Later, Applied Dexterity http://applieddexterity.com/about/history) was formed to support the RAVEN community, and also the development of the RAVEN-III platform was started. In 2019, there are 16 RAVEN sites worldwide within the cutting-edge research of surgical robotics.

The research platform is fully open-source, consists of two 3 DoF positioning arms, with 4 DoF attachable instruments—similar to the da Vinci. Like the DVRK, the system is Linux-based, and uses ROS interface for programming (Hannaford et al., 2013).

RECENT TRENDS IN AUTOMATING SURGERY

Currently, partial (or conditional) automation is the most intensively research domain of surgical automation. The workflow of RAMIS interventions often contain subtask elements, that are time-consuming and monotonous for the surgeon to complete, such as knot-tying or blunt dissection. Partial automation—the automation of such subtasks—may reduce the cognitive load and fatigue on the human surgeon, making possible them to pay more attention on the critical parts of the operation (D. Á. Nagy, Rudas & Haidegger, 2018).

The technological advancements of the last few years in the domain of deep learning or mechatronics offer a rising potential on the research of surgical subtask automation (Karoly, Kuti & Galambos, 2018). Several of those are under intensive development or implemented already by research groups worldwide (Table 2.).

Table 2. List of surgical subtasks from the aspect of automation. FRS: Fundamentals of Robotic Surgery (Florida Hospital Nicholson Center, Celebration, FL, STAR: Smart Tissue Autonomous Robot)

Subtask	Platform	Sensor Integration	Experimental Environment	Complexity	Clinical Relevance	Reference
shape cutting	DVRK	stereo camera	gauze patch, FRS Dome	medium	high	(Murali et al., 2015)
suturing	DVRK	stereo camera	silicone, foam, FRS Dome	high	high	(Sen et al., 2016)
palpation	DVRK	force sensor	special silicone phantom, FRS Dome	medium	medium	(Garg et al., 2016; Nichols & Okamura, 2013)
tumor palpation and resection	DVRK, RAVEN	force sensor	special silicone phantom, FRS Dome	high	medium	(McKinley et al., 2016)
debridement	DVRK	stereo camera	tiny objects	medium	high	(Murali et al., 2015; Seita et al., 2017)
bowel anastomosis	STAR	3D camera	porcine bowel	high	high	(Shademan et al., 2016)
blunt dissection	DVRK	stereo camera	sandwich-like silicone phantom	medium	high	(Elek et al., 2017; D. Á. Nagy, Nagy, et al., 2018)
tissue retraction	DVRK	stereo camera	silicone phantom	low	high	(T. D. Nagy, Takács, Rudas & Haidegger, 2018)
peg transfer	DVRK	stereo camera	training phantom	medium	low	(T. D. Nagy & Haidegger, 2018)

One of the first successful projects in the domain—the work of UC Berkeley AUTOLAB and Center for Automation and Learning for Medical Robotics (CAL-MR)—was presented (Fig. 5.) (Murali et al., 2015), with not one, but two surgical subtasks completed autonomously in phantom environment, using the DVRK: multilateral (using at least two arms) debridement (Fig. 6. a–b) and shape cutting (Fig. 6. c). In this work, the learning by observation approach was used: human motion patterns were recorded and segmented, and then those patterns were used to generate robot trajectories during autonomous execution. The motion segments were, e.g., in case of debridement: motion, penetration, grasping, retraction and cutting. In order to autonomously execute the motion segments, a state-machine was compiled for each subtask. The state machine required parameters for the motion segments for execution, e.g., the height of lifting motion in case of retraction. The parameters were determined empirically, using binary search methodology. After each motion segment was parameterized and tested individually, the whole state machine was tested again, and the parameters updated, if necessary (Fig. 5.). The positions of the targets were estimated by computer vision, two pairs of stereo cameras were used to observe the field of operation. The debridement targets and the circle for cutting were detected in the images, and robot's trajectories were translated based on the 3D coordinates of the detected objects. To evaluate the autonomous execution, autonomous debridement was executed 10 times with 5 targets each, and shape-cutting were performed 20 times. The repeatability of the subtasks (ratio of successful trials) was respectively 96% and 70%.

Another work of the same research group (McKinley et al., 2016), aimed at autonomous multilateral tumor resection based on palpation in phantom environment (Fig. 6. d). To achieve the completion of this series of subtask, custom-built instruments were designed, attachable to end effector of the da Vinci: a palpation probe, a scalpel and a fluid injector. Also, a state machine was compiled to execute the series of subtasks: scan the phantom by palpation and localize the hard inclusion, making of the incision, removal of the inclusion (debridement) and fluid injection to close the wound. To evaluate the performance of the system, 10 end-to-end trials were performed, with overall 50% success rate. In 2 of the trials, the tumor location was estimated incorrectly, another 2 times retraction failed, and in 1 trial the tumor was not fully resected, which shows the challenges given the complexity of any surgical procedure. The authors asserted that the performance could be enhanced by visual feedback and are planned to include computer vision features in the future.

Suturing is probably the most intensively researched subtask of RAMIS, it occurs quite frequently in the workflow of surgical interventions, yet extremely time-consuming for the surgeon, and challenging for automation. Suturing has two, highly difficult aspects: needle guidance through the tissue on a given trajectory, and also thread manipulation, especially during knot-tying (Gao et al., 2014; Vedula et al., 2016). In (Sen et al., 2016), another work of the UC Berkeley Automation Lab, a solution for the former one is presented. To precisely grasp the needle, a positioning adapter (Suture Needle Angular Positioner—SNAP) was designed, that itself achieved a 3-folds error reduction in needle pose. The needle position was estimated using camera image, the needle size, trajectory, and control parameters using were optimized using sequential convex programming. During the trials, the system was able to complete 86% of attempted suture throws successfully.

Figure 5. The learning by observation process used to compile state machines for autonomous subtask execution.
Image credit: (Murali et al., 2015)

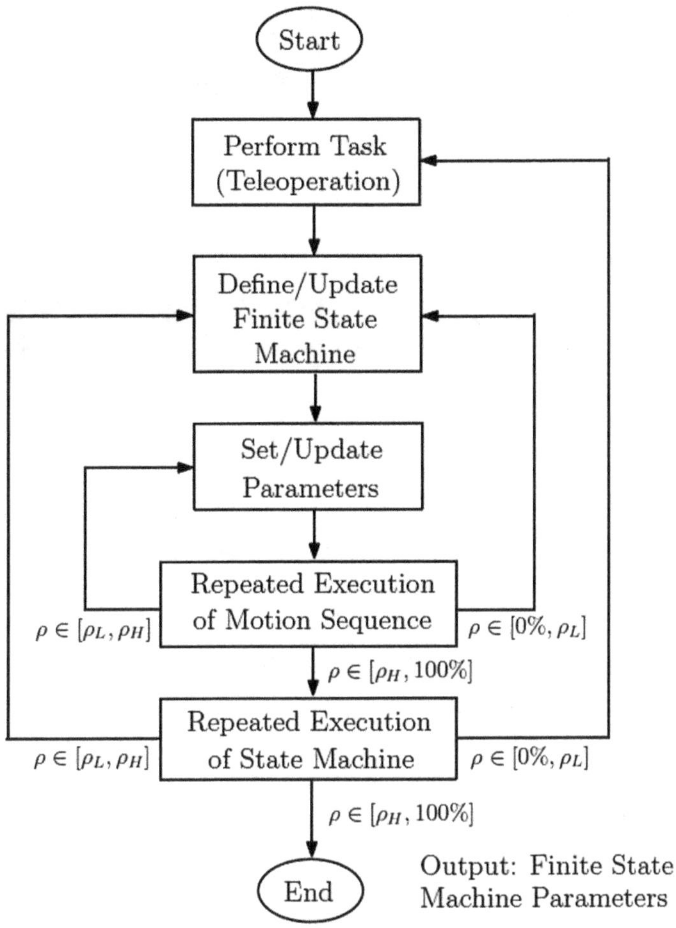

Implementation of autonomous subtasks on cable-driven robots, like the da Vinci or the RAVEN can be challenging due to the their inherent non-linearities. The inaccurate robot positioning causes no issue in teleoperation, as the human surgeon, who follows the tool position on the endoscopic camera stream, is part of the control loop. However, in the case of the automation of position-critical subtasks those inaccuracies can easily cause failures. In (Seita et al., 2017) a two-phase calibration method was presented, to decrease position errors of cable-driven surgical robots, using deep neural network and random forest techniques. By precise calibration, the debridement subtask was automated with 94.5% success rate (6. e).

Figure 6. Recently automated surgical subtasks. a-c) Multilateral cutting, d) tumor palpation and e) resection, debridement.
Image credit: (McKinley et al., 2016; Murali et al., 2015; Seita et al., 2017)

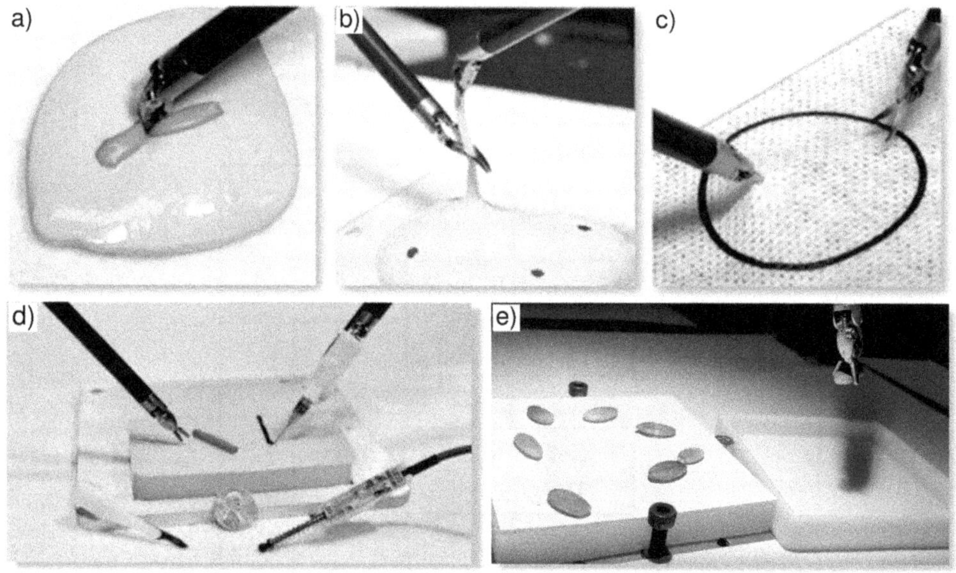

Blunt dissection automated in phantom environment was presented in (Elek et al., 2017). Blunt dissection is used to separate loosely connected layers of tissue, without harming sensitive components in between with sharp object (Fig. 7.). This autonomous approach is also relies on stereo camera, and built by the software framework developed by the authors (T. D. Nagy & Haidegger, 2018). This modular, ROS-based framework contains a motion library; a set of parameterizable surgemes, universal to a various subtasks. Additionally to blunt dissection, this framework was also used to implement soft tissue retraction (T. D. Nagy, Takács, et al., 2018) and a surgical training exercise, peg transfer (T. D. Nagy & Haidegger, 2018).

SOLUTIONS AND RECOMMENDATIONS

With the growing number of robotic applications, new challenges also arise, for which autonomous robot capabilities may be able to provide a solution (Kaplan, Nichols & Okamura, 2016). Nevertheless, the widening adoption of service robots in the surgical domain also increases the incidence of malfunctions, potential injuries and damage. As a consequence, litigation fears are escalating for companies developing new types of robots Parallel, urgency is growing to have international

Figure 7. Motion primitives of the blunt dissection subtask. a) The surgical instrument (large needle driver) moves to the dissection target; b) the robot pushes the instrument into the phantom; c) the instrument is opened; d) the robot pulls out the instrument. *Image credit: (Elek et al., 2017)*

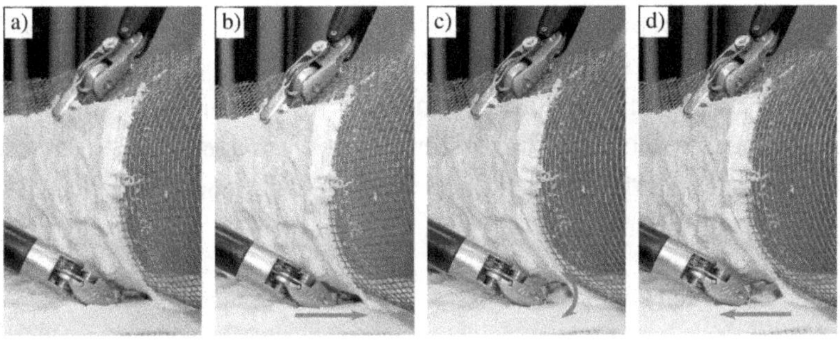

safety standards published to allow new robots to get certified in a transparent way to operate in complex, real-world scenarios.

The market has well understood the need for rigorous development methods and stringent testing. Representing this, the funding of the surgical robotics projects has risen significantly (to the extremes of Auris Health, which collected $733 million venture capital money for development and early sales). There is a stronger need nowadays towards incremental advancement in the field, which is better suited for international standardization as well.

FUTURE RESEARCH DIRECTIONS

CIS and RAMIS within still requires a lot of fundamental research to deliver reliable solutions to the most urging open medical needs. It is remarkable that over 1 m RAMIS interventions were performed in 2018, yet this is still a fraction of the global annual 300 m surgeries. The rise of robotic MD training is obviously leading to the wider spread of robotic procedures, and will eventually resolve currently existing resistance among surgeons (Sándor et al., 2010). Given the popularity of open source initiatives (such as ROS-Industrial), software development pathways have become streamlined, successful FDA clearances have been backed by toolkits such as 3D Slicer (https://www.slicer.org/) and PLUS (Lasso et al., 2014).

As it is already happening, smaller scale robots, such as capsules, natural orifice and nano-size robots represent a rising future domain (Forbrigger et al., 2019; Li, Esteban-Fernández de Ávila, Gao, Zhang & Wang, 2017; Ma, Song, Chiu & Li, 2019), however, these will fundamentally change the clinical workflow, therefore cannot be handled together with the current surgical robots regarding standardization. Already the locomotion capabilities of an active capsule robot require significant autonomous functions from the system, as such are now being tested by many companies, and human clinical trials are underway.

In the meanwhile, capturing the essence of human surgery remains a significant challenge for computer systems. Ontologies and surgical process models are believed to serve efficiently the current automation needs for certain more simple sub-tasks at LoA 3 and presumably LoA 4 (Olszewska et al., 2018). Most of the publicly disclosed projects are in the stage of *Technology Readiness Level* (TRL) 4–6, in the research and development phase.

CONCLUSION

We are witnessing the age of robotics and AI, and even the most challenging domains, such as human surgery, are receiving substantial support and uptake by autonomous capabilities. It is evident that safety has the key role in the field of medical robotics, and therefore current research and development efforts mostly address just elements of the complex challenge. In this chapter, the autonomy of surgical robots was reviewed, providing an insight into the current standardization landscape and offering some practical tools for system capability assessment. Lower and middle level of autonomy system prototypes and development projects were reviewed which aim at autonomous subtask level execution. Current major efforts, trends and possible future directions were presented

ACKNOWLEDGMENT

This work was partially supported by ACMIT (Austrian Center for Medical Innovation and Technology), which is funded within the scope of the COMET (Competence Centers for Excellent Technologies) program of the Austrian Government. Partial support of this work comes from the Hungarian State and the European Union under the EFOP-3.6.1-16-2016-00010 project. T. D. Nagy and T. Haidegger are supported through the New National Excellence Program of the Ministry of Human Capacities. T. Haidegger is a Bolyai Fellow of the Hungarian Academy of Sciences. The grammatical finalization of the chapter was supported by the V4+ACARDC – CLOUD AND VIRTUAL SPACES grant.

REFERENCES

Chen, S.-Y. (2016). *Study on situation awareness for medical robots*. National Taipei University of Technology. [Technical Report]

Chen, Z., Deguet, A., Taylor, R. H., & Kazanzides, P. (2017). Software Architecture of the Da Vinci Research Kit. *Proc. of the IEEE International Conference on Robotic Computing (IRC)*, 180–187. 10.1109/IRC.2017.69

Elek, R., Nagy, T. D., Nagy, D. Á., Garamvölgyi, T., Takács, B. & Galambos, P., … Haidegger, T. (2017). Towards surgical subtask automation—Blunt dissection. *Proc. Of the IEEE 21st International Conference on Intelligent Engineering Systems*, 253–258.

Elek, R., Nagy, T. D., Nagy, D. Á., Kronreif, G., Rudas, I. J., & Haidegger, T. (2016). Recent Trends in Automating Robotic Surgery. In A. Szakál (Ed.), *Proc. Of the 20th IEEE Jubilee International Conference on Intelligent Engineering Systems* (pp. 27–32). Budapest: IEEE.

Fei, B., Ng, W. S., Chauhan, S. & Kwoh, C. K. (2001). *The safety issues of medical robotics*. Academic Press.

Forbrigger, C., Lim, A., Onaizah, O., Salmanipour, S., Looi, T., Drake, J., & Diller, E. D. (2019). Cable-Less, Magnetically Driven Forceps for Minimally Invasive Surgery. *IEEE Robotics and Automation Letters*, *4*(2), 1202–1207. doi:10.1109/LRA.2019.2894504

Gao, Y., Vedula, S. S., Reiley, C. E., Ahmidi, N., Varadarajan, B. & Lin, H. C. … Yuh, D. D. (2014). JHU-ISI gesture and skill assessment working set (JIGSAWS): A surgical activity dataset for human motion modeling. *Proc. of the MICCAI Workshop: M2CAI, 3*. Retrieved from https://cirl.lcsr.jhu.edu/wp-content/uploads/2015/11/JIGSAWS.pdf

Garg, A., Sen, S., Kapadia, R., Jen, Y., McKinley, S., Miller, L., & Goldberg, K. (2016). Tumor localization using automated palpation with Gaussian Process Adaptive Sampling. *Proc. of the 2016 IEEE International Conference on Automation Science and Engineering (CASE)*, 194–200. 10.1109/COASE.2016.7743380

Haidegger, T. (2019). Autonomy for Surgical Robots: Concepts and Paradigms. *IEEE Trans. on Medical Robotics and Bionics*, *1*(2), 65–76. doi:10.1109/TMRB.2019.2913282

Haidegger, T. (2019). Probabilistic Method to Improve the Accuracy of Computer-Integrated Surgical Systems. Acta Polytechnica Hungarica. *Special Issue on Platforms for Medical Robotics Research, 16*(8). doi:10.12700/APH.16.8.2019.8.8

Hannaford, B., Rosen, J., Friedman, D. W., King, H., Roan, P., Cheng, L., ... White, L. (2013). Raven-II: An Open Platform for Surgical Robotics Research. *IEEE Transactions on Biomedical Engineering, 60*(4), 954–959. doi:10.1109/TBME.2012.2228858 PMID:23204264

History – Applied Dexterity. (n.d.). Retrieved April 30, 2019, from http://applieddexterity.com/about/history/

Huang, H.-M. (2008). *Autonomy Levels for Unmanned Systems (ALFUS)*. Academic Press.

Kaber, D. B., & Endsley, M. R. (2004). The effects of level of automation and adaptive automation on human performance, situation awareness and workload in a dynamic control task. *Theoretical Issues in Ergonomics Science, 5*(2), 113–153. doi:10.1080/1463922021000054335

Kaplan, K. E., Nichols, K. A., & Okamura, A. M. (2016). Toward human-robot collaboration in surgery: Performance assessment of human and robotic agents in an inclusion segmentation task. *2016 IEEE International Conference on Robotics and Automation (ICRA)*, 723–729. 10.1109/ICRA.2016.7487199

Karoly, A. I., Kuti, J. & Galambos, P. (2018). Unsupervised real-time classification of cycle stages in collaborative robot applications. *2018 IEEE 16th World Symposium on Applied Machine Intelligence and Informatics (SAMI)*. 10.1109/SAMI.2018.8323994

Kazanzides, P. (2005). *Open Source Software Libraries for Computer Integrated Surgery*. Retrieved from https://cisst.org/wp-content/uploads/2016/03/YR_8_Open_Source_Software.pdf

Kazanzides, P., Chen, Z., Deguet, A., Fischer, G. S., Taylor, R. H., & DiMaio, S. P. (2014). An open-source research kit for the da Vinci® Surgical System. *Proc. of the IEEE International Conference on Robotics and Automation*, 6434–6439. 10.1109/ICRA.2014.6907809

Lasso, A., Heffter, T., Rankin, A., Pinter, C., Ungi, T., & Fichtinger, G. (2014). PLUS: Open-Source Toolkit for Ultrasound-Guided Intervention Systems. *IEEE Transactions on Biomedical Engineering, 61*(10), 2527–2537. doi:10.1109/TBME.2014.2322864 PMID:24833412

Li, J., Esteban-Fernández de Ávila, B., Gao, W., Zhang, L. & Wang, J. (2017). Micro/nanorobots for biomedicine: Delivery, surgery, sensing, and detoxification. *Science Robotics, 2*(4). doi:10.1126cirobotics.aam6431

Ma, X., Song, C., Chiu, P. W., & Li, Z. (2019). Autonomous Flexible Endoscope for Minimally Invasive Surgery With Enhanced Safety. *IEEE Robotics and Automation Letters*, *4*(3), 2607–2613. doi:10.1109/LRA.2019.2895273

MacKenzie, L., Ibbotson, J. A., Cao, C. G. L., & Lomax, A. J. (2001). Hierarchical decomposition of laparoscopic surgery: A human factors approach to investigating the operating room environment. *Minimally Invasive Therapy & Allied Technologies*, *10*(3), 121–127. doi:10.1080/136457001753192222 PMID:16754003

Márton, L., Szántó, Z., Haidegger, T., Galambos, P., & Kövecses, J. (2017). Internet—Based Bilateral Teleoperation Using a Revised Time—Domain Passivity Controller. *Acta Polytechnica Hungarica*, 27–45.

McKinley, S., Garg, A., Sen, S., Gealy, D. V., McKinley, J., Jen, Y. & Goldberg, K. (2016). *Autonomous Multilateral Surgical Tumor Resection with Interchangeable Instrument Mounts and Fluid Injection Device*. Academic Press.

Mohamed, A., Hossny, M., Nahavandi, S., Dalvand, M., & Asadi, H. (2018). Towards Trusted Autonomous Surgical Robots. *Proc. of the IEEE International Conference on Systems, Man, and Cybernetics (SMC)*, 4083–4088.

Murali, A., Sen, S., Kehoe, B., Garg, A., McFarland, S., & Patil, S., … Goldberg, K. (2015). Learning by observation for surgical subtasks: Multilateral cutting of 3D viscoelastic and 2D Orthotropic Tissue Phantoms. *Proc. of the IEEE International Conference on Robotics and Automation*, 1202–1209. 10.1109/ICRA.2015.7139344

Nagy, D. Á., Nagy, T. D., Elek, R., Rudas, I. J., & Haidegger, T. (2018). Ontology-Based Surgical Subtask Automation, Automating Blunt Dissection. *Journal of Medical Robotics Research*, *3*(3). doi:10.1142/S2424905X18410052

Nagy, D. Á., Rudas, I. J. & Haidegger, T. (2018). OntoFlow, a software tool for surgical workflow recording. *2018 IEEE 16th World Symposium on Applied Machine Intelligence and Informatics (SAMI)*, 119–124. 10.1109/SAMI.2018.8323998

Nagy, Tamas D. & Haidegger, T. (2019). A DVRK-based Framework for Surgical Subtask Automation. Acta Polytechnica Hungarica. *Special Issue on Platforms for Medical Robotics Research*, *16*(8). doi:10.12700/APH.16.8.2019.8.5

Nagy, T. D., Takács, M., Rudas, I. J., & Haidegger, T. (2018). Surgical Subtask Automation—Soft Tissue Retraction. *Proc. of the 16th IEEE World Symposium on Applied Machine Intelligence and Informatics*, 55–60.

Nagyné Elek, R., & Haidegger, T. (2019). Robot-Assisted Minimally Invasive Surgical Skill Assessment—Manual and Automated Platforms. Acta Polytechnica Hungarica. *Special Issue on Platforms for Medical Robotics Research, 16*(8). doi:10.12700/APH.16.8.2019.8.9

Nichols, K. A., & Okamura, A. M. (2013). Autonomous robotic palpation: Machine learning techniques to identify hard inclusions in soft tissues. *Proc. of the 2013 IEEE International Conference on Robotics and Automation*, 4384–4389. 10.1109/ICRA.2013.6631198

Olszewska, J. I., Houghtaling, M., Gonçalves, P., Haidegger, T., Fabiano, N., Carbonera, J. L., … Prestes, E. (2018). *Robotic Ontological Standard Development Life Cycle*. Presented at the IEEE ICRA 2018 WELCARO workshop. Retrieved from http://real.mtak.hu/86419/

Richter, F., Orosco, R. K. & Yip, M. C. (2019). *Open-Sourced Reinforcement Learning Environments for Surgical Robotics*. Retrieved from http://arxiv.org/abs/1903.02090

Rosen, J. (2013). Surgical Robotics. In *Medical Devices* (pp. 63–97). Surgical and Imaging-Guided Technologies.

Seita, D., Krishnan, S., Fox, R., McKinley, S., Canny, J. & Goldberg, K. (2017). Fast and Reliable Autonomous Surgical Debridement with Cable-Driven Robots Using a Two-Phase Calibration Procedure. *Proc. of the 2018 IEEE International Conference on Robotics and Automation (ICRA)*, 6651–6658. Retrieved from http://arxiv.org/abs/1709.06668

Sen, S., Garg, A., Gealy, D. V., McKinley, S., Jen, Y., & Goldberg, K. (2016). Automating multi-throw multilateral surgical suturing with a mechanical needle guide and sequential convex optimization. *Proc. of the IEEE International Conference on Robotics and Automation*, 4178–4185. 10.1109/ICRA.2016.7487611

Shademan, A., Decker, R. S., Opfermann, J., Leonard, S., Krieger, A., & Kim, P. C. W. (2016). Supervised autonomous robotic soft tissue surgery. *Science Translational Medicine*, *8*(337), 337ra64. doi:10.1126citranslmed.aad9398 PMID:27147588

Takács, Á., Drexler, D. A., Galambos, P., Rudas, I. J., & Haidegger, T. (2018). Assessment and Standardization of Autonomous Vehicles. *Proc. of the 2018 IEEE 22nd International Conference on Intelligent Engineering Systems (INES)*, 185–192. 10.1109/INES.2018.8523899

Takács, Á., Nagy, D. Á., Rudas, I. J., & Haidegger, T. (2016). Origins of Surgical Robotics: From Space to the Operating Room. *Acta Polytechnica Hungarica*, *13*(1), 13–30.

Takacs, A., Rudas, I., Bosl, D., & Haidegger, T. (2018). Highly Automated Vehicles and Self-Driving Cars [Industry Tutorial]. *IEEE Robotics & Automation Magazine*, *25*(4), 106–112. doi:10.1109/MRA.2018.2874301

Vedula, S. S., Malpani, A. O., Tao, L., Chen, G., Gao, Y., Poddar, P., ... Chen, C. C. (2016). Analysis of the Structure of Surgical Activity for a Suturing and Knot-Tying Task. *PLoS One*, *11*(3), e0149174. doi:10.1371/journal.pone.0149174 PMID:26950551

Wei, G. Q., Arbter, K., & Hirzinger, G. (1997). Real-time visual servoing for laparoscopic surgery. Controlling robot motion with color image segmentation. *IEEE Engineering in Medicine and Biology Magazine*, *16*(1), 40–45. doi:10.1109/51.566151 PMID:9058581

Yamamoto, T., Vagvolgyi, B., Balaji, K., Whitcomb, L. L., & Okamura, A. M. (2009). Tissue property estimation and graphical display for teleoperated robot-assisted surgery. *Proc. of the 2009 IEEE International Conference on Robotics and Automation (ICRA)*, 4239–4245. 10.1109/ROBOT.2009.5152674

Yang, G.-Z., Cambias, J., Cleary, K., Daimler, E., Drake, J., Dupont, P. E., … Taylor, R. H. (2017). Medical robotics—Regulatory, ethical, and legal considerations for increasing levels of autonomy. *Science Robotics*, *2*(4). doi:10.1126cirobotics.aam8638

ENDNOTES

1 https://ec.europa.eu/digital-single-market/en/news/attitudes-towards-impact-digitisation-and-automation-daily-life

2 https://transmitter.ieee.org/ai/

Compilation of References

Abedi, M., Moghaddam, M. M., & Firoozabadi, S. M. P. (2014). A full bio-inspired bipedal gait locomotion system. *2014 2nd RSI/ISM International Conference on Robotics and Mechatronics, ICRoM 2014*. 10.1109/ICRoM.2014.6990897

Acosta-Mejia, C. A. (1999). Improved *p* charts to monitor process quality. *IIE Transactions*, *31*(6), 509–516. doi:10.1080/07408179908969854

Aertbeliën, E., & De Schutter, J. (2014, September). eTaSL/eTC: A constraint-based task specification language and robot controller using expression graphs. In *2014 IEEE/RSJ International Conference on Intelligent Robots and Systems* (pp. 1540-1546). IEEE. 10.1109/IROS.2014.6942760

Aggarwal, L., Aggarwal, K., & Urbanic, R. J. (2014). Use of artificial neural networks for the development of an inverse kinematic solution and visual identification of singularity zone(s). *Procedia CIRP*, *17*, 812–817. doi:10.1016/j.procir.2014.01.107

Ahn, C. K., & Lee, M. C. (2000). An Off-Line Automatic Teaching by Vision Information for Robotic Assembly Task. *Proceedings of IEEE International Conference on Industrial Electronics, Control and Instrumentation*, 2171–2176.

Almusawi, A. R. J., Dülger, L. C., & Kapucu, S. (2016), A new artificial neural network approach in solving inverse kinematics of robotic arm (Denso VP6242). Computational Intelligence and Neuroscience, 2016(5720163).

Alves, T., D'Carvalho, M. C., & Gonçalves, R. S. (2018). Controle "Assist-As-Needed" em Estruturas Robóticas Atuadas por Cabos para reabilitação das Articulações do Corpo Humano. ENEBI 2018, Águas de Lindóia, SP, Brazil. (in Portuguese)

Alves, T., D'Carvalho, M.C., & Gonçalves, R.S. (2019). Assist-As-Needed Control in A Cable-Actuated Robot for Human Joints Rehabilitation. *Journal of Mechanical Engineering and Biomechanics*.

Ambler, A. P., & Popplestone, R. J. (1975). Inferring the positions of bodies from specified spatial relationships. *Artificial Intelligence*, *6*(2), 157–174. doi:10.1016/0004-3702(75)90007-7

Anam, K., & Al-Jumaily, A. A. (2012). Active exoskeleton control systems: State of the art. *Procedia Engineering*, *41*, 988–994. doi:10.1016/j.proeng.2012.07.273

Arisumi, H., Miossec, S., Chardonnet, J. R., & Yokoi, K. (2008, September). Dynamic lifting by whole body motion of humanoid robots. In *Intelligent Robots and Systems, 2008. IROS 2008. IEEE/RSJ International Conference on* (pp. 668-675). IEEE. 10.1109/IROS.2008.4651195

Avrin, G. (2017). *Modélisation du contrôle moteur humain lors de tâches rythmiques hybrides et application à la commande de robots anthropomorph*. Université Paris-Saclay.

Avrin, G., Makarov, M., Rodriguez-Ayerbe, P., & Siegler, I. A. (2016). Particle Swarm Optimization of Matsuoka's oscillator parameters in human-like control of rhythmic movements. *Proceedings of the American Control Conference*. 10.1109/ACC.2016.7524938

Bang, D.H., & Shin, W.S. (2016). Effects of robot-assisted gait training on spatiotemporal gait parameters and balance in patients with chronic stroke: A randomized controlled pilot trial. *NeuroRehabilitation*, *38*(4), 343–349.

Barber, C. B., David, P., Dobkin, D. P., & Huhdanpaa, H. (1996). The Quickhull Algorithm for Convex Hulls. *ACM Transactions on Mathematical Software*, *22*(4), 469–483. doi:10.1145/235815.235821

Barbosa, A. M., Carvalho, J. C. M., & Gonçalves, R. S. (2018). Cable-driven lower limb rehabilitation robot. *Journal of the Brazilian Society of Mechanical Sciences and Engineering*, *40*(5), 245. doi:10.100740430-018-1172-y

Barrett, L., & Narayanan, S. (2008). Learning All Optimal Policies with Multiple Criteria. In *Proceedings of the 25th International Conference on Machine Learning (ICML-2008)* (pp. 41-47). ACM. 10.1145/1390156.1390162

Bartels, G., Kresse, I., & Beetz, M. (2013, October). Constraint-based movement representation grounded in geometric features. In *2013 13th IEEE-RAS International Conference on Humanoid Robots (Humanoids)* (pp. 547-554). IEEE. 10.1109/HUMANOIDS.2013.7030027

Barto, A. G., Steven, J., Bradtke, S. J., & Singh, S. P. (1995). Learning to act using real-time dynamic programming. *Artificial Intelligence*, *72*(1-2), 81–138. doi:10.1016/0004-3702(94)00011-O

Basl, J., & Doucek, P. (2019). A Metamodel for Evaluating Enterprise Readiness in the Context of Industry 4.0. *Information*, *10*(89), 1–13.

Bastos, P., Lopes, P. I., & Pires, L. (2014). Application of data mining in a maintenance system for failure prediction. In Safety, Reliability and Risk Analysis: Beyond the Horizon. Taylor & Francis Group.

Batista, R. F. (2017). *Sistema dinâmico híbrido aplicado a modelo mecânico bípede antropomórfico*. University of Campinas.

Beebe, J. A., & Lang, C. E. (2009). Active Range of Motion predicts Upper Extremity Function Three months post-stroke. *Stroke*, *40*(5), 1772–1779. doi:10.1161/STROKEAHA.108.536763 PMID:19265051

Beyl, P. (2009). Safe and compliant guidance in robot-assisted gait rehabilitation using proxy-based sliding mode control. *IEEE 11th International Conference on Rehabilitation Robotics*. 10.1109/ICORR.2009.5209505

Biggs, G., & MacDonald, B. (2003, December). A survey of robot programming systems. In *Proceedings of the Australasian conference on robotics and automation* (pp. 1-3). Academic Press.

Boschetti, G., Carbone, G., & Passarini, C. (2019). Cable Failure Operation Strategy for a Rehabilitation Cable-Driven Robot. *Robotics*, *8*(1), 17. doi:10.3390/robotics8010017

Bosch, T., van Eck, J., Knitel, K., & de Looze, M. (2016). The effects of a passive exoskeleton on muscle activity, discomfort and endurance time in forward bending work. *Applied Ergonomics*, *54*, 212–217. doi:10.1016/j.apergo.2015.12.003 PMID:26851481

Bouteille, D. (2000). *Les automatismes programmables*. Toulouse: Cepadues.

Bruce, A., Hanrahan, S., Vaughan, K., Mackinnon, L., & Pandy, M. G. (1998). The Biophysical Foundations of Human Movement. *Medicine &amp Science in Sports & Exercise*. doi:10.1097/00005768-199801000-00025

Bruce, J. (2013, August 21). *Home Automation with Raspberry Pi and Arduino*. Retrieved from http://www.makeuseof.com/tag/how-to-build-home-automation-system-raspberry-pi-and-arduino

Brum, M. (2016, December 22). *Introduction to Educational Robotics*. Retrieved from http://www.educacional.com.br/upload/dados/materialapoio/124590001/8214768/Rob%C3%B3tica%20 Educativa.pdf

Bruzzone, L. (2004). *Meccanica applicata alle macchine*. Genova: Università degli Studi di Genova.

Buitrago Salazar, G. D. (2018). *Concepção de Diferentes Estruturas para Exoesqueletos de Membro Inferior baseado no Estudo Dinâmico e Utilização de Biomateriais Concepção de Diferentes Estruturas para Exoesqueletos de Membro Inferior baseado no Estudo Dinâmico e Utilização de Biomateria*. University of Campinas.

Burris, M. (2019, May 17). *Choosing Between Stepper Motors or Servo Motors*. Retrieved from https://www.lifewire.com/stepper-motor-vs-servo-motors-selecting-a-motor-818841

Buschmann, T., Ewald, A., Ulbrich, H., & Buschges, A. (2012). Event-based walking control - From neurobiology to biped robots. In *2012 IEEE/RSJ International Conference on Intelligent Robots and Systems* (pp. 1793–1800). IEEE. 10.1109/IROS.2012.6385783

Cafolla, D., Russo, M., & Carbone, G. (2019). CUBE, a Cable-driven Device for Limb Rehabilitation. *Journal of Bionic Engineering*, *16*(2).

Cafolla, D., Russo, M., & Carbone, G. (2019). CUBE, a Cable-driven Device for Limb Rehabilitation. *Journal of Bionics Engineering*, *16*, 492. https://doi.org/10.1007/s42235-019-0040-5

Carbone, G., & Ceccarelli, M. (2016). *Sistema a cavi per assistenza motoria* [Cable driven system for motion assistance]. Patent Application n.102016000038975.

Carbone, G., Gherman, B., Ulinici, J., Vaida, C., & Pisla, D. (2017). Design Issues for an Inherently Safe Robotic Rehabilitation Device. In *Mechanisms and Machine Science 49: Advances in Service and Industrial Robotics* (Vol. 49, pp. 1025–1032). Springer. doi:10.1007/978-3-319-61276-8_110

Carlos, L.F., Jesus, H.B., & Alanis, A.Y. (2018). Inverse kinematics of mobile manipulators based on differential evolution. *International Journal of Advanced Robotic Systems, 15*(1).

Caudell, T. P., & Mizell, D. W. (1992). Augmented reality: an application of heads-up display technology to manual manufacturing processes. *Proceedings of the 25th Hawaii International Conference on System Sciences, 2*(1), 659. 10.1109/HICSS.1992.183317

Ceccarelli, M., Pîslă, D., & Graur, F. (2010). Design and operation issues for parallel robotic devices in the rehabilitation of stroke patients. *International Journal of Mechanics and Control, 11*(1), 51–57.

Ceccarelli, M., & Romdhane, L. (2010). Design Issues for Human-Machine Platform Interface in Cable Based Parallel Manipulators for Physiotherapy Applications. *Journal of Zhejiang University. Science A, 11*(4), 231–239. doi:10.1631/jzus.A1000027

Chakraborty, A. (2016). Importance of PDCA cycle for SMEs. *SSRG International Journal of Mechanical Engineering, 3*(5), 30-34.

Chen, J., Zhang, X., & Zhu, L. (2014). *Kinematics analysis and three-dimensional simulation of the rehabilitation lower extremity exoskeleton robot.* arXiv preprint arXiv:1401.6517

Chen, S.-Y. (2016). *Study on situation awareness for medical robots.* National Taipei University of Technology. [Technical Report]

Chen, Z., Deguet, A., Taylor, R. H., & Kazanzides, P. (2017). Software Architecture of the Da Vinci Research Kit. *Proc. of the IEEE International Conference on Robotic Computing (IRC)*, 180–187. 10.1109/IRC.2017.69

Chestnutt, J., Lau, M., Cheung, G., Kuffner, J., Hodgins, J., & Kanade, T. (2005). Footstep Planning for the Honda ASIMO Humanoid. In *Proceedings of the 2005 IEEE International Conference on Robotics and Automation* (pp. 629–634). IEEE. 10.1109/ROBOT.2005.1570188

Cicala, C., Massaro, A., Velardi, L., Senesi, G. S., & Valentini, A. (2014). Self-assembled pillar-like structures in nanodiamond layers by pulsed spray technique. *ACS Applied Materials & Interfaces, 6*(23), 21101–21109. doi:10.1021/am505974d PMID:25402729

Colombo, R., Sterpi, I., Mazzone, A., Delconte, C., & Pisano, F. (2012). Taking a Lesson from Patients' Recovery Strategies to Optimize Training During Robot-Aided Rehabilitation. *IEEE Transactions on Neural Systems and Rehabilitation Engineering, 20*(3), 276–285. doi:10.1109/TNSRE.2012.2195679 PMID:22623406

Corke, P. (1996). A robotics toolbox for MATLAB. *IEEE Robotics & Automation Magazine, 3*(1), 24–32. doi:10.1109/100.486658

Côté, G. (2003). *Analyse et conception de mécanismes parallèles actionnés par cables* (M.Sc. dissertation). Université Laval, Quebec. (in French)

Cunedioglu, Y., Mugan, A., & Akcay, H. (2006). Frequency domain analysis of model order reduction techniques. *Finite Elements in Analysis and Design*, *42*(5), 367–403. doi:10.1016/j.finel.2005.08.005

Cyberdyne. (2019). *Cyberdyne HAL*. Retrieved from https://www.cyberdyne.jp/english/products/HAL/

Davison, E. J. (1966). A method for simplifying linear dynamic systems. *IEEE Transactions on Automatic Control*, *11*(1), 93–101. doi:10.1109/TAC.1966.1098264

De Schutter, J., De Laet, T., Rutgeerts, J., Decré, W., Smits, R., Aertbeliën, E., ... Bruyninckx, H. (2007). Constraint-based task specification and estimation for sensor-based robot systems in the presence of geometric uncertainty. *The International Journal of Robotics Research*, *26*(5), 433–455. doi:10.1177/027836490707809107

Dipietro, L., Krebs, H. I., Fasoli, S. E., Volpe, B. T., Stein, J., Bever, C., & Hogan, N. (2007). Changing Motor Synergies in Chronic Stroke. *Journal of Neurophysiology*, *98*(2), 757–768. doi:10.1152/jn.01295.2006 PMID:17553941

Duka, A. V. (2014). Neural network based inverse kinematics solution for trajectory tracking of a robotic arm. *Procedia Technology*, *12*, 20–27. doi:10.1016/j.protcy.2013.12.451

Duka, A. V. (2015). ANFIS Based Solution to the Inverse Kinematics of a 3DOF Planar Manipulator. *Procedia Technology*, *19*, 526–533. doi:10.1016/j.protcy.2015.02.075

Dumitru, N., Copilusi, C., Geonea, I., Tarnita, D., & Dumitrache, I. (2015). Dynamic Analysis of an Exoskeleton New Ankle Joint Mechanism. In *New Trends in Mechanism and Machine Science* (pp. 709–717). Springer International Publishing. doi:10.1007/978-3-319-09411-3_75

Duret, C., Grosmaire, A.-G., & Krebs, H. I. (2019). Robot-Assisted Therapy in Upper Extremity Hemiparesis: Overview fo an Evidence-Based Approach. *Frontiers in Neurology*, *10*, 412. doi:10.3389/fneur.2019.00412 PMID:31068898

Dzahir, M. A. M., & Yamamoto, S.-I. (2014). Recent Trends in Lower-Limb Robotic Rehabilitation Orthosis: Control Scheme and Strategy for Pneumatic Muscle Actuated Gait Trainers. *Robotics*, *3*(2), 120–148. doi:10.3390/robotics3020120

Elek, R., Nagy, T. D., Nagy, D. Á., Garamvölgyi, T., Takács, B. & Galambos, P., … Haidegger, T. (2017). Towards surgical subtask automation—Blunt dissection. *Proc. Of the IEEE 21st International Conference on Intelligent Engineering Systems*, 253–258.

Elek, R., Nagy, T. D., Nagy, D. Á., Kronreif, G., Rudas, I. J., & Haidegger, T. (2016). Recent Trends in Automating Robotic Surgery. In A. Szakál (Ed.), *Proc. Of the 20th IEEE Jubilee International Conference on Intelligent Engineering Systems* (pp. 27–32). Budapest: IEEE.

Elliott, G., Sawicki, G. S., Marecki, A., & Herr, H. (2013). The biomechanics and energetics of human running using an elastic knee exoskeleton. In *Rehabilitation Robotics (ICORR), 2013 IEEE International Conference on*, (pp. 1-6). IEEE. 10.1109/ICORR.2013.6650418

Elsa, Hocoma Armeo Boom - Elsa. (2019). Disponível em: http://www.elsa.web.tr/tr/urun/robotik-rehabilitasyon/hocoma-armeo-boom

Endo, G., Nakanishi, J., Morimoto, J., & Cheng, G. (2005). Experimental Studies of a Neural Oscillator for Biped Locomotion with QRIO. In *Proceedings of the 2005 IEEE International Conference on Robotics and Automation* (pp. 596–602). IEEE. 10.1109/ROBOT.2005.1570183

Favi, C., Germani, M., & Marconi, M. (2017). 4M approach for a comprehensive analysis and improvement of manual assembly lines. *Procedia Manufacturing*, *11*(1), 1510–1518. doi:10.1016/j.promfg.2017.07.283

Fei, B., Ng, W. S., Chauhan, S. & Kwoh, C. K. (2001). *The safety issues of medical robotics*. Academic Press.

Forbrigger, C., Lim, A., Onaizah, O., Salmanipour, S., Looi, T., Drake, J., & Diller, E. D. (2019). Cable-Less, Magnetically Driven Forceps for Minimally Invasive Surgery. *IEEE Robotics and Automation Letters*, *4*(2), 1202–1207. doi:10.1109/LRA.2019.2894504

Fouad, R. H. & Mukattash, A. (2010). *Statistical process control tools: a practical guide for Jordanian industrial organizations*. Academic Press.

Frascati Manual. (2015). *The Measurement of Scientific, Technological and Innovation Activities-Guidelines for Collecting and Reporting Data on Research and Experimental Development*. OECD.

Fryman, J., & Matthias, B. (2012, May). Safety of industrial robots: From conventional to collaborative applications. In *ROBOTIK 2012; 7th German Conference on Robotics* (pp. 1-5). VDE.

Gan, Y., Dai, X., & Li, D. (2013). Off-line programming techniques for multi-robot Cooperation System. *International Journal of Advanced Robotic Systems*, *10*(7), 1–17. doi:10.5772/56506

Gao, Y. (2006, November). *Research on Average Reward. Reinforcement Learning*. Algorithms, National Laboratory for Novel Software Technology, Nanjing University. Retrieved from http://lamda.nju.edu.cn/conf/MLA06/files/Gao.Y.pdf

Gao, Y., Vedula, S. S., Reiley, C. E., Ahmidi, N., Varadarajan, B. & Lin, H. C. … Yuh, D. D. (2014). JHU-ISI gesture and skill assessment working set (JIGSAWS): A surgical activity dataset for human motion modeling. *Proc. of the MICCAI Workshop: M2CAI, 3*. Retrieved from https://cirl.lcsr.jhu.edu/wp-content/uploads/2015/11/JIGSAWS.pdf

Garg, A., Sen, S., Kapadia, R., Jen, Y., McKinley, S., Miller, L., & Goldberg, K. (2016). Tumor localization using automated palpation with Gaussian Process Adaptive Sampling. *Proc. of the 2016 IEEE International Conference on Automation Science and Engineering (CASE)*, 194–200. 10.1109/COASE.2016.7743380

Gassert, R., & Dietz, V. (2018). Rehabilitation robots for the treatment of sensorimotor deficits: A neurophysiological perspective. *Journal of Neuroengineering and Rehabilitation*, *15*(1), 46. doi:10.118612984-018-0383-x PMID:29866106

Gates, D.H., Walters, L.S., Cowley, J., Wilken, J.M., & Resnik, L. (2016). Range of Motion Requirements for Upper-Limb Activities of Daily Living. *Am J Occup Ther, 70*(1). doi: . doi:10.5014/ajot.2016.015487

Gates. (2017, January 3). *Timing belt theory*. Retrieved from http://www.gatesmectrol.com/mectrol/downloads/download_common.cfm?file=belt_theory06sm.pdf&folder=brochure

Ge, D. F., Takeuchi, Y., & Asakawa, N. (1993). Automation of Polishing Work by an Industrial Robot, – 2nd report, Automatic Generation of Collision- Free Polishing Path –. [in Japanese]. *Transactions of the Japan Society of Mechanical Engineers*, *59*(561), 1574–1580. doi:10.1299/kikaic.59.1574

Giurgiutiu, V., & Lyshevski, S. E. (2009). *Micromechatronics: Modeling, Analysis, and Design with Matlab*. CRC Press.

Gomes, M. A., Siqueira, A. A. G., & Gobbo, R. G. (2011). Improving the parameters of neural oscillators to generate joint trajectories of an exoskeleton for lower limbs. In *2011 9th IEEE International Conference on Control and Automation (ICCA)* (pp. 286–291). IEEE. 10.1109/ICCA.2011.6138064

Gonçalves, R. S., & Carvalho, J. C. M. (2014). Robot modeling for physical rehabilitation. In *Robotics: Concepts, Methodologies, Tools, and Applications*. IGI Global. . doi:10.4018/978-1-4666-4607-0.ch058

Gonçalves, R. S., & Carvalho, J. C. M. (2010). Desenvolvimento de uma estrutura robótica paralela atuada por cabos para reabilitação dos movimentos do ombro. *VI Congresso Nacional de Engenharia Mecânica - CONEM 2010*. (in Portuguese)

Gonçalves, R. S., Carvalho, J. C. M., & Lobato, F. S. (2016). Design of a robotic device actuated by cables for human lower limb rehabilitation using self-adaptive differential evolution and robust optimization. *Bioscience Journal*, *32*(6). doi:10.14393/BJ-v32n1a2016-32436

Gonçalves, R. S., Carvalho, J. C. M., Ribeiro, J. F., & Salim, V. V. (2015). *Cable-Driven Robot for Upper and Lower Limbs Rehabilitation. In Handbook of Research on Advancements in Robotics and Mechatronics* (pp. 284–315). IGI Global. doi:10.4018/978-1-4666-7387-8.ch011

Gonçalves, R. S., Carvalho, J. C. M., Rodrigues, A. A. O., & Barbosa, A. M. (2013). Cable-Driven Parallel Manipulator for Lower Limb Rehabilitation. *Applied Mechanics and Materials*, *459*, 535–542. doi:10.4028/www.scientific.net/AMM.459.535

Gonçalves, R. S., & Krebs, H. I. (2017). MIT-Skywalker: Considerations on the Desing of a Body Weight Support System. *Journal of Neuroengineering and Rehabilitation*, *14*(1), 88. doi:10.118612984-017-0302-6 PMID:28877750

Gonçalves, R. S., Soares, G., & Carvalho, J. C. M. (2019). Conceptual design of a rehabilitation device based on cam-follower and crank-rocker mechanisms hand actioned. *Journal of the Brazilian Society of Mechanical Sciences and Engineering*, *41*(7), 277. doi:10.100740430-019-1772-1

Gopura, R. A. R. C., & Kiguchi, K. (2009). Mechanical designs of active upper-limb exoskeleton robots. *11th Int. Conf. on Rehabilitation Robotics*, 178-187.

Gordo, N., & Ferreira, J. (2012). *Machine Elements*. São Paulo: Telecurso 2000 Profissionalizante.

Gou, H., Wang, J., Wu, H., Wang, C., Yan, L., & Xiao, J. (2014). Design of information acquisition and control system for the exoskeleton robot. *Journal of Electrical and Computer Engineering*, *2014*, 2. doi:10.1155/2014/309814

Haage, M., Piperagkas, G., Papadopoulos, C., Mariolis, I., Malec, J., Bekiroglu, Y., ... Tzovaras, D. (2017). Teaching assembly by demonstration using advanced human robot interaction and a knowledge integration framework. *Procedia Manufacturing*, *11*, 164–173. doi:10.1016/j.promfg.2017.07.221

Habib, M. K. (2007). Mechatronics - A unifying interdisciplinary and intelligent engineering science paradigm. IEEE Indsutrial Electronics Magazine, 2, 2 – 24.

Habib, M. K., Liu, G. L., Watanabe, K., & Izumi, K. (2007). Bipedal Locomotion Control via CPGs with Coupled Nonlinear Oscillators. In *Mechatronics, 2007 IEEE International Conference on* (pp. 1–6). IEEE. 10.1109/ICMECH.2007.4280021

Habib, M. K., Watanabe, K., & Izumi, K. (2009). Biped locomotion using CPG with sensory interaction. *IEEE International Symposium on Industrial Electronics*, 1452–1457. 10.1109/ISIE.2009.5219063

Haidegger, T. (2019). Autonomy for Surgical Robots: Concepts and Paradigms. *IEEE Trans. on Medical Robotics and Bionics*, *1*(2), 65–76. doi:10.1109/TMRB.2019.2913282

Haidegger, T. (2019). Probabilistic Method to Improve the Accuracy of Computer-Integrated Surgical Systems. Acta Polytechnica Hungarica. *Special Issue on Platforms for Medical Robotics Research*, *16*(8). doi:10.12700/APH.16.8.2019.8.8

Handa, H. (2009). Solving multi-objective reinforcement learning problems by EDA-RL -, acquisition of various strategies, In *Proceedings of the Ninth International Conference on Intelligent Systems Design and Applications* (pp. 426-431). Pisa, Italy: IEEE 10.1109/ISDA.2009.92

Hannaford, B., Rosen, J., Friedman, D. W., King, H., Roan, P., Cheng, L., ... White, L. (2013). Raven-II: An Open Platform for Surgical Robotics Research. *IEEE Transactions on Biomedical Engineering*, *60*(4), 954–959. doi:10.1109/TBME.2012.2228858 PMID:23204264

Harada, K., Kajita, S., Kanehiro, F., Fujiwara, K., Kaneko, K., Yokoi, K., & Hirukawa, H. (2007). Real-time planning of humanoid robot's gait for force-controlled manipulation. *Mechatronics, IEEE/ASME Transactions on, 12*(1), 53-62

Hatem, S. M., Saussez, G., Della Faille, M., & ... . (2016). Rehabilitation of Motor Function after Stroke: A Multiple Systematic Review Focused on Techniques to Stimulate Upper Extremity Recovery. *Frontiers in Human Neuroscience, 10*(September), 1–22. PMID:27679565

Heer, C. (2016, August 18). *Survey: 1.3 million industrial robots to enter service by 2018.* Retrieved from http://www.ifr.org/news/ifr-press-release/survey-13-million-industrial-robots-to-enter-service-by-2018-799/

Heliot, R., Azevedo, C., & Espiau, B. (2012). Functional Rehabilitation: Coordination of Artificial and Natural Controllers. Rehabilitation Robotics. doi:10.5772/5160

Herrmann, M. (2015). *RL 16: Model-based RL and Multi-Objective Reinforcement Learning.* University of Edinburgh, School of Informatics. Retrieved from http://www.inf.ed.ac.uk/teaching/courses/rl/slides15/rl16.pdf

Hiller, M., Hirsch, K., Bruckmann, T., Brandt, T., & Schramm, D. (2009). Common aspects in methods for the design of mechatronic systems - Applications in automotive and robotic systems. *XII International Symposium on Dynamic Problems of Mechanics – DINAME 2009.*

Hiraoka, K., Yoshida, M., & Mishima, T. (2009). Parallel reinforcement learning for weighted multi-criteria model with adaptive margin. *Cognitive Neurodynamics, 3*(1), 17–24. doi:10.100711571-008-9066-9 PMID:19003453

Hirschhorn, A. D., Lockhart, J. W., & Breckenridge, J. D. (2015). Can a physical activity monitor provide a valid measure of arm elevation angle? A study to assess agreement between the SenseWear Mini Armband and the universal goniometer. *BMC Musculoskeletal Disorders,* 16–46. PMID:25886361

History – Applied Dexterity. (n.d.). Retrieved April 30, 2019, from http://applieddexterity.com/about/history/

Hoang, G., Min, J. H., Lee, G. M., Jun, B. H., Kim, H. K., & Kim, S. B. (2014, October), Omni-directional walking control for a six-legged robot using differential kinematics algorithm, In *Proceedings of 2014 14th International Conference on Control, Automation and Systems (ICCAS 2014)* (pp. 1163–1168). Academic Press. 10.1109/ICCAS.2014.6987735

Hochreiter, S., & Schmidhuber, J. (1997). Long short-term memory. *Neural Computation, 9*(8), 1735–1780. doi:10.1162/neco.1997.9.8.1735 PMID:9377276

Hocoma. (2019a). *Armeo® Therapy Concept.* Retrieved from https://products.iisartonline.org/products/24/marketing/BRO_Armeo_Therapy_Concept_130225_en.pdf

Hocoma. (2019b). Retrieved from http://www.hocoma.com/

Hogan, N. (1985). Impedance control: an approach to manipulation: Part I- Theory, Part II-Implementation, Part III - Applications. *ASME Journal of Dynamic Systems. Measurement and Control, 107*(1), 1–24. doi:10.1115/1.3140702

Hreljac, A., & Marshall, R. N. (2000). Algorithms to determine event timing during normal walking using kinematic data. *Journal of Biomechanics*, *33*(6), 783–786. doi:10.1016/S0021-9290(00)00014-2 PMID:10808002

Huang, H.-M. (2008). *Autonomy Levels for Unmanned Systems (ALFUS)*. Academic Press.

Hu, J., & Tsai, M. (2008). Design of robust stabilization and fault diagnosis for an auto-balancing two-wheeled cart. *Advanced Robotics*, *22*(2-3), 319–338. doi:10.1163/156855308X292600

Hussain, A., Budhota, A., & Contu, S. (2017). Quantitative assessment of motor functions post-stroke: Responsiveness of upper-extremity robotic measures and its task dependence. *IEEE International Conference on Rehabilitation Robotics*, 1037–1042. 10.1109/ICORR.2017.8009386

Ijspeert, A. J. (2008). Central pattern generators for locomotion control in animals and robots: A review. *Neural Networks*, *21*(4), 642–653. doi:10.1016/j.neunet.2008.03.014 PMID:18555958

Ikeda, R., & Horie, R. (2016). Control of bipedal locomotion with a neural oscillator-based brain-computer interface. *Journal of Information and Communication Technology*, *15*(2), 19–37.

InMotion. (2018). *InMotion ARM™ Interactive Therapy System - Bionik Labs*. Retrieved from http://bionikusa.com/healthcarereform/upper-extremity-rehabilitiation/inmotion2-arm/

InMotion. (2019). *Helps Traumatic Brain Injury Patient's - Bionik Labs*. Accessed Feb 13, 2019, <https://www.bioniklabs.com/products/inmotion-arm>

Ionescu, T.G. (2003). Terminology for the Mechanism and Machine Science. *Mech. and Mach. Theory*, (38), 819-825.

Isermann, R. (2003). *Mechatronic systems fundamentals*. Springer.

ISO Standards. (2016). Robots and robotic devices. *Collaborative robots*, PD ISO/TS 15066:2016.

Jäger, A., Moll, C., Som, O., Zanker, C., Kinkel, S., & Lichtner, R. (2015). *Analysis of the impact of robotic systems on employment in the European Union. Final report*. Luxembourg: Publications Office of the European Union.

Janardhanan, S. (2013). *Model Order Reduction and Controller Design Techniques*. Academic Press.

Jatsun, S., & Yatsun, A. (2018). Investigation of Human Cargo Handling in Industrial Exoskeleton. In *Proceedings - 2018 Global Smart Industry Conference*. IEEE.

Jatsun, S., Savin, S., & Yatsun, A. (2016, October). Study of controlled motion of an exoskeleton performing obstacle avoidance during a single support walking phase. In *System Theory, Control and Computing (ICSTCC), 2016 20th International Conference on* (pp. 113-118). IEEE.

Jatsun, S., Savin, S., & Yatsun, A. (n.d.). Modelling characteristics of human-robot interaction in an exoskeleton system with elastic elements. *Lecture Notes in Computer Science, 11097*, 85.

Jezernik, S., Colombo, G., & Morari, M. (2004). Automatic Gait-Pattern Adaptation Algorithms for Rehabilitation with A 4-Dof Robotic Orthosis. IEEE Transactions on Robotics and Automation, 20(3), 574-582.

Jezernik, S. (2003). Robotic orthosis lokomat: A rehabilitation and research tool. Neuromodulation: Technology at the Neural Interface. *Wiley Online Library*, *6*(2), 108–115.

Kaber, D. B., & Endsley, M. R. (2004). The effects of level of automation and adaptive automation on human performance, situation awareness and workload in a dynamic control task. *Theoretical Issues in Ergonomics Science*, *5*(2), 113–153. doi:10.1080/1463922021000054335

Kahn, L. E., Averbuch, M., & Rymer, W. Z. (2001). *Comparison of Robot-Assisted Reaching to Free Reaching in Promoting Recovery from Chronic Stroke*. National Institute on Disability and Rehabilitation Research.

Kahn, L. E., Zygman, M. L., Rymer, W. Z., & Reinkensmeyer, D. J. (2006). Robot-assisted reaching exercise promotes arm movement recovery in chronic hemiparetic stroke: A randomized controlled pilot study. *Journal of Neuroengineering and Rehabilitation*, 1–13. PMID:16790067

Kajita, S., Kanehiro, F., Kaneko, K., Fujiwara, K., Harada, K., Yokoi, K., & Hirukawa, H. (2003, September). Biped walking pattern generation by using preview control of zero-moment point. In *Robotics and Automation, 2003. Proceedings. ICRA'03. IEEE International Conference on* (Vol. 2, pp. 1620-1626). IEEE. 10.1109/ROBOT.2003.1241826

Kajita, S., Kanehiro, F., Kaneko, K., Yokoi, K., & Hirukawa, H. (2001). The 3D Linear Inverted Pendulum Mode: A simple modeling for a biped walking pattern generation. In *Intelligent Robots and Systems, 2001. Proceedings. 2001 IEEE/RSJ International Conference on* (Vol. 1, pp. 239-246). IEEE.

Kamisetty, K. V., & McDermott, K. J. (1992). Development of a CAD/CAM Robotic Translator for Programming the IBM 7535 SCARA Robot off-line. *Computers in Industry*, *20*(2), 219–228. doi:10.1016/0166-3615(92)90056-S

Kapandji, A. I. (2007). The Physiology of the Joints: Vol. 1. *The Upper Limb* (6th ed.). Churchill Livingstone.

Kapandji, A. I. (2010). The Physiology of the Joints: Vol. 2. *Lower Limb* (6th ed.). Churchill Livingstone.

Kaplan, K. E., Nichols, K. A., & Okamura, A. M. (2016). Toward human-robot collaboration in surgery: Performance assessment of human and robotic agents in an inclusion segmentation task. *2016 IEEE International Conference on Robotics and Automation (ICRA)*, 723–729. 10.1109/ICRA.2016.7487199

Karoly, A. I., Kuti, J. & Galambos, P. (2018). Unsupervised real-time classification of cycle stages in collaborative robot applications. *2018 IEEE 16th World Symposium on Applied Machine Intelligence and Informatics (SAMI)*. 10.1109/SAMI.2018.8323994

Kazanzides, P. (2005). *Open Source Software Libraries for Computer Integrated Surgery*. Retrieved from https://cisst.org/wp-content/uploads/2016/03/YR_8_Open_Source_Software.pdf

Kazanzides, P., Chen, Z., Deguet, A., Fischer, G. S., Taylor, R. H., & DiMaio, S. P. (2014). An open-source research kit for the da Vinci® Surgical System. *Proc. of the IEEE International Conference on Robotics and Automation*, 6434–6439. 10.1109/ICRA.2014.6907809

Kemerich, P., Piovesan, M., Bertoletti, L., Altmeyer, L., & HohmVorpagel, T. (2013, January-April). *Glass Fiber: Characterization, Disposal and Environmental Impact Generated.* Retrieved from https://periodicos.ufsm.br/reget/article/viewFile/7590/pdf

Khabbazi, M. R., Yusof Ismail, M. D., Ismail, N., & Mousavi, S. A. (2010). Modeling of traceability information system for material flow control data. *Australian Journal of Basic and Applied Sciences*, *4*(2), 208–216.

Khosravi, M., & Taghirad, H. D. (2011). Dynamic analysis and control of cable driven robots with elastic cables. Transactions – Canadian Society for Mechanical Engineering 35(4), 543-557. doi:10.1139/tcsme-2011-0033

Kiguchi, K., Iwami, K., Yasuda, M., Watanabe, K., & Fukuda, T. (2003). An exoskeletal robot for human shoulder joint motion assist. *IEEE/ASME Transactions on Mechatronics*, *8*(1), 125–135. doi:10.1109/TMECH.2003.809168

Kim, Y., Kim, S. H., & Kwak, Y. (2005). Dynamic analysis of a nonholonomic two-wheeled inverted pendulum robot. *Journal of Intelligent & Robotic Systems*, *44*(1), 25–46. doi:10.100710846-005-9022-4

Koker, R. (2013, February). A genetic algorithm approach to a neural network-based inverse kinematics solution of robotic manipulators based on error minimization. *Information Sciences*, *222*, 528–543. doi:10.1016/j.ins.2012.07.051

Koker, R., Oz, C., Çakar, T., & Ekiz, H. (2004, December). A study of neural network based inverse kinematics solution for a three-joint robot. *Robotics and Autonomous Systems*, *49*(3), 227–234. doi:10.1016/j.robot.2004.09.010

Konda, T., Tensyo, S., & Yamaguchi, T. (2002). LC-Learning: Phased Method for Average Reward Reinforcement Learning - Preliminary Results. In *Proceedings of 7th Pacific Rim International Conference on Artificial Intelligence (PRICAI2002): PRICAI2002 Trends in Artificial Intelligence, Lecture notes in Artificial Intelligence 2417* (pp. 208-217). Berlin, Germany: Springer. 10.1007/3-540-45683-X_24

Krabben, T., & Prange, G. B. (2012). Influence of gravity compensation training on synergistic movement patterns of the upper extremity after stroke, a pilot study. *Journal of Neuroengineering and Rehabilitation*, 9–44. PMID:22824488

Krebs, H.I., Bruce V., and Hogan N. (2009). A working model of stroke recovery from rehabilitation robotics practitioners. *Journal of NeuroEngineering and Rehabilitation, 6*(1).

Krebs, H.I., Finley, M.A., Dipietro, L., Ohlhof F, J., Whitall, J. & Bever, C.T. (2004). *Does MIT-MANUS upper extremity robot testing create a learning effect in healthy adults?* Academic Press.

Krebs, H. I., Palazzolo, J. J., Dipietro, L., Ferraro, M., Krol, J., Rannekleiv, K., ... Hogan, N. (2003). Rehabilitation Robotics: Performance-Based Progressive Robot-Assisted Therapy. *Autonomous Robots, Springer*, *15*(1), 7–20. doi:10.1023/A:1024494031121

Krenek, J., Kuca, K., Krejcar, O., Blazek, P., & Jun, D. (2016). Application of artificial neural networks in condition based predictive maintenance. In D. Król, L. Madeyski, & N. Nguyen (Eds.), *Recent Developments in Intelligent Information and Database Systems. Studies in Computational Intelligence* (Vol. 642). Cham: Springer. doi:10.1007/978-3-319-31277-4_7

Kumar, D., & Nagar, S. K. (2014). Model reduction by extended minimal degree optimal Hankel norm approximation. *Applied Mathematical Modelling*, *38*(11-12), 2922–2933. doi:10.1016/j.apm.2013.11.012

Kushida, D., Nakamura, M., Goto, S., & Kyura, N. (2001). Human Direct Teaching of Industrial Articulated Robot Arms based on Force-Free Control. *Artificial Life and Robotics*, *5*(1), 26–32. doi:10.1007/BF02481317

Kuteken, R. S., Batista, R. F., & Rosario, J. M. (2017). Cognitively inspired artificial bipedal humanoid gait generation. *2017 International Work Conference on Bio-Inspired Intelligence: Intelligent Systems for Biodiversity Conservation, IWOBI 2017 - Proceedings*. 10.1109/IWOBI.2017.7985518

Kuznetsov, A. N., Rybalko, N. V., Daminov, V. D., & Luft, A. R. (2013). Early Poststroke Rehabilitation Using a Robotic Tilt-Table Stepper and Functional Electrical Stimulation. Stroke Research and Treatment. doi:10.1155/2013/946056

Laribi, M. A., Carbone, G., & Zeghloul, S. (2019). On the Optimal Design of Cable Driven Parallel Robot with a Prescribed Workspace for Upper Limb Rehabilitation Tasks. *Journal of Bionics Engineering*, *16*, 503. doi:10.100742235-019-0041-4

Lasso, A., Heffter, T., Rankin, A., Pinter, C., Ungi, T., & Fichtinger, G. (2014). PLUS: Open-Source Toolkit for Ultrasound-Guided Intervention Systems. *IEEE Transactions on Biomedical Engineering*, *61*(10), 2527–2537. doi:10.1109/TBME.2014.2322864 PMID:24833412

Latombe, J. C. (1991). *Robot Motion Planning*. Kluwer. doi:10.1007/978-1-4615-4022-9

Li, J., Esteban-Fernández de Ávila, B., Gao, W., Zhang, L. & Wang, J. (2017). Micro/nanorobots for biomedicine: Delivery, surgery, sensing, and detoxification. *Science Robotics, 2*(4). doi:10.1126cirobotics.aam6431

Li, Q., & Zhang, H. H. (2019, May). Model Reduction of A Real Time Multidisciplinary Mechatronic System. In *20th International Conference on Research and Education in Mechatronics (REM 2019)*. University of Applied Sciences Upper Austria. 10.1109/REM.2019.8744102

Li, Yan, Qian, Wu, Wu, & Men. (2015). Review on Lower Extremity Exoskeleton Robot. *Open Automation and Control Systems Journal*, *7*, 441–453.

Li, Z., Vanderborght, B., Tsagarakis, N. G., & Caldwell, D. G. (2013). Quasi-straightened knee walking for the humanoid robot. Cognitive Systems Monographs. doi:10.1007/978-3-642-36368-9_9

Linhart, A., & Skogestad, S. (2012). An Aggregation Model Reduction Method for One-Dimensional Distributed Systems. *AIChE Journal. American Institute of Chemical Engineers*, *58*(5), 1524–1537. doi:10.1002/aic.12688

Li, Q. (2015). *Large Scale Modeling, Model Reduction and Control Design for A Real-time Mechatronic System*. Purdue University.

Liu, C. J., Fan, Z., Seo, K., Tan, X. B., & Goodman, E. D. (2012). Synthesis of Matsuoka-Based Neuron Oscillator Models in Locomotion Control of Robots. In *2012 Third Global Congress on Intelligent Systems* (pp. 342–347). IEEE. 10.1109/GCIS.2012.99

Liu, C., Wang, D., Goodman, E. D., & Chen, Q. (2016). Adaptive walking control of biped robots using online trajectory generation method based on neural oscillators. *Journal of Bionics Engineering*, *13*(4), 572–584. doi:10.1016/S1672-6529(16)60329-3

Liu, C., Xu, X., & Hu, D. (2015). Multiobjective Reinforcement Learning: A Comprehensive Overview. *IEEE Transactions on Systems, Man, and Cybernetics. Systems*, *45*(3), 385–398. doi:10.1109/TSMC.2014.2358639

Liu, G. L., Habib, M. K., Watanabe, K., & Izumi, K. (2007). The Design of Central Pattern Generators Based on the Matsuoka Oscillator to Generate Rhythmic Human-Like Movement for Biped Robots. *Journal of Advanced Computational Intelligence and Intelligent Informatics*, *11*(8), 946–955. doi:10.20965/jaciii.2007.p0946

Liu, G. L., Habib, M. K., Watanabe, K., & Izumi, K. (2008a). Central pattern generators based on Matsuoka oscillators for the locomotion of biped robots. *Artificial Life and Robotics*, *12*(1–2), 264–269. doi:10.100710015-007-0479-z

Lizotte, D. J., Bowling, M., & Murphy, S. A. (2012). Linear fitted-q iteration with multiplereward functions. *Journal of Machine Learning Research*, *13*, 3253–3295. PMID:23741197

Lotlikar, T., Kankapurkar, R., Parekar, A., & Mohite, A. (2013). Comparative study of barcode, QR-code and RFID system. *International Journal of Computer Technology & Applications*, *4*(5), 817–821.

Lu, Q., & Tian, J. (2015). Research on Walking Gait of Biped Robot Based on a Modified CPG Model. *Mathematical Problems in Engineering*, *2015*, 1–9. doi:10.1155/2015/584954

Maciejasz, P., Eschweiler, J., Gerlach-Hahn, K., Jansen-Troy, A., & Leonhardt, S. (2014). A survey on robotic devices for upper limb rehabilitation. *Journal of Neuroengineering and Rehabilitation*, *11*(3), 1–29. PMID:24401110

MacKenzie, L., Ibbotson, J. A., Cao, C. G. L., & Lomax, A. J. (2001). Hierarchical decomposition of laparoscopic surgery: A human factors approach to investigating the operating room environment. *Minimally Invasive Therapy & Allied Technologies*, *10*(3), 121–127. doi:10.1080/136457001753192222 PMID:16754003

Maeda, Y., Fujiwara, T., & Ito, H. (2014, September), Robot control using high dimensional neural networks. In *Proceedings of SICE Annual Conference 2014 -International Conference on Instrumentation, Control, Information Technology and System Integration-* (pp. 738–743). Academic Press. 10.1109/SICE.2014.6935220

Maeda, Y., Ishido, N., Kikuchi, H., & Arai, T. (2002). Teaching of Grasp/Graspless Manipulation for Industrial Robots by Human Demonstration. *Proceedings IEEE/RSJ International Conference on Intelligent Robots and Systems*, 1523–1528. 10.1109/IRDS.2002.1043971

Mahadevan, S. (1996). Average Reward Reinforcement learning: Foundations, Algorithms, and Empirical Results. *Machine Learning*, *22*(1-3), 159–196. doi:10.1007/BF00114727

Mahler, J., Krishnan, S., Laskey, M., & Sen, S. (2014). Learning accurate kinematic control of cable-driven surgical robots using data cleaning and Gaussian process regression. *IEEE International Conference on Automation Science and Engineering*. 10.1109/CoASE.2014.6899377

Major, K. A., Major, Z. Z., & Carbone, G. (2008). Ranges of motion as basis for robot-assisted post- stroke rehabilitation. *Human & Veterinary Medicine - International Journal of the Bioflux Society*, *8*(4), 192–196.

Major, K. A., Major, Z. Z., Carbone, G., Pîslă, A., Vaida, C., Gherman, B., & Pîslă, D. (2016). Ranges of motion as basis for robot-assisted post-stroke rehabilitation. *J Bioflux Soc Hum Vet Med*, *8*(4), 192–196.

Makris, S., Tsarouchi, P., Surdilovic, D., & Krüger, J. (2014). Intuitive dual arm robot programming for assembly operations. *CIRP Annals*, *63*(1), 13–16. doi:10.1016/j.cirp.2014.03.017

Malvezzi, F. (2008, February 27). *Evaluation of kinematic performance of a tridimensional parallel mechanism*. Retrieved from http://www.teses.usp.br/teses/disponiveis/3/3151/tde-08012008-085930/pt-br.php

Manoonpong, P., Geng, T., Kulvicius, T., Porr, B., & Wörgötter, F. (2007). Adaptive, Fast Walking in a Biped Robot under Neuronal Control and Learning. *PLoS Computational Biology*, *3*(7), 1305–1320. doi:10.1371/journal.pcbi.0030134 PMID:17630828

Mansard, N. (2012, May). A dedicated solver for fast operational-space inverse dynamics. In *2012 IEEE International Conference on Robotics and Automation* (pp. 4943-4949). IEEE. 10.1109/ICRA.2012.6224851

Mansard, N., & Chaumette, F. (2007). Task sequencing for high-level sensor-based control. *IEEE Transactions on Robotics*, *23*(1), 60–72. doi:10.1109/TRO.2006.889487

Mansard, N., Khatib, O., & Kheddar, A. (2009). A unified approach to integrate unilateral constraints in the stack of tasks. *IEEE Transactions on Robotics*, *25*(3), 670–685. doi:10.1109/TRO.2009.2020345

Mao, Y., Jin, X., Gera Dutta, G., Scholz, J. P., & Agrawal, S. K. (2015). Human Movement Training with a Cable Driven ARm EXoskeleton (CAREX). *IEEE Transactions on Neural Systems and Rehabilitation Engineering*, *23*(1), 84–92. doi:10.1109/TNSRE.2014.2329018 PMID:24919202

Mao, Z., & Hsia, T. C. (1997, January). Obstacle avoidance inverse kinematics solution of redundant robots by neural networks. *Robotica*, *15*(1), 3–10. doi:10.1017/S0263574797000027

Martelli, D., Vannetti, F., Cortese, M., Tropea, P., Giovacchini, F., Micera, S., ... Vitiello, N. (2014). The effects on biomechanics of walking and balance recovery in a novel pelvis exoskeleton during zero-torque control. *Robotica*, *32*(08), 1317–1330. doi:10.1017/S0263574714001568

Martin, E. M., & Cazorla, M. (2019). *Rehabilitation Technology: Assistance from Hospital to Home*. Hindawi Computational Intelligence and Neuroscience. doi:10.1155/2019/1431509

Márton, L., Szántó, Z., Haidegger, T., Galambos, P., & Kövecses, J. (2017). Internet—Based Bilateral Teleoperation Using a Revised Time—Domain Passivity Controller. *Acta Polytechnica Hungarica*, 27–45.

Masiero, S., Armani, M., Ferlini, G., Rosati, G., & Rossi, A. (2014a). Randomized trial of a robotic assistive device for the upper extremity during early inpatient stroke rehabilitation. *Neurorehabilitation and Neural Repair*, *28*(4), 377–386. doi:10.1177/1545968313513073 PMID:24316679

Masiero, S., Armani, M., & Rosati, G. (2011). Upper-limb robot-assisted therapy in rehabilitation of acute stroke patients: Focused review and results of new randomized controlled trial. *Journal of Rehabilitation Research and Development*, *48*(4), 355–366. doi:10.1682/JRRD.2010.04.0063 PMID:21674388

Masiero, S., Celia, A., Rosati, G., & Armani, M. (2007). Robotic-assisted rehabilitation of the upper limb after acute stroke. *Archives of Physical Medicine and Rehabilitation*, *88*(2), 142–149. doi:10.1016/j.apmr.2006.10.032 PMID:17270510

Masiero, S., Poli, P., Armani, M., Ferlini, G., Rizziello, R., & Rosati, G. (2014c). Robotic upper limb rehabilitation after acute stroke by nerebot: Evaluation of treatment costs. *BioMed Research International*. PMID:24967345

Masiero, S., Poli, P., Rosati, G., Zanotto, D., Iosa, M., Paolucci, S., & Morone, G. (2014b). The value of robotic systems in stroke rehabilitation. *Expert Review of Medical Devices*, *11*(2), 187–198. doi:10.1586/17434440.2014.882766 PMID:24479445

Massaro, A., Galiano, A., Meuli, G., & Massari, S. F. (2018). Overview and application of enabling technologies oriented on energy routing monitoring, on network installation and on predictive maintenance. *International Journal of Artificial Intelligence and Applications*, *9*(2), 1–20. doi:10.5121/ijaia.2018.9201

Massaro, A., Maritati, V., Galiano, A., Birardi, V., & Pellicani, L. (2018). ESB platform integrating KNIME data mining tool oriented on Industry 4.0 based on artificial neural network predictive maintenance. *International Journal of Artificial Intelligence and Applications*, *9*(3), 1–17. doi:10.5121/ijaia.2018.9301

Massaro, A., Maritati, V., Savino, N., Galiano, A., Convertini, D., De Fonte, E., & Di Muro, M. (2018). A study of a health resources management platform integrating neural networks and DSS telemedicine for homecare assistance. *Information*, *9*(176), 1–20.

Massaro, A., Meuli, G., Savino, N., & Galiano, A. (2018). A precision agriculture DSS based on sensor threshold management for irrigation field. Signal & Image Processing *International Journal*, *9*(6), 39–58.

Massaro, A., Vitti, V., & Galiano, A. (2018). Automatic image processing engine oriented on quality control of electronic boards. *Signal and Image Processing: an International Journal*, *9*(2), 1–14. doi:10.5121ipij.2018.9201

MathWorks. (2018). *Stateflow - MATLAB & Simulink*. Author.

Matthaiakis, S. A., Dimoulas, K., Athanasatos, A., Mparis, K., Dimitrakopoulos, G., Gkournelos, C., ... Angione, G. (2017). Flexible programming tool enabling synergy between human and robot. *Procedia Manufacturing*, *11*, 431–440. doi:10.1016/j.promfg.2017.07.131

Ma, X., Song, C., Chiu, P. W., & Li, Z. (2019). Autonomous Flexible Endoscope for Minimally Invasive Surgery With Enhanced Safety. *IEEE Robotics and Automation Letters*, *4*(3), 2607–2613. doi:10.1109/LRA.2019.2895273

Mayhew, D., Bachrach, B., Rymer, W. Z., & Beer, R. F. (2005). Development of the MACARM-a novel cable robot for upper limb neurorehabilitation. *Proc. of ICORR 9th International Conference on Rehabilitation Robotics*, 299-302. 10.1109/ICORR.2005.1501106

Mccomb, G. (2011). *Robot builder's bonanza*. TAB.

McKinley, S., Garg, A., Sen, S., Gealy, D. V., McKinley, J., Jen, Y. & Goldberg, K. (2016). *Autonomous Multilateral Surgical Tumor Resection with Interchangeable Instrument Mounts and Fluid Injection Device*. Academic Press.

Medical Devices. (2015). *Recognized essential principles of safety and performance of medical devices*. BS ISO 16142-2.

Mehrholz, J. (2012). Electromechanical and robot-assisted arm training for improving generic activities of daily living, arm function, and arm muscle strength after stroke. *Cochrane Database of Systematic Reviews*, *6*, CD006876. PMID:22696362

Moffaert, K. V., & Nowe, A. (2014). Multi-Objective Reinforcement Learning using Sets of Pareto Dominating Policies. *Journal of Machine Learning Research*, *15*, 3663–3692.

Mohamed, A., Hossny, M., Nahavandi, S., Dalvand, M., & Asadi, H. (2018). Towards Trusted Autonomous Surgical Robots. *Proc. of the IEEE International Conference on Systems, Man, and Cybernetics (SMC)*, 4083–4088.

Mooney, L. M., Rouse, E. J., & Herr, H. M. (2014). Autonomous exoskeleton reduces metabolic cost of walking. In *Engineering in Medicine and Biology Society (EMBC), 2014 36th Annual International Conference of the IEEE*, (pp. 3065-3068). IEEE. 10.1109/EMBC.2014.6944270

Moore, K. L., & Dalley, A. F. (2009). *Clinically oriented anatomy* (6th ed.). Philadelphia: Lippincott Williams & Wilkins.

Murali, A., Sen, S., Kehoe, B., Garg, A., McFarland, S., & Patil, S., … Goldberg, K. (2015). Learning by observation for surgical subtasks: Multilateral cutting of 3D viscoelastic and 2D Orthotropic Tissue Phantoms. *Proc. of the IEEE International Conference on Robotics and Automation*, 1202–1209. 10.1109/ICRA.2015.7139344

Nadas, I., Vaida, C., Gherman, B., Pisla, D., & Carbone, G. (2017). Considerations for designing robotic upper limb rehabilitation devices. *AIP Conference Proceedings*, *1917*, 030005. doi:10.1063/1.5018278

Nagata, F., & Watanabe, K. (2002, September), Learning of contact motion using a neural network and its application for force control. In *Proceedings of the 4th Asian Control Conference (ASCC2002)* (pp. 420–424). Academic Press.

Nagata, F., Inoue, S., Fujii, S., Otsuka, A., Watanabe, K., & Habib, M. K. (2015, August). *Learning of inverse kinematics using a neural network and its application to kinematic control of position-based servo motor*. Paper presented at The 2015 World Congress on Advances in Aeronautics, Nano, Bio, Robotics, and Energy (ANBRE15).

Nagata, F., Otsuka, A., Sakakibara, K., Watanabe, K., & Habib, M. K. (2013, September), Experiment Systems Using Three Types of Motors for Biomimetic Machine Research. In *Proceedings of SICE Annual Conference 2013 -International Conference on Instrumentation, Control, Information Technology and System Integration* (pp. 2711–2717). Academic Press.

Nagata, F., Yamane, Y., Okada, Y., Kusano, T., Watanabe, K. & Habib, M. K. (2018). Development of Post Processor Approach for an Industrial Robot FANUC R2000iC. *Journal of Artificial Life and Robotics*, *23*(2), 186-191.

Nagata, F., Hayashi, S., Nagatmi, T., Otsuka, A., & Watanabe, K. (2016). Application of Fuzzy Reasoning and Neural Network to Feed Rate Control of a Machining Robot. *International Journal of Applied Electromagnetics and Mechanics*, *52*(3-4), 897–905. doi:10.3233/JAE-162217

Nagata, F., Kusumoto, Y., & Watanabe, K. (2009). Intelligent Machining System for the Artistic Design of Wooden Paint Rollers. *Robotics and Computer-integrated Manufacturing*, *25*(3), 680–688. doi:10.1016/j.rcim.2008.05.001

Nagata, F., & Watanabe, K. (2016). Neural network-based inverse kinematics for Motoman HS20 and its efficient learning method. *Journal of the Institute of Industrial Applications Engineers*, *4*(4), 166–171. doi:10.12792/JIIAE.4.166

Nagata, F., Watanabe, K., & Izumi, K. (2001). Furniture Polishing Robot using a Trajectory Generator based on Cutter Location Data. *Proceedings of 2001 IEEE International Conference on Robotics and Automation*, 319–324. 10.1109/ROBOT.2001.1620978

Nagata, F., Watanabe, K., & Kiguchi, K. (2006). *Joystick Teaching System for Industrial Robots using Fuzzy Compliance Control. In Industrial Robotics: Theory, Modelling and Control* (pp. 799–812). INTECH.

Nagy, D. Á., Rudas, I. J. & Haidegger, T. (2018). OntoFlow, a software tool for surgical workflow recording. *2018 IEEE 16th World Symposium on Applied Machine Intelligence and Informatics (SAMI)*, 119–124. 10.1109/SAMI.2018.8323998

Nagy, D. Á., Nagy, T. D., Elek, R., Rudas, I. J., & Haidegger, T. (2018). Ontology-Based Surgical Subtask Automation, Automating Blunt Dissection. *Journal of Medical Robotics Research*, *3*(3). doi:10.1142/S2424905X18410052

Nagyné Elek, R., & Haidegger, T. (2019). Robot-Assisted Minimally Invasive Surgical Skill Assessment—Manual and Automated Platforms. Acta Polytechnica Hungarica. *Special Issue on Platforms for Medical Robotics Research*, *16*(8). doi:10.12700/APH.16.8.2019.8.9

Nagy, T. D., Takács, M., Rudas, I. J., & Haidegger, T. (2018). Surgical Subtask Automation—Soft Tissue Retraction. *Proc. of the 16th IEEE World Symposium on Applied Machine Intelligence and Informatics*, 55–60.

Nagy, Tamas D. & Haidegger, T. (2019). A DVRK-based Framework for Surgical Subtask Automation. Acta Polytechnica Hungarica. *Special Issue on Platforms for Medical Robotics Research*, *16*(8). doi:10.12700/APH.16.8.2019.8.5

Nasrallah, D., Michalska, H., & Angeles, J. (2007). Controllability and posture control of a wheeled pendulum moving on an inclined plane. *IEEE Transactions on Robotics*, *23*(3), 564–577. doi:10.1109/TRO.2007.898953

Natarajan, S., & Tadepalli, P. (2005). Dynamic preferences in multi-criteria reinforcement learning. In *Proceedings of the 22nd international conference on machine learning (ICML-2005)* (pp. 601-60). ACM

Nef, T., & Riener, R. (2005). ARMin: Design of a novel arm rehabilitation robot. *IEEE 9th Int. Conf. Rehabilitation Robotics - ICORR2005*, 57–60.

Neto, P., Pires, J. N., & Moreira, A. P. (2010). CAD based Off-Line Robot Programming. *Proceedings of IEEE International Conference on Robotics Automation and Mechatronics*, 516–521.

Nichols, K. A., & Okamura, A. M. (2013). Autonomous robotic palpation: Machine learning techniques to identify hard inclusions in soft tissues. *Proc. of the 2013 IEEE International Conference on Robotics and Automation*, 4384–4389. 10.1109/ICRA.2013.6631198

Novak, D. E., & Riener, R. (2015). Control Strategies and Artificial Intelligence in Rehabilitation Robotics. *AI Magazine*, *36*(4), 23. doi:10.1609/aimag.v36i4.2614

Novak-Marcincin, J., Barna, J., Janak, M., & Novakova-Marcincinova, L. (2013). Augmented reality aided manufacturing. *Procedia Computer Science*, *25*(1), 23–31. doi:10.1016/j.procs.2013.11.004

Nunes, W. M., Ribeiro, J. F., Carvalho, J. C. M., & Gonçalves, R. S. (2011a). Kinematics modeling and workspace analysis of a cable-based parallel manipulator for shoulder rehabilitation. *21th Brasilian Congress on Mechanical Engineering – COBEM 2011*.

Nunes, W. M., Ribeiro, J. F., Carvalho, J. C. M., & Gonçalves, R. S. (2011b). Workspace Analysis of a Cable-Base Parallel Manipulator for Rehabilitation of Shoulder Movements. *Proceedings of the XIV International Symposium on Dynamic Problems of Mechanics*.

Nunes, W. M., Rodrigues, L. A. O., Oliveira, L. P., Ribeiro, J. F., Carvalho, J. C. M., & Gonçalves, R. S. (2011c). Cable-Based Parallel Manipulator for Rehabilitation of Shoulder and Elbow Movements. *International Conference on Rehabilitation Robotics*. 10.1109/ICORR.2011.5975503

O'Connor, C. M., Thorpe, S. K., O'Malley, M. J., & Vaughan, C. L. (2007). Automatic detection of gait events using kinematic data. *Gait & Posture*, *25*(3), 469–474. doi:10.1016/j.gaitpost.2006.05.016 PMID:16876414

OECD. (2017). *The Next Production Revolution: Implications for Governments and Business*. Paris: OECD Publishing.

Olszewska, J. I., Houghtaling, M., Gonçalves, P., Haidegger, T., Fabiano, N., Carbonera, J. L., … Prestes, E. (2018). *Robotic Ontological Standard Development Life Cycle*. Presented at the IEEE ICRA 2018 WELCARO workshop. Retrieved from http://real.mtak.hu/86419/

O'Sullivan, Nugent, & van der Vorm. (2015). *Standards for the Safety of Exoskeletons Used by Industrial Workers Performing Manual Handling Activities*. A Contribution from the Robo-Mate Project to their Future Development.

Özdemir, V., & Hekim, N. (2018). Birth of Industry 5.0: Making Sense of Big Data with Artificial Intelligence, 'The Internet of Things" and Next-Generation Technology Policy'. *OMICS: A Journal of Integrative Biology*, *22*(1), 65–76. doi:10.1089/omi.2017.0194 PMID:29293405

Oztemel, E., & Gursev, S. (2018). Literature Review of Industry 4.0 and related Technologies. *Journal of Intelligent Manufacturing*, *30*(176), 1–56.

Pan, Z., Polden, J., Larkin, N., Van Duin, S., & Norrish, J. (2010, June). Recent progress on programming methods for industrial robots. In *ISR 2010 (41st International Symposium on Robotics) and ROBOTIK 2010 (6th German Conference on Robotics)* (pp. 1-8). VDE.

Paraskevopoulos, P. N. (1980). Padé-type order reduction of two-dimensional systems. *IEEE Transactions on Circuits and Systems*, *CAS-27*(5), 413–416. doi:10.1109/TCS.1980.1084833

Pedersen, M. R., Nalpantidis, L., Andersen, R. S., Schou, C., Bøgh, S., Krüger, V., & Madsen, O. (2016). Robot skills for manufacturing: From concept to industrial deployment. *Robotics and Computer-integrated Manufacturing, 37*, 282–291. doi:10.1016/j.rcim.2015.04.002

Pennycott, A., Wyss, D., Vallery, H. K., Marganka, V., & Riener, R. (2012). Towards more effective robotic gait training for stroke rehabilitation: a review. Journal of NeuroEngineering and Rehabilitation, 9(65).

Perzylo, A., Somani, N., Profanter, S., Kessler, I., Rickert, M., & Knoll, A. (2016, October). *Intuitive instruction of industrial robots: Semantic process descriptions for small lot production. In 2016 IEEE/RSJ international conference on intelligent robots and systems (iros)* (pp. 2293–2300). IEEE.

Pinder, J. M. (2016). *Multi-Objective Reinforcement Learning Framework for Unknown Stochastic & Uncertain Environments* (Unpublished doctoral dissertation). Retrieved from http://usir.salford.ac.uk/id/eprint/39978/ 2/John%20Pinder%20PhD%20Thesis%20Complete.pdf

Pons, J. L. (2008). *Wearable Robots: Biomechatronic Exoskeletons*. West Sussex, UK: John Wiley and Sons Ltd. doi:10.1002/9780470987667

Prange, G. B. (2009) An explorative Study into Changes in Reach Performance After Gravity Compensation Training in Chronic Stroke Patients. *IEEE 11th International Conference on Rehabilitation Robotics*.

Prange, G. B., Kottink, A. I. R., Buurke, J. H., Eckhardt, M. M. E. M., van Keulen-Rouweler, B. J., Ribbers, G. M., & Rietman, J. S. (2015). The effect of Arm Support combined with rehabilitation games on upper-extremity function in subacute stroke: A randomized controlled trial. *Neurorehabilitation and Neural Repair, 29*(2), 174–182. doi:10.1177/1545968314535985 PMID:24878589

Prange, G. B., Kottink, A. I. R., Buurke, J. H., & Rietman, J. S. (2013). Application of arm support training in sub-acute stroke rehabilitation: first results on effectiveness and user experiences. *IEEE International Conference on Rehabilitation Robotics*. 10.1109/ICORR.2013.6650470

Pratt, G., & Manzo, J. (2013). The DARPA robotics challenge. *IEEE Robotics & Automation Magazine, 20*(2), 10–12. doi:10.1109/MRA.2013.2255424

Pristovani, R. D., Henfri, B. E., Sanggar, D., & Dadet, P. (2018). Walking strategy model based on zero moment point with single inverted pendulum approach in "T-FLoW" humanoid robot. In *Proceedings - 2017 2nd International Conferences on Information Technology, Information Systems and Electrical Engineering, ICITISEE 2017*. 10.1109/ICITISEE.2017.8285498

Puerta Barrera, J. F. (2017). *Proposta de arquitetura de controle para prótese robótica de membro inferior*. University of Campinas; doi:10.1901/jaba.2012.45-839

Puterman, M. L. (1994). *Markov Decision Processes: Discrete Stochastic Dynamic Programming*. Hoboken, NJ: John Wiley & Sons, Inc. doi:10.1002/9780470316887

Quinones, J. (2017, January 20). *Applying acceleration and deceleration profiles to bipolar stepper motors*. Retrieved from http://www.ti.com/lit/an/slyt482/slyt482.pdf

Reinkensmeyer, D. J., & Boninger, M. L. (2011). *Technologies and combination therapies for enhancing movement training for people with a disability. J Neuroeng Rehab.*

Reprap. (2017, January 27). *Choosing Belts and Pulleys.* Retrieved from http://reprap.org/wiki/Choosing_Belts_and_Pulleys

Richter, F., Orosco, R. K. & Yip, M. C. (2019). *Open-Sourced Reinforcement Learning Environments for Surgical Robotics.* Retrieved from http://arxiv.org/abs/1903.02090

Righettini, P. (2010). *Progettazione funzionale di sistemi meccatronci: Introduzione al modulo.* Bergamo: Università degli Studi di Bergamo.

Rivest, J. F., Soille, P., & Beucher, S. (1993). Morphological gradients. *Journal of Electronic Imaging, 2*(1), 326–336.

Roijers, D. M., Whiteson, S., Vamplew, P., & Dazeley, R. (2015). Why Multi-objective Reinforcement Learning? In *Proceedings of the 12th European Workshop on Reinforcement Learning (EWRL 2015),* (pp. 1-2). Academic Press.

Roijers, D. M., Vamplew, P., Whiteson, S., & Dazeley, R. (2013). A Survey of Multi-Objective Sequential Decision-Making. *Journal of Artificial Intelligence Research, 48,* 67–113. doi:10.1613/jair.3987

Rosário, J. M., Suekichi Kuteken, R., & Izquierdo Cordoba, L. M. (2019). *Proposal of Methodology of a Bipedal Humanoid Gait Generation Based on Cognitive Algorithm.* doi:10.1007/978-3-030-13835-6_13

Rosario, J. M. (2010). *Robótica Industrial I Modelagem, Utilização e Programação.* Sao Paulo, Brazil: Editora Baraúna.

Rosati, G. (2010). The place of robotics in post-stroke rehabilitation. *Expert Review of Medical Devices, 7*(6), 753–758. doi:10.1586/erd.10.49 PMID:21050086

Rosati, G., Masiero, S., & Rossi, A. (2017). "On the Use of Cable-Driven Robots in Early Inpatient Stroke Rehabilitation." Advances in Italian mechanism science. *International Journal of Mechanics and Control, 18*(2), 551–558.

Rosen, J. (2013). Surgical Robotics. In *Medical Devices* (pp. 63–97). Surgical and Imaging-Guided Technologies.

Rubmann, M., Lorenz, M., Gerbert, P., Waldner, M., Justus, J. & Harnish, M. (2015). *Industry 4.0: the future of productivity and growth in manufacturing industries.* The Boston Consulting Group Report.

Samson, C., Espiau, B., & Borgne, M. L. (1991). *Robot control: the task function approach.* Oxford University Press, Inc.

Santos, C. P., Alves, N., & Moreno, J. C. (2017). Biped Locomotion Control through a Biomimetic CPG-based Controller. *Journal of Intelligent & Robotic Systems*, *85*(1), 47–70. doi:10.100710846-016-0407-3

Satoh, K., & Yamaguchi, T. (2006). *Preparing various policies for interactive reinforcement learning*. Paper presented at the meeting of SICE-ICASE International Joint Conference 2006.

Schmidt, R. A. (2018). Motor control and learning: a behavioral emphasis (6th ed.). Human Kinetics Publishers.

Scioni, E., Decré, W., Aertbeliën, E., De Schutter, J., Tirmizi, A., & Witters, M. (2017). Fast and Intuitive Robot Programming: a Constraint-based Approach. ICRA.

Segovia, D., Mendoza, M., Mendoza, E., & González, E. (2015). Augmented reality as a tool for production and quality monitoring. *Procedia Computer Science*, *75*(1), 291–300. doi:10.1016/j.procs.2015.12.250

Seita, D., Krishnan, S., Fox, R., McKinley, S., Canny, J. & Goldberg, K. (2017). Fast and Reliable Autonomous Surgical Debridement with Cable-Driven Robots Using a Two-Phase Calibration Procedure. *Proc. of the 2018 IEEE International Conference on Robotics and Automation (ICRA)*, 6651–6658. Retrieved from http://arxiv.org/abs/1709.06668

Sellaouti, R., Stasse, O., Kajita, S., Yokoi, K., & Kheddar, A. (2006, October). Faster and smoother walking of humanoid HRP-2 with passive toe joints. In *Intelligent Robots and Systems, 2006 IEEE/RSJ International Conference on* (pp. 4909-4914). IEEE.

Sen, S., Garg, A., Gealy, D. V., McKinley, S., Jen, Y., & Goldberg, K. (2016). Automating multi-throw multilateral surgical suturing with a mechanical needle guide and sequential convex optimization. *Proc. of the IEEE International Conference on Robotics and Automation*, 4178–4185. 10.1109/ICRA.2016.7487611

Sentis, L., & Khatib, O. (2006, May). A whole-body control framework for humanoids operating in human environments. In *Proceedings 2006 IEEE International Conference on Robotics and Automation, 2006. ICRA 2006.* (pp. 2641-2648). IEEE. 10.1109/ROBOT.2006.1642100

Shademan, A., Decker, R. S., Opfermann, J., Leonard, S., Krieger, A., & Kim, P. C. W. (2016). Supervised autonomous robotic soft tissue surgery. *Science Translational Medicine*, *8*(337), 337ra64. doi:10.1126citranslmed.aad9398 PMID:27147588

Shamaei, K., Cenciarini, M., Adams, A., Gregorczyk, K. N., Schiffman, J. M., & Dollar, A. M. (2014). Design and evaluation of a quasi-passive knee exoskeleton for investigation of motor adaptation in lower extremity joints. *Biomedical Engineering, IEEE Transactions on*, *61*(6), 1809–1821. doi:10.1109/TBME.2014.2307698 PMID:24845291

Siegler, I. A., Bardy, B. G., & Warren, W. H. (2010). Passive vs. active control of rhythmic ball bouncing: The role of visual information. *Journal of Experimental Psychology. Human Perception and Performance*, *36*(3), 729–750. doi:10.1037/a0016462 PMID:20515200

Siltanen, S. (2012). *Theory and applications of marker-based augmented reality*. JULKAISIJA – UTGIVARE – Publisher.

Skobelev, P. O., & Borovik, S. Y. (2017). On the Way from Industry 4.0 to Industry 5.0: from Digital Manufacturing to Digital Society. *International Scientific Journal Industry 4.0, 2*(6), 207-311.

Stefano, M., Patrizia, P., Mario, A., Ferlini, G., Rizzelli, R., & Rosati, G. (2014). Robotic Upper Limb Rehabilitation after Acute Stroke by NeReBot: Evaluation of Treatment Costs. *BioMed Research International, 2014*, 1–5. doi:10.1155/2014/265634 PMID:24967345

Stein, J. (2012). Robotics in rehabilitation: technology as destiny. *Am J Phys Med Rehabil, 91*(11), S199–S203.

Stienen, A. H. A. (2009). *Development of novel devices for upper-extremity rehabilitation* (Ph.D. dissertation). University of Twente.

Stienen, A. H. A. (2007). Freebal: dedicated gravity compensation for the upper extremities. *Proceedings of the 2007 IEEE 10th International Conference on Rehabilitation Robotics*. 10.1109/ICORR.2007.4428517

Stock Drive Products. (2017, January 27). In *Handbook of timing belts and pulleys*. Retrieved from: http://www.sdp-si.com/d265/html/d265t003.html

Suekichi, R. (2017). *Geração de Padrões de Marcha para Sistemas Robóticos Bípedes Humanoides*. University of Campinas.

Sugimoto, N., & Morimoto, J. (2011). Phase-dependent trajectory optimization for CPG-based biped walking using path integral reinforcement learning. In *2011 11th IEEE-RAS International Conference on Humanoid Robots* (pp. 255–260). IEEE. 10.1109/Humanoids.2011.6100908

Sugitani, Y., Kanjo, Y., & Murayama, M. (1996). Systemization with CAD/CAM Welding Robots for Bridge Fabrication. *Proceedings of 4th International Workshop on Advanced Motion Control*, 80–85. 10.1109/AMC.1996.509384

Sugita, S., Itaya, T., & Takeuchi, Y. (2003). Development of Robot Teaching Support Devices to Automate Deburring and Finishing Works in Casting. *International Journal of Advanced Manufacturing Technology, 23*(3/4), 183–189.

Surdilovic, D., Zhang, J., & Bernhardt, R. R. (2007). STRING-MAN: Wire-robot technology for safe, flexible and human-friendly gait rehabilitation. *IEEE 10th Int. Conf. on Rehabilitation Robotics*, 446-453.

Tadepalli, P., & Ok, D. (1998). Model-Based Average Reward Reinforcement Learning. *Artificial Intelligence, 100*(1-2), 177–224. doi:10.1016/S0004-3702(98)00002-2

Tajmajer, T. (2017). *Multi-objective deep Q-learning with subsumption architecture*. Retrieved from https://arxiv.org/pdf/1704.06676v1.pdf

Takács, Á., Drexler, D. A., Galambos, P., Rudas, I. J., & Haidegger, T. (2018). Assessment and Standardization of Autonomous Vehicles. *Proc. of the 2018 IEEE 22nd International Conference on Intelligent Engineering Systems (INES)*, 185–192. 10.1109/INES.2018.8523899

Takács, Á., Nagy, D. Á., Rudas, I. J., & Haidegger, T. (2016). Origins of Surgical Robotics: From Space to the Operating Room. *Acta Polytechnica Hungarica, 13*(1), 13–30.

Takacs, A., Rudas, I., Bosl, D., & Haidegger, T. (2018). Highly Automated Vehicles and Self-Driving Cars [Industry Tutorial]. *IEEE Robotics & Automation Magazine*, *25*(4), 106–112. doi:10.1109/MRA.2018.2874301

Takahashi, K., Domen, K., Sakamoto, T., Toshima, M., Otaka, Y., Seto, M., ... Hachisuka, K. (2016). Efficacy of Upper Extremity Robotic Therapy in Subacute Poststroke Hemiplegia. *Stroke*, *45*(5), 1385–1388. doi:10.1161/STROKEAHA.115.012520 PMID:27006452

Takebayashi, T., Takahashi, K., Amano, S., Uchiyama, Y., Gosho, M., Domen, K., & Hachisuka, K. (2018). Assessment of the Efficacy of ReoGo-J Robotic Training Against Other Rehabilitation Therapies for Upper-Limb Hemiplegia After Stroke: Protocol for a Randomized Controlled Trial. *Frontiers in Neurology*, *9*, 730. doi:10.3389/fneur.2018.00730 PMID:30210446

Takeuchi, N., & Izumi, S.-I. (2013). Rehabilitation with Poststroke Motor Recovery: A Review with a Focus on Neural Plasticity. Stroke Research and Treatment. doi:10.1155/2013/128641

Tavolieri, C. (2008). *Design of a cable-based parallel manipulator for rehabilitation applications* (Ph.D. dissertation). University of Cassino, Italy, and INRIA, France.

Tejomurtula, S., & Kak, S. (1999). Inverse kinematics in robotics using neural networks. *Information Sciences*, *116*(2-4), 147–164. doi:10.1016/S0020-0255(98)10098-1

Thelen, E., & Smith, L. B. (2009). Dynamic Systems Theories. In Handbook of Child Psychology. Academic Press. doi:10.1002/9780470147658.chpsy0106

Tirmizi, A., De Cat, B., Janssen, K., Pane, Y., Leconte, P., & Witters, M. (2019, May). User-Friendly Programming of Flexible Assembly Applications with Collaborative Robots. In *2019 20th International Conference on Research and Education in Mechatronics (REM)* (pp. 1-7). IEEE. 10.1109/REM.2019.8744135

Tsitsiklis, J. N. (2007). NP-Hardness of checking the unichain condition in average cost MDPs. *Operations Research Letters*, *35*(3), 319–323. doi:10.1016/j.orl.2006.06.005

Tucan, P., Vaida, C., Plitea, N., Pisla, A., Carbone, G., & Pisla, D. (2019). Risk-Based Assessment Engineering of a Parallel Robot Used in Post-Stroke Upper Limb Rehabilitation. *Sustainability*, *11*, 2893.

Tyromotion. (n.d.). *Diego*. Retrieved from https://tyromotion.com/en/produkte/diego/

Vaida, C., Carbone, G., Major, K., Major, Z., Plitea, N., & Pisla, D. (2017). ACTA TECHNICA NAPOCENSIS. *Applied Mathematics. Mechanical Engineering*, *60*, 91–102.

Vaidya, S., Ambad, P., & Bhosle, S. (2018). Industry 4.0 –A Glimpse. *Procedia Manufacturing*, *20*(1), 233–238. doi:10.1016/j.promfg.2018.02.034

van der Schaft, A., & Schumacher, H. (2000). *An introduction to hybrid dynamical systems* (Vol. 251). London: Springer London; doi:10.1007/BFb0109998

Vanthienen, D., Klotzbu, M., De Schutter, J., De Laet, T., & Bruyninckx, H. (2013, November). Rapid application development of constrained-based task modelling and execution using domain specific languages. In *2013 IEEE/RSJ International Conference on Intelligent Robots and Systems*(pp. 1860-1866). IEEE. 10.1109/IROS.2013.6696602

Vaughan, C. L., Davis, B. L., Christopher, L., & Connor, J. C. O. (2005). Dynamics of human gait. *European Journal of Physics*.

Vedula, S. S., Malpani, A. O., Tao, L., Chen, G., Gao, Y., Poddar, P., ... Chen, C. C. (2016). Analysis of the Structure of Surgical Activity for a Suturing and Knot-Tying Task. *PLoS One*, *11*(3), e0149174. doi:10.1371/journal.pone.0149174 PMID:26950551

Verrelst, B., Stasse, O., Yokoi, K., & Vanderborght, B. (2006, December). Dynamically stepping over obstacles by the humanoid robot HRP-2. In *Humanoid Robots, 2006 6th IEEE-RAS International Conference on* (pp. 117-123). IEEE.

Viteckova, S., Kutilek, P., & Jirina, M. (2013). Wearable lower limb robotics: A review. *Biocybernetics and Biomedical Engineering*, *33*(2), 96–105. doi:10.1016/j.bbe.2013.03.005

Vora, A., Wu, H., Wang, C., Qian, Y., Shaver, G., Motevalli, V., . . . Zhang, H. (2014). Development of a SIL, HIL and Vehicle Test-bench for Model-based Design and Validation of Hybrid Powertrain Control Strategies (SAE 2014-01-1906). *SAE Annual Meeting*.

Wang, Z., Li, L., & Wang, X. (2011). Modification Algorithm on Routh-Pade Model Reduction of Interval Systems. In ICIC 2011, LNCS 6838 (pp. 701–704). Springer. doi:10.1007/978-3-642-24728-6_94

Wang, H. (2009). Comparison of *p* control charts for low defective rate. *Computational Statistics & Data Analysis*, *53*(12), 4210–4220. doi:10.1016/j.csda.2009.05.024 PMID:20161085

Wang, Y., Yang, Y., Liu, W., & Wang, R. (2015). Structure Design and Dynamic Model Analysis of Multi-degree-of-freedom Exoskeleton. In *2015 International Conference on Mechatronics, Electronic, Industrial and Control Engineering (MEIC-15)*. Atlantis Press. 10.2991/meic-15.2015.342

Warren, W. H. (2006). The dynamics of perception and action. *Psychological Review*, *113*(2), 358–389. doi:10.1037/0033-295X.113.2.358 PMID:16637765

Warren, W. H., & Fajen, B. R. (2008). Behavioral dynamics of visually guided locomotion. *Understanding Complex Systems*, 45–75. doi:10.1007/978-3-540-74479-5_3

Wei, G. Q., Arbter, K., & Hirzinger, G. (1997). Real-time visual servoing for laparoscopic surgery. Controlling robot motion with color image segmentation. *IEEE Engineering in Medicine and Biology Magazine*, *16*(1), 40–45. doi:10.1109/51.566151 PMID:9058581

Wentzel, M. (2016, January 22). *Fourth Industrial Revolution: how Brazil can prepare for the Economy of the Future*. Retrieved from http://www.bbc.com/portuguese/noticias/2016/01/160122_quarta_revolucao_industrial_mw_ab

Wiering, M. A., Withagen, M., & Drugan, M. M. (2014). Model-based multiobjective reinforcement learning. In *Proceedings of the IEEE Symposium on Adaptive Dynamic Programming and Reinforcement Learning* (pp. 1-6). IEEE.

Winter, D. A. (2009). *Biomechanics and Motor Control of Human Movement*. Hoboken, NJ: John Wiley & Sons; doi:10.1002/9780470549148

Wittenberg, G. (1995). Developments in offline programming: An overview. *Industrial Robot: An International Journal*, *22*(3), 21–23. doi:10.1108/EUM0000000004186

Wolbrecht, E. T., Chan, V., Reinkensmeyer, D. J., & Bobrow, J. E. (2008). Optimizing compliant, modelbased robotic assistance to promote neurorehabilitation. *IEEE Transactions on Neural Systems and Rehabilitation Engineering*, *16*(3), 286–297. doi:10.1109/TNSRE.2008.918389 PMID:18586608

Xia, Y., & Wang, J. (2001, February). A dual neural network for kinematic control of redundant robot manipulators. *IEEE Transactions on Systems, Man, and Cybernetics. Part B, Cybernetics*, *31*(1), 147–154. doi:10.1109/3477.907574 PMID:18244777

Xia, Z., Deng, H., Zhang, X., Weng, S., Gan, Y., & Xiong, J. (2017). A central pattern generator approach to footstep transition for biped navigation. *International Journal of Advanced Robotic Systems*, *14*(1), 1–9. doi:10.1177/1729881416682708

Xu, W., Mu, Z., Liu, T., & Liang, B. (2017). A modified modal method for solving the mission-oriented inverse kinematics of hyper-redundant space manipulators for on-orbit servicing. *Acta Astronautica*, *139*, 54–66. doi:10.1016/j.actaastro.2017.06.015

Yadav, P. K., & Prajapati, N. L. (2012). One Dimensional Routh Pade Approximantsfor Two Dimensional Systems using Genetic Algorithm. *International Journal of Advanced and Innovative Research*, 576-580.

Yamamoto, T., Vagvolgyi, B., Balaji, K., Whitcomb, L. L., & Okamura, A. M. (2009). Tissue property estimation and graphical display for teleoperated robot-assisted surgery. *Proc. of the 2009 IEEE International Conference on Robotics and Automation (ICRA)*, 4239–4245. 10.1109/ROBOT.2009.5152674

Yang, G.-Z., Cambias, J., Cleary, K., Daimler, E., Drake, J., Dupont, P. E., … Taylor, R. H. (2017). Medical robotics—Regulatory, ethical, and legal considerations for increasing levels of autonomy. *Science Robotics*, *2*(4). doi:10.1126cirobotics.aam8638

Yang, S., Gao, Y., Bo, A., Wang, H., & Chen, X. (2016). Efficient Average Reward Reinforcement Learning Using Constant Shifting Values. *Proceedings of the Thirtieth AAAI Conference on Artificial Intelligence (AAAI-16)*, 2258-2264

Yoshida, E., Esteves, C., Sakaguchi, T., Laumond, J. P., & Yokoi, K. (2006, October). Smooth collision avoidance: Practical issues in dynamic humanoid motion. In *Intelligent Robots and Systems, 2006 IEEE/RSJ International Conference on* (pp. 827-832). IEEE.

Yoshitake, S., Nagata, F., Otsuka, A., Watanabe, K., & Habib, M. K. (2012). Proposal and Implementation of CAM System for Industrial Robot RV1A. *Proceedings of the 17th International Symposium on Artificial Life and Robotics*, 158–161.

Yu, S., Han, C., & Cho, I. (2014). Design considerations of a lower limb exoskeleton system to assist walking and load-carrying of infantry soldiers. *Applied Bionics and Biomechanics*, *11*(3), 119–134. doi:10.1155/2014/585837

Yu, W., & Chen, H. (2018). Analog Controller Design for Mechatronic Rotary Inverted Pendulum (Part 1). *2018 ASEE Mid-Atlantic Section Spring Conference*.

Zanchettin, A. M., & Rocco, P. (2016, October). Robust constraint-based control of robot manipulators: An application to a visual aided grasping task. In *2016 IEEE/RSJ International Conference on Intelligent Robots and Systems (IROS)* (pp. 3634-3639). IEEE. 10.1109/IROS.2016.7759535

Zhang, H. H., Nawrocki, R., & Li, Q. (2018, July). On Basics and Applications of Multidisciplinary Engineering and Technology Education. *3rd International Contemporary Educational Research Congress*.

Zhang, X., Xiao, L., & Kan, J. (2015). Degradation prediction model based on a neural network with dynamic windows. *Sensors (Basel)*, *15*(1), 6996–7015. doi:10.3390150306996 PMID:25806873

Zhao, Y., Zhang, W., Ge, W., & Li, S. (2013). Finite Element Simulation of Soldier Lower Extremity Exoskeleton. *Journal of Multimedia*, *8*(6), 705–711. doi:10.4304/jmm.8.6.705-711

Zhou, Z.; Guo, H.; Wang, Y.; Zhu, Z.; Wu, J. & Liu, X. (2018, August). Inverse kinematics solution for robotic manipulator based on extreme learning machine and sequential mutation genetic algorithm. *International Journal of Advance Robotic Systems*.

About the Contributors

Maki Habib obtained his Doctor of Engineering Sciences in Intelligent and Autonomous Robot from the University of Tsukuba, Japan. He was a senior research scientist with GMD, Japan, leading the Telecooperation group, Associate Professor with Monash University and lead the Mechatronics Engineering Programme. He was appointed as a full Professor of Robotics and Mechatronics at Swinburne University. Then, he was an invited Professor at KAIST, Korea, Visiting Professor at Saga University, Japan, and now a full Professor at AUC. His main area of research is focusing on human adaptive and friendly Mechatronics, autonomous navigation, service robots and humanitarian demining, telecooperation, distributed teleoperation and collaborative control, wireless sensor networks and ambient intelligence, biomemtic, bioinspiration and biomedical robots.

* * *

Thiago Alves graduated in Mechatronics Engineering at the Federal University of Uberlândia (2014) and master's degree in Mechanical Engineering at the Federal University of Uberlândia (2019). Currently studying for a PhD in Mechanical Engineering at the Federal University of Uberlândia. Main fields are on Robotics, Cable-Driven Robots and Rehabilitation.

Darío Amaya Hurtado is a Professor at the Nueva Granada Military University (Bogotá, Colombia), Department of Mechatronic Engineering. Responsible for the Virtual Applications Research Group (GAV), working with emphasis in the following thematic areas: mechatronics, automation and robotics, control and systems and communications engineering. He holds a degree in Electronic Engineer, Antonio Nariño University (Bogotá, Colombia, 1995), specialization diploma in Industrial Process Control, University of Los Andes (Bogotá, Colombia, 2000), Master in Teleinformatics, District University Francisco José de Caldas (Bogotá, Colombia 2007), and Doctorate in Mechanical Engineering, University of Campinas (Campinas, Brazil, 2011).

Giuseppe Carbone has got the Master cum laude at University of Cassino (Italy) where he also completed the PhD studies being a Key Member of LARM (Laboratory of Robotics and Mechatronics) for about 20 years. In 2015-2017 he has been Senior Lecturer at Sheffield Hallam University (UK) and member of the executive board of Sheffield Robotics. Since 2019 he is Associate Professor at DIMEG, University of Calabria, Italy. Prof. Carbone has received several Awards including three IFToMM "Young Delegate" Awards, two JSPS Awards in Japan. His research interests cover aspects of Mechanics of Manipulation and Grasp, Mechanics of Robots, Mechanics of Machinery with more than 300 published papers and over 10 patent submissions. He edited/co-edited four books with Springer International Publisher. He has been participating or coordinating more than 20 research projects at national and international level including 7th European Framework and Horizon 2020. Currently, he is Deputy Chair of IFToMM Technical Committee on Robotics and Mechatronics, Deputy Chair of the Youth Committee of the Society of Bionics and Biomechanics, Treasurer of the IFToMM Italy Society.

Marcos Carnevale is a Mechanical Design Engineer at Electrolux Italy; Mechanical Engineer, Federal University of Rio de Janeiro (2017).

Marco Ceccarelli received PhD degree in 1987 at the University La Sapienza of Rome. He was Professor in Cassino University and since March 2019 he is Professor at University of Rome Tor Vergata. He is Director of LARM2: Laboratory of Robot Mechatronics. He is Scientific Editor of Springer Book Series on History of MMS and on Mechanism and Machine Science, and Associate Editor of several Journals. He wrote the book 'Fundamentals of Mechanics of Robotic Manipulation' in 2004. In 2008 he co-authored a book on Mechanisms Design in Spanish. In 2010 he co-authored a book A brief illustrated history of machines and mechanisms. He received Degree of Doctor Honoris Causa from foreigner Universities and ASME Historian Award. He is ASME fellow. He was elected Secretary-General of IFToMM in 2004-2007 and IFToMM President in 2008-2011 and 2016-2019. His research interests cover aspects of Mechanism Design, Mechanics and Design of Robots, and History of Mechanical Engineering. He is author/co-author of several other books and papers, which are published in conference proceedings and international journals. More info in http://larmlaboratory.net.

Rogério Gonçalves received the mechanical engineering degree from the Federal University of Uberlândia in 2004, Master degree in 2006 and Ph.D. at the same University in 2009. In 2016 was visiting scientist at The Eric P. and Evelyn E. Newman Laboratory for Biomechanics and Human Rehabilitation, Massachusetts Institute of Technology – MIT, Department of Mechanical Engineering. He joined the School of Mechanical Engineering of Federal University of Uberlândia, Brazil, in 2008, where he is currently Professor. He is member of RoboCup Brazil and ABCM (Brazilian Society of Engineering and Mechanical Sciences). He is author or co-author of about 100 papers, which have been presented in national and international conferences or published in national and international journals. His research interest includes kinematics and dynamics of serial and parallel structure, stiffness, cable-driven parallel structure, mobile robots, biorobotics, bioengineering, rehabilitation robots and humanoid robots.

Tamás Haidegger received his MSc degrees from the Budapest University of Technology and Economics (BME) in Electrical Engineering and Biomedical Engineering in 2006 and 2008, and PhD in 2011, respectively. His main field of research is control/teleoperation of surgical robots, image-guided therapy and supportive medical technologies. Currently, he is associate professor at Óbuda University, serving as the director of the University Research, Innovation and Service Center, and as the technical lead of medical robotics research at the Antal Bejczy Center for Intelligent Robotics. Besides, he is a research area manager at the Austrian Center of Medical Innovation and Technology (ACMIT), working on minimally invasive surgical simulation and training, medical robotics and usability/workflow assessment through ontologies. Tamás is the co-founder of a university spin-off—HandInScan—focusing on objective hand hygiene control in the medical environment. He is an active member of various other professional organizations, including the IEEE Robotics an Automation Society, IEEE SMC, IEEE EMBC and euRobotics aisbl. He is a national delegate to an ISO/IEC standardization committee focusing on the safety and performance of medical robots. He has been maintaining a professional blog on medical robotic technologies for 10 years: surgrob.blogspot.com.

Yoshimichi Honma received his M.E. degree from Toyohashi University of Technology, Japan, in 1994. He worked at National Institute of Technology (KOSEN), Nara College as research associate in 1997 and is currently an lecturer from 2000. His research interests include learning support system, network management, computer security and computer architecture.

Jean Hoyos is an associate research engineer in Human-Robot Collaboration at Flanders Make. His focus is on Industry 4.0 applications. He did his masters from Ecole Polytechnique de Bruxelles in Mechatronics Engineering.

Yoshihiro Ichikawa got Doctor of Engineering Degree from The University of Electro-Communications, Japan, in 12014. He joined the University of Tsukuba from 2014 to 2017 as a postdoctoral researcher. He moved to the National Institute of

Technology (KOSEN), Nara College as an assistant professor in 2017. His research interests include multiagent system, artificial intelligence, and human-computer interaction.

Luis Izquierdo-Córdoba is a PhD candidate in Mechanical Engineering (University of Campinas, 2017), with a master's degree in Mechatronic Engineering from the University of Brasília (2014, Brazil), and a bachelor's degree in Electronic Engineering from the District University Francisco José de Caldas (2010, Colombia). Currently, his main work focused on robotics and automation, linear and nonlinear control systems, and artificial intelligence algorithms.

Karel Janssen obtained in 2011 his Master of Science degree in Electronics Engineering. During the last 8 years, he was active in the research department of Flanders Make, Belgium. His main research interests are in the areas of computer vision and software development, for robotics and the vehicle-of-the-future.

Sergey Jatsun has a Scientific degree: Doctor of technical sciences Academic title: Professor In 1979 Jatsun Sergey has received PhD degree in Mishkole technical University in Hungary. In 1990 received degree Doctor science in a field of Dynamics and durability of machines devices and apparatus in Riga, Latvia. In 1990 he defended his doctoral dissertation: "Theory and Synthesis of Vibratory, Technological Processes and Vibromachines for the Processing of Dry Materials". He is author of more than 250 publications and has more than 100 patents on inventions. He is a member of the International Academy of Ecology and Safety and was awarded the token "High Achiever of Invention and Rationalization" and "USSR's Inventor". He has the silver and golden medal of exhibition of achievements of national economy of USSR and Russian. He awarded the title of Honored worker of science of the Russian Federation. Jatsun S.F. is a member of the editorial council of the journals: "Kursk State University News" and "Mechanism and Machine Theory" (Germany), "Izvestiya SWSU". He has prepared more than 40 PhD students.

Patricia Leconte is an electromechanical engineer with a biomedical background. Presently, she is working as a research engineer working on industrial and collaborative robots at Flanders Make. The purpose is to bring new technologies on the work floor in manufacturing industries. She accomplished her PhD in 2017 at the Université catholique de Louvain. During her thesis she brought new insides on how stroke patient perform specific types of movements (discrete vs rhythmic movements) and developed a new robotic assist-as-needed rehabilitation program for rhythmic movements.

Qiong Li received her PhD degree in Mechanical Engineering Technology at Purdue University in 2015.

Alessandro Massaro received the Laurea degree in electronic engineering and the Ph.D. degree in telecommunication engineering from the Università Politecnica delle Marche, Ancona, Italy, in 2001 and 2004, respectively.From 2004 to 2006 he worked as Research Scientist (post-doc) in the Department of Electromagnetism and Bioengineering at Università Politecnica delle Marche. In 2006, he spent one year in Research and Development at medical and industrial optics industry(endoscope design and optical systems). He worked for two years with National Nanotechnology Laboratory of CNR-INFM, Università del Salento, as principal investigator. He is currently team leader in Robotics Lab. 1 of the Center of Bio-Molecular Nanotechnology of Italian Institute Technology (IIT), Arnesano, Lecce, Italy. His research interests are in the design and modeling of photonic band gap circuits, in the development of computer aided design (CAD) tools in the area of integrated optics, MEMS technology and robotic systems, and smart material implementation. Actually he is member of the European Microwave Association (EuMA) and of IEEE (IEEE Senior Member). Actually he is R&D chief of Dyrecta Lab srl and he is Associate Professor in Electronic and in Experimetal Physics.

João Maurício Rosário is a professor at the State University of Campinas (UNICAMP), Department of Mechanical Design, responsible for the Laboratory of Integrated Automation, Robotics and Intelligent Systems, working in the area of Control and Automation Engineering, with emphasis in the following thematic areas: mechatronics, automation and Robotics, control, modeling and Intelligent Systems. He holds a degree in Mechanical Engineering, UNICAMP (1981), Specialization Diploma in Industrial Automation: Automated Production - Universite de Nancy I (1986), Master in Mechanical Engineering, UNICAMP (1984), Doctorate in Automation Robotics Specialty - Ecole Centrale des Arts et Manufactures (Paris) (1990), post-doctorate in the institutions: Geestach Technological Institute - GKSS, Germany (1992), SUPELEC, France (2005), and ISEP, Portugal (2006). CENTRALE-SUPELEC (2015-2016 and 2017-2019). (Source: Lattes Curriculum)

Shota Nagahama graduated the Faculty of Advanced Engineering, National Institute of Technology (KOSEN), Nara College, Japan and received his B.A. degree from The National Institution for Academic Degrees and University Evaluation, Japan in 2018. He has been a graduate student of the Graduate School of Science and Technology, Nara Institute of Science and Technology, Japan since 2018. His research interests include reinforcement learning, recommendation system, graph theory, parallel algorithm and distributed algorithm.

Fusaomi Nagata received the B.E. degree from the Department of Electronic Engineering at Kyushu Institute of Technology in 1985, and the D.E. degree from the Faculty of Engineering Systems and Technology at Saga University in 1999. He was a research engineer with Kyushu Matsushita Electric Co. from 1985 to 1988, and a special researcher with Fukuoka Industrial Technology Centre from 1988 to 2006. From 2006 to 2016, he was a professor at the Department of Mechanical Engineering, Faculty of Engineering, Tokyo University of Science, Yamaguchi (TUSY), Japan. After the TUSY become a public university (SOCU) supported by Sanyo-Onoda City in 2016, He has been a dean of the faculty of Engineering, SOCU. His current research interests include deep convolutional neural networks (DCNNs) and support vector machines (SVMs) for visual inspection, intelligent control for industrial robots and its application to machining process. Up to now, a design and training application for DCNNs & SVMs has been developed for students and engineers and applied to the anomaly detection of resin molded articles. A robot sander, mold polishing robot, desktop NC machine tool with compliance control capability, machining robot with robotic CAM system, and 3D printer-like data interface for the machining robot have been also developed for related industrial fields to smartly and dexterously machine wood material, aluminum PET bottle mold, LED lens mold, foamed polystyrene, and so on. Research results: http://www.rs.tusy.ac.jp/nagata/research_2016.htm

Tamás Dániel Nagy earned his bachelor's degree in Molecular Bionics in 2014, and master's degree in Computer Science Engineering in 2016 at the University of Szeged. During his university studies he joined the Noise Research Group, where he worked on physiological measurements, signal processing and those application in telemedicine, furthermore he was employed as a demonstrator on laboratory practices. After graduation he was hired as a technical assistant by Department of Software Engineering at the University of Szeged for six months, where he worked in cooperation with Noise Research Group on projects related to medical signals and telemonitoring. Since September 2016 he is a PhD student of the Doctoral School of Applied Informatics and Applied Mathematics at Óbuda University, he is currently working on the analysis and low level automation of movement patterns in robot surgery interventions.

Asad Tirmizi received the Aeronautical Engineering degree in 2008 from the National University of Sciences and Technology. From December 2008 to August 2010, he was a Research Engineer at Institute of Avionics and Aeronautics, Air University, Islamabad, Pakistan. He received the M.S. degree in Electrical Engineering (Controls) in Sweden. From 2012 to 2016 he was a Ph.D. student at the University of Siena, During the Ph.D. his focus has been on the perception and wearability of haptic devices. Since 2016 he is associated with Flanders Make, where he works on collaborative robots and their role in flexible assembly systems.

Maarten Witters did his MS and PhD in Mechanical engineering from KU Leuven in 2010. Since then he is associated with Flanders Make. He is currently the corelab manager for ProductionS. It is a division of the organization that looks into innovation in assembly systems to bring flexibility and agility in production.

Tomohiro Yamaguchi received his M.E. degree from Osaka University, Japan, in 1987. He joined Mitsubishi Electric Corporation in 1987 and moved to Matsushita Electric Industrial in 1988. He worked at Osaka University from 1991 to 1998 as a research associate and got Doctor of Engineering Degree from Osaka University in 1996. He moved to Nara National College of Technology as associate professor in 1998 and is currently a professor from 2007. His research interests include multi-objective reinforcement learning, creative learning, autonomous learning agent, human-agent interaction, learning support system, human learning process and mastery process, interactive recommender system and music information retrieval. He is a member of The Japanese Society for Artificial Intelligence and The Society of Instrument and Control Engineers, Japan.

Wangling Yu is an assistant professor in the Electrical & Computer Engineering Technology Department of the Purdue University Northwest. He was a test engineer of a government contractor over 15 years, providing technical leadership in the certification, testing and evaluation of custom integrated security systems used by the United States government. He received his PhD degree in Electrical Engineering from the City University of New York in 1992, specializing in control theory and electronic technology.

H. Henry Zhang is a full professor in Mechanical Engineering Technology in the College of Technology at Purdue University. He is also director of Center for Technology Development. He received his PhD degree in Mechanical Engineering at University of Michigan-Ann Arbor in 1996. He also received degrees in Hydraulic Control, Neural Networking, and Electrical & Computer Engineering. Before he joined Purdue in 2007, he was a senior specialist of automatic transmission engineering in Chrysler Technology Center with 13 years of automotive industry experience. He established Multidisciplinary Design Laboratory to support his research interests in Mechatronics modeling, design and manufacturing (including high value added consumer products, industrial robotic systems, MEMS devices, and medical instrumentation) and vehicle powertrain/automatic transmission design and manufacturing.

Index

H

I

K

L

M

N

O

P

R

S

T

V

W

Z

www.ingramcontent.com/pod-product-compliance
Lightning Source LLC
LaVergne TN
LVHW081347300926
849161LV00008B/2087
* 9 7 8 1 7 9 9 8 1 3 8 2 8 *